# Chemical Symmetry Breaking

# Chemical Symmetry Breaking

Editor

**Rui Tamura**

MDPI • Basel • Beijing • Wuhan • Barcelona • Belgrade • Manchester • Tokyo • Cluj • Tianjin

*Editor*
Rui Tamura
Kyoto University
Japan

*Editorial Office*
MDPI
St. Alban-Anlage 66
4052 Basel, Switzerland

This is a reprint of articles from the Special Issue published online in the open access journal *Symmetry* (ISSN 2073-8994) (available at: https://www.mdpi.com/journal/symmetry/special_issues/Chemical_Symmetry_Breaking).

For citation purposes, cite each article independently as indicated on the article page online and as indicated below:

LastName, A.A.; LastName, B.B.; LastName, C.C. Article Title. *Journal Name* **Year**, *Volume Number*, Page Range.

**ISBN 978-3-0365-1130-6 (Hbk)**
**ISBN 978-3-0365-1131-3 (PDF)**

© 2021 by the authors. Articles in this book are Open Access and distributed under the Creative Commons Attribution (CC BY) license, which allows users to download, copy and build upon published articles, as long as the author and publisher are properly credited, which ensures maximum dissemination and a wider impact of our publications.

The book as a whole is distributed by MDPI under the terms and conditions of the Creative Commons license CC BY-NC-ND.

# Contents

**About the Editor** .................................................. vii

**Preface to "Chemical Symmetry Breaking"** .................................................. ix

**Shuichi Sato, Yoshiaki Uchida and Rui Tamura**
Spin Symmetry Breaking: Superparamagnetic and Spin Glass-Like Behavior Observed in Rod-Like Liquid Crystalline Organic Compounds Contacting Nitroxide Radical Spins
Reprinted from: *Symmetry* **2020**, *12*, 1910, doi:10.3390/sym12111910 .................................................. 1

**Yoichi Takanishi**
Chiral Symmetry Breaking in Liquid Crystals: Appearance of Ferroelectricity and Antiferroelectricity
Reprinted from: *Symmetry* **2020**, *12*, 1900, doi:10.3390/sym12111900 .................................................. 31

**Gérard Coquerel and Marine Hoquante**
Spontaneous and Controlled Macroscopic Chiral Symmetry Breaking by Means of Crystallization
Reprinted from: *Symmetry* **2020**, *12*, 1796, doi:10.3390/sym12111796 .................................................. 47

**Naohiro Uemura, Seiya Toyoda, Waku Shimizu, Yasushi Yoshida, Takashi Mino and Masami Sakamoto**
Absolute Asymmetric Synthesis Involving Chiral Symmetry Breaking in Diels–Alder Reaction
Reprinted from: *Symmetry* **2020**, *12*, 910, doi:10.3390/sym12060910 .................................................. 65

**Daichi Kitagawa, Christopher J. Bardeen and Seiya Kobatake**
Symmetry Breaking and Photomechanical Behavior of Photochromic Organic Crystals
Reprinted from: *Symmetry* **2020**, *12*, 1478, doi:10.3390/sym12091478 .................................................. 81

**Toshikazu Ono and Yoshio Hisaeda**
Vapochromism of Organic Crystals Based on Macrocyclic Compounds and Inclusion Complexes
Reprinted from: *Symmetry* **2020**, *12*, 1903, doi:10.3390/sym12111903 .................................................. 97

**Shotaro Hayashi**
Elastic Organic Crystals of $\pi$-Conjugated Molecules: New Concept for Materials Chemistry
Reprinted from: *Symmetry* **2020**, *12*, 2022, doi:10.3390/sym12122022 .................................................. 109

**Michiya Fujiki**
Resonance in Chirogenesis and Photochirogenesis: Colloidal Polymers Meet Chiral Optofluidics
Reprinted from: *Symmetry* **2021**, *13*, 199, doi:10.3390/sym13020199 .................................................. 129

**Yoshitane Imai**
Generation of Circularly Polarized Luminescence by Symmetry Breaking
Reprinted from: *Symmetry* **2020**, *12*, 1786, doi:10.3390/sym12111786 .................................................. 181

**Fuyuki Ito**
Photochemical Methods for the Real-Time Observation of Phase Transition Processes upon Crystallization
Reprinted from: *Symmetry* **2020**, *12*, 1726, doi:10.3390/sym12101726 .................................................. 195

**Yoshiyuki Kageyama**
Robust Dynamics of Synthetic Molecular Systems as a Consequence of Broken Symmetry
Reprinted from: *Symmetry* **2020**, *12*, 1688, doi:10.3390/sym12101688 . . . . . . . . . . . . . . . . . . **209**

**Hirofumi Sakuma, Izumi Ojima, Motoichi Ohtsu and Hiroyuki Ochiai**
Off-Shell Quantum Fields to Connect Dressed Photons with Cosmology
Reprinted from: *Symmetry* **2020**, *12*, 1244, doi:10.3390/sym12081244 . . . . . . . . . . . . . . . . . . **225**

# About the Editor

**Rui Tamura** (Professor Emeritus of Kyoto University) became a professor of the Department of Interdisciplinary Environment, Graduate School of Human and Environmental Studies at Kyoto University in 2002, from which he retired in 2018. His research fields cover synthetic and structural organic chemistry, organic crystal and liquid crystal chemistry, as well as colloidal, magnetic and chiral chemistry. His current research interests focus on the discovery of novel complexity phenomena occurring upon the phase transition of organic crystals and in liquid crystals under nonequilibrium conditions. It should be stressed that the observed unusual phenomena cannot be reproduced in ordinary equilibrium systems. To date, two novel phenomena have been discovered: "preferential enrichment" is a unique spontaneous enantiomeric resolution phenomenon observed upon the recrystallization of certain types of racemic organic crystals and cocrystals; while the "magneto-LC effect" refers to the generation of unique superparamagnetic domains in liquid crystalline phases of organic nitroxide radical compounds in low magnetic fields. Professor Tamura completed his Ph.D. at the Department of Chemistry, Faculty of Science, at Kyoto University in 1980, and took two post-doctoral research fellowships at the Department of Chemistry at Colorado State university (1980–1981) and Princeton University (1982) in the USA. He has undertaken faculty positions in chemistry at the National Defense Academy (1983–1987), Ehime University (1988–1994), Hokkaido University (1995–1996) and Kyoto University (1997–2018) in Japan.

# Preface to "Chemical Symmetry Breaking"

Nowadays, it is well recognized that a concept of the nonlinear complexity theory governs a variety of dynamic phenomena observed in all fields of science; the fluctuation in a nonequilibrium state induces a phase transition from a chaotic or dissipative state to another one to trigger symmetry breaking, and eventually, the nonlinear amplification of fluctuation leads to dissymmetric circumstances. A typical example is the birth of the universe by the cosmic inflation followed by the Big Bang starting from a quantum fluctuation, which finally led to the selection of "matter" against "anti-matter" in the universe, according to a violation of the charge conjugation parity (CP) symmetry. Thus, symmetry breaking has been playing a primordial role in physics, chemistry, life science, economics and so on.

This Special Issue focuses on the various chemical and physical phenomena that originate from symmetry breaking occurring upon a thermally or photochemically induced phase transition in the organic condensed phases, such as metastable liquid crystals, crystals, amorphous solids, and colloidal polymer materials under nonequilibrium conditions, including an experimental and theoretical report on the connection of dressed photons with cosmology.

**Rui Tamura**
*Editor*

*Review*

# Spin Symmetry Breaking: Superparamagnetic and Spin Glass-Like Behavior Observed in Rod-Like Liquid Crystalline Organic Compounds Contacting Nitroxide Radical Spins

Shuichi Sato [1,*], Yoshiaki Uchida [2,*] and Rui Tamura [3,*]

1. Department of Physics, Osaka Dental University, Hirakata, Osaka 573-1121, Japan
2. Graduate School of Engineering Science, Osaka University, Toyonaka, Osaka 560-8531, Japan
3. Graduate School of Human and Environmental Studies, Kyoto University, Kyoto 606-8501, Japan
* Correspondence: shuichi-s@cc.osaka-dent.ac.jp (S.S.); yuchida@cheng.es.osaka-u.ac.jp (Y.U.); tamura.rui.45x@st.kyoto-u.ac.jp (R.T.); Tel.: +81-90-2997-3881 (S.S.); +81-6-6850-6256 (Y.U.); +81-77-577-1337 (R.T.)

Received: 24 October 2020; Accepted: 16 November 2020; Published: 20 November 2020

**Abstract:** Liquid crystalline (LC) organic radicals were expected to show a novel non-linear magnetic response to external magnetic and electric fields due to their coherent collective molecular motion. We have found that a series of chiral and achiral all-organic LC radicals having one or two five-membered cyclic nitroxide radical (PROXYL) units in the core position and, thereby, with a negative dielectric anisotropy exhibit spin glass (SG)-like superparamagnetic features, such as a magnetic hysteresis (referred to as 'positive magneto-LC effect'), and thermal and impurity effects during a heating and cooling cycle in weak magnetic fields. Furthermore, for the first time, a nonlinear magneto-electric (ME) effect has been detected with respect to one of the LC radicals showing a ferroelectric (chiral Smectic C) phase. The mechanism of the positive magneto-LC effect is proposed and discussed by comparison of our experimental results with the well-known magnetic properties of SG materials and on the basis of the experimental results of a nonlinear ME effect. A recent theoretical study by means of molecular dynamic simulation and density functional theory calculations suggesting the high possibility of conservation of the memory of spin-spin interactions between magnetic moments owing to the ceaseless molecular contacts in the LC and isotropic states is briefly mentioned as well.

**Keywords:** spin symmetry breaking; magnetic liquid crystals; magneto-LC effect; nitroxide radicals; superparamagnetic domain; spin glass state

## 1. Introduction

Following the birth and establishment of 'Einsteinian general theory of relativity' and 'quantum theory and mechanics' at the beginning of the 20th century [1], the 'complexity theory' developed rapidly since the 1970s is recognized as one of the paradigm shifts or innovations in science in the same century [2–4]. Currently, a concept of the nonlinear and nonequilibrium (or out of equilibrium) complexity theory is known to govern a variety of dynamic behaviors observed in both natural and social sciences [5–9]. In the nonequilibrium complexity system, symmetry breaking takes place easily during transition from a chaotic or dissipative state to another one. The fluctuation in a nonequilibrium state induces a phase transition to trigger the symmetry breaking, and, eventually, the nonlinear amplification of fluctuation leads to dissymmetric circumstances [5–10]. The most dramatic and important event is the birth of the universe by the cosmic inflation followed by the Big Bang starting from a quantum fluctuation, according to the 'uncertainty principle of energy and time' as a hypothetical explanation [1]. After the selection of 'matter' against 'anti-matter' resulting from the CP (charge conjugation parity symmetry) violation in the universe, the important and familiar events on the earth include many body interactions of multiple elements responsible for the nonlinearity [11]. The intermolecular interactions can create nonequilibrium objects such as cells [12], bubbles [13], and metastable crystals [14], which may have strong links to the origin of selected chirality of life. Thus, symmetry breaking has been playing a primordial role in physics, chemistry, life science, and so forth.

Liquid crystals, which are defined as a thermal mesophase between crystalline and isotropic phases, and, hence, can be regarded as high-temperature polymorphs of crystals, are unique soft materials that combine anisotropy and fluidity. From a different point of view, liquid crystalline (LC) phases are considered to be a sort of complexity system consisting of nonequilibrium dynamic states due to the molecular motion and the coherent collective properties of molecules in the LC state [8]. Therefore, they are so sensitive to external stimuli, such as electric or magnetic field, heat, light, temperature, pressure, and added chiral dopants, that the LC superstructure can easily be altered and the change can be controlled [15,16]. In this context, it seems challenging and promising to develop organic soft materials that can show a novel complexity phenomenon in response to various external stimuli by making use of the LC environment.

In this connection, magnetic LC compounds had attracted great interest as soft materials to enhance the effect of magnetic fields on the electric and optical properties of liquid crystals. They were anticipated to exhibit unique magnetic interactions and, thereby, unconventional magneto-electric [17–20] or magneto-optical properties in the LC state [21–23]. However, there had been no prominent study on these interesting subjects in the 20th century because the majority of magnetic liquid crystals examined were of a highly viscous transition (d-block or f-block) metal-containing metallomesogens (Figure 1) [24,25], which were not suitable for investigation of the swift molecular motion and reorientation in the LC phases at moderate temperatures in weak magnetic fields. At present, no appreciable intermolecular ferromagnetic or superparamagnetic interaction (in other words, spin symmetry breaking) has been observed in the LC state of d-block or f-block metallomesogens.

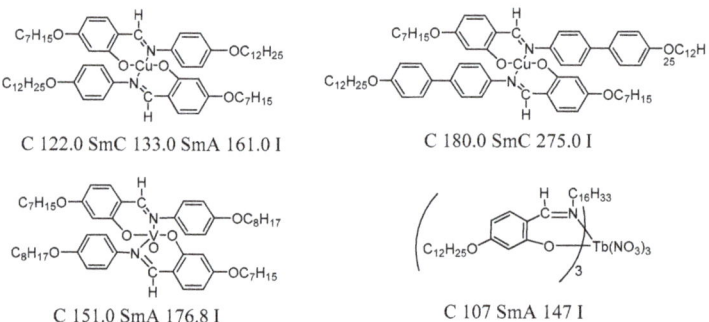

**Figure 1.** Representative metallomesogens and their LC transition temperatures.

Only a few all-organic LC radical compounds were prepared until 2003 because it was believed that the geometry and bulkiness of the radical-stabilizing substituents were detrimental to the stability of liquid crystals, which requires molecular linearity or planarity [25–27]. Although several rod-like organic LC compounds with a stable cyclic nitroxide unit as the spin source were prepared (Figure 2), their molecular structures were limited to those containing a nitroxyl group in the terminal alkyl chain, away from the rigid core, and, thereby, allowed the free rotation of the nitroxyl moiety inside the molecule, resulting in a very small dielectric anisotropy ($\Delta\varepsilon$) of the whole molecule [25–27].

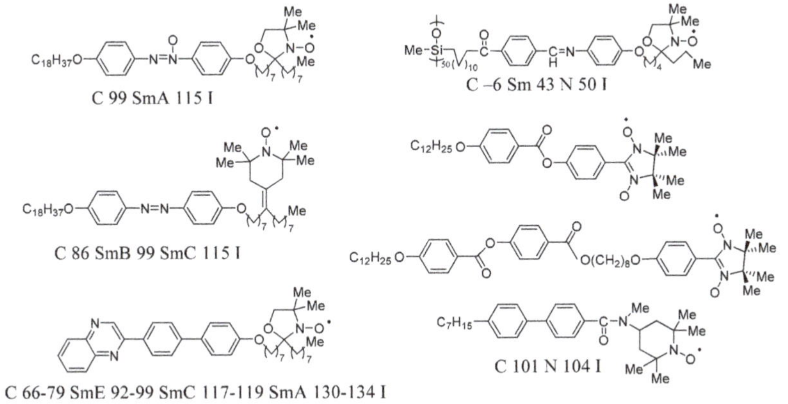

**Figure 2.** Rod-like LC organic mono-radicals synthesized before 2004 and their LC transition temperatures [25–27].

With this situation in mind, since 2004, we have designed and synthesized a series of chiral and achiral all-organic LC nitroxide radicals such as **1–5** having one or two PROXYL (2,2,5,5-tetrassubstituted-1-pyrrolidinyloxyl) units in the core position and, thereby, with a negative dielectric anisotropy ($\Delta\varepsilon < 0$) (Figure 3) [28–44]. Their LC phases and magnetic properties have been fully characterized. Consequently, we could observe a magnetic hysteretic behavior between the heating and cooling processes for all of these LC compounds by a SQUID magnetic susceptibility measurement. The molar magnetic susceptibility ($\chi_M$) always increased at the crystal-to-LC phase transition and this $\chi_M$ increase was always preserved during the cooling process. This unique magnetic hysteretic behavior was referred to as a 'positive magneto-LC effect' [34–36]. Measurement of the magnetic field ($H$) dependence of molar magnetization ($M$) indicates that the observed magnetic hysteresis corresponds to the emergence of superparamagnetic behavior, that is, the partial formation of ferromagnetic domains in the LC phases. These experimental results were contrary to the general knowledge that the possibility of a ferromagnetic LC material had been considered unrealistic due to the inaccessibility of long-range spin-spin interactions between rotating molecules in the LC state at ambient or higher temperatures [26]. However, since the positive magneto-LC effect is observed only for specific PROXYL radicals showing an LC phase (Figure 3), it is suggested that the breaking of time reversal symmetry of radical spins is likely to have a strong connection with the partially broken space symmetry of LC phases of these compounds. Thus, the observed magnetic behavior reminds us of a spin glass state. In fact, for example, considerable enhancement of a positive magneto-LC effect or superparamagnetic behavior was noted by adding a small amount (5 to 20 mol%) of racemic *cis*-diastereomers ($R^*,R^*$)-**3** [a 1:1 mixture of ($R,R$)-**3** and ($S,S$)-**3**] to meso ($R,S$)-**3** as an organic impurity [42]. Such thermal and impurity effects are known to be commonly observed for metallic spin glass materials.

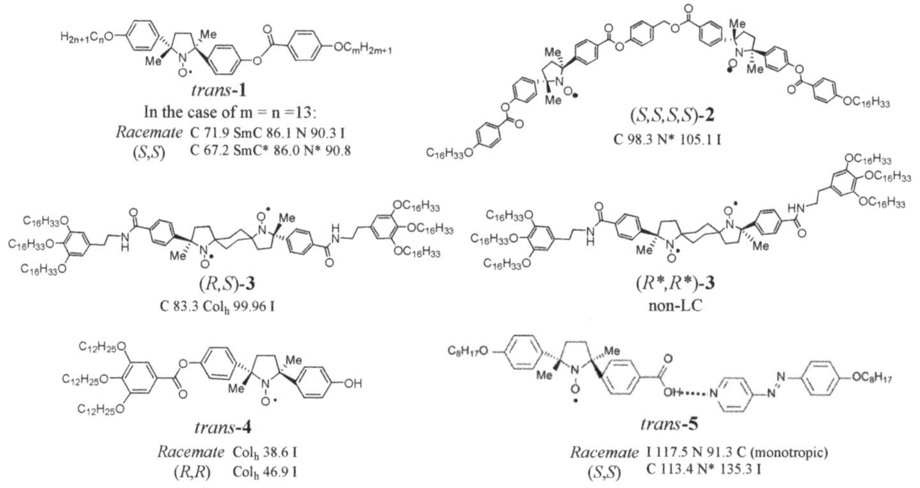

**Figure 3.** Representative PROXYL-based organic radicals synthesized since 2004 and their LC transition temperatures. Rod-like monoradicals trans-**1** [28–40] and biradical (S,S,S,S)-**2** [41], discotic diradical meso (R,S)-**3**, and the non-LC racemic diastereomers (R*,R*)-**3** [42], discotic monoradical trans-**4** [44], and rod-like hydrogen-bonded monoradical trans-**5** [43]. The LC temperatures refer to the heating process, except for monotropic racemice trans-**5**. These compounds except (R*,R*)-**3** showed a positive magneto-LC effect. The individual magnetic properties for **1–4** are discussed in Section 4.

Here, it should be emphasized that the positive magneto-LC effect ($\chi_M$ increase) (i) proved to be independent of the magnetic field-induced molecular reorientation because of the too small molecular magnetic anisotropy ($\Delta\chi$) [45] and (ii) did not result from the contamination by extrinsic magnetic metal or metal ion impurities, as confirmed by the quantitative analysis using inductively coupled plasma atomic emission spectroscopy (ICP-AES) as well as the observation of thermal and impurity effects [42].

Thus far, there have been two theoretical studies to contribute to the elucidation of the mechanism of positive magneto-LC effect. By the quantitative analysis of angular dependence of $g$-values and $\Delta H_{pp}$ (line-width) of EPR spectra and DFT calculations of spin density distribution in the interacting molecules based on the crystal structure of an analogous compound, Vorobiev et al. revealed that an intermolecular spin polarization mechanism operating between neighboring radical molecules rather than the direct through-space interactions between the paramagnetic centers contributes to the occurrence of the positive magneto-LC effect [46]. Furthermore, quite recently, by means of molecular dynamic (MD) simulation and density functional theory (DFT) calculations, Uchida et al. has revealed that the positive magneto-LC effect can originate from the conservation of the memory of spin-spin interactions between magnetic moments, owing to the ceaseless molecular contacts in the LC and isotropic states. They incorporated the molecular mobility effects [44] into the quantitative Thouless-Anderson-Palmer approach, which was proposed to solve the Edwards-Anderson model for spin-glasses with inhomogeneous magnetic interactions [47].

In addition, it is noteworthy that the nonlinear magneto-electric (ME) effect was, for the first time, detected for organic LC materials, i.e., (S,S)-**1** (m = n = 13) showing a ferroelectric LC [chiral SmC (SmC*)] phase [37–39]. By applying an external DC electric field (between +25 V and −25 V) to the (S,S)-**1** confined in a rubbed surface-stabilized liquid-crystal thin (4 μm thick) sandwich cell, the hysteresis loops were observed for the experimental EPR parameters such as the $g$-value, $\Delta H_{pp}$, and $\chi_{rel}$ (relative paramagnetic susceptibility). These experimental results suggest that the interactions between the electric polarization and the magnetic moment in the magnetic LC phase are likely to contribute to the enhancement of a positive magneto-LC effect in weak magnetic fields. In fact, the highly enantiomerically-enriched compounds (*S*,*S*)-**1** always exhibited a larger positive magneto-LC effect than the corresponding racemic ones *trans*-**1** [35,36].

Due to the latest comprehensive review as well as database-like article regarding the preparation, characterization, and magnetic properties of a wide array of all-organic LC radicals synthesized, thus, far, an excellent one entitled "Liquid crystalline derivatives of heterocyclic radicals" by Kaszyński is strongly recommended to refer to Reference [27].

In Section 2, the fundamentals of super-para-magnetism are briefly mentioned. In Section 3, the typical properties of canonical spin glasses composed of diluted magnetic alloys and the non-canonical spin glass-like behavior of the other systems are summarized. In Section 4, the selected important experimental results featuring the magnetic properties of positive magneto-LC effect observed for LC nitroxide radical compounds are summarized. In Section 5, the mechanism of positive magneto-LC effect is proposed and discussed on the basis of the magnetism of the spin glass state and the ME effect.

## 2. Super-Para-Magnetism

Super-para-magnetism is a paramagnetic behavior of magnetic domains or clusters. The collective magnetic spins in a domain are aligned parallel to the external magnetic field, and there is no remanent magnetization of the sample, similarly to paramagnets. However, the spins are ferromagnetically-ordered inside each domain. To behave like paramagnets, domains of ferromagnetically-ordered moments should readily respond to the applied field (Figure 4). Such a situation can be realized in domains with a rotational degree of freedom or domains with a sufficiently low energy barrier of magnetic anisotropy (i.e., coercive force).

The net magnetization of super-para-magnets are expressed by the equation below.

$$M = N\mu L(\mu) = N\mu \times \left(\coth\frac{\mu H}{k_B T} - \frac{k_B T}{\mu H}\right) \quad (1)$$

$N$ denotes a number of domains, $\mu$ does the magnetization summed in a domain, and $k_B$ is the Boltzmann factor. $L(\mu)$ is called Langevin function. In the case that there is distribution in $\mu$, the net magnetization should be estimated by integration over the distribution function $P(\mu)$ [48].

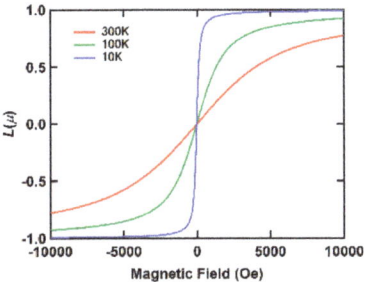

**Figure 4.** Temperature variation in the magnetic field dependence of Langevin function $L(\mu)$ shown in Equation (1) for $\mu = 1000\ \mu_B$, where $\mu_B$ is the Bohr magneton.

Here, it is worth noting that the size effect of the domain is a very important factor, since the anisotropic energy in a domain is expressed by the product of the anisotropic coefficient ($K$) and volume size ($V$) of the domain. The $V$ value, which satisfies $KV < T$, makes the magnetic moments paramagnetic. As the $V$ becomes larger, the anisotropic energy of a domain also increases so that it turns to the coercive force at $KV > T$.

## 3. Spin Glass

### 3.1. What Is a Spin Glass?—General Experimental Features

Spin glass (SG) is a frozen state of spins without long-range periodicity [49]. In the SG, although the long-time average of spins at each site has a non-zero value as seen in a conventional, magnetically-ordered state, both the magnitude and direction are random at each site (Figure 5). The name SG comes from the glass transition in which the atomic positions take random values.

If the SG state is expressed in a thermodynamic way with a spin ($S_i$) of the $i$-th site, the local spontaneous magnetization ($m_i$) has a non-zero value (Equation (2)), even though the site averaged value ($m$) is zero (Equation (3)).

$$m_i = \langle S_i \rangle \neq 0 \tag{2}$$

$$m = \frac{1}{N} \sum_i^N m_i = 0 \tag{3}$$

Here, $N$ denotes the total number of spins. To distinguish the SG state from a paramagnetic state or a long range ordered state, the combination of $m$ and the second order moment ($q$) with respect to the thermal average of $S_i$ is often used (Equation (4)).

$$q = \frac{1}{N} \sum_i^N \langle S_i^2 \rangle \neq 0 \tag{4}$$

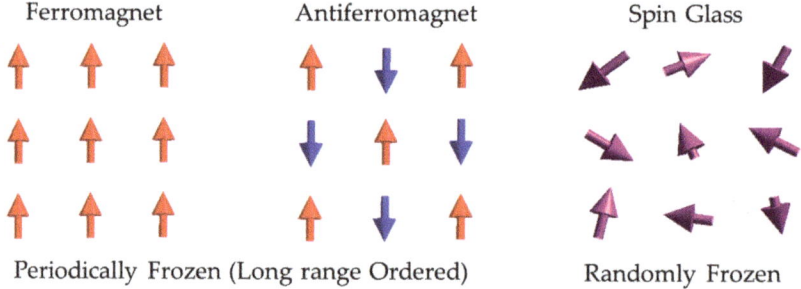

**Figure 5.** An image of the SG state compared to conventional long-range magnetic ordering.

Historically, the first SG state was found in dilute magnetic alloys such as AuFe and CuMn [50,51]. Such systems are called canonical SGs, which are the most studied and established system in the SG theory. Canonical SGs consist of a random distribution of local magnetic impurities in nonmagnetic metal hosts. The features in which canonical SGs show experimentally can be summarized as follows.

1. Hysteresis: The emergence of hysteresis of magnetic susceptibility ($\chi$) under a different temperature control process (Figure 6 [51]).
2. Impurity effect: Non-linear or non-monotonic increase of $\chi$ with the increment of an amount of the magnetic impurity (Figure 6 [51]).
3. Long-time scale dynamics.
4. AC magnetic susceptibility sensitive to the frequency of the external magnetic fields near the spin freezing temperature.
5. A broad peak in magnetic specific heat around the spin freezing temperature.

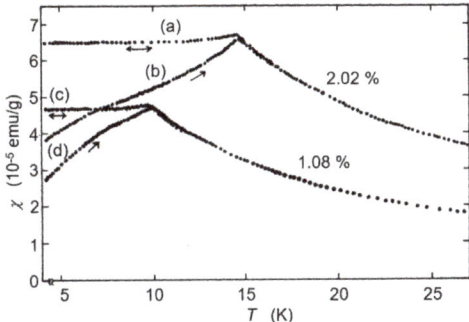

**Figure 6.** Temperature dependence of magnetic susceptibility for the SG alloy CuMn containing the magnetic impurity Mn of 1.08- and 2.02-at.%. After zero field cooling (ZFC, $H < 0.05$ G), the initial susceptibilities (b,d) increased with rising temperature at a field of 5.90 G. The susceptibilities (a,c) were obtained at the same magnetic field, which was applied above $T_{SG}$ before cooling the samples. Data cited from Ref. [51].

The features 1–3 observed for LC nitroxide radicals are described in Section 4. For feature 4, we do not discuss it because of the difficulty in the measurement of AC magnetic susceptibility for LC nitroxide radicals at high temperatures. Likewise, we cannot discuss the feature 5 concerning the LC nitroxide radicals at high temperatures because the contribution by molecular motion is dominant in the specific heat data for these compounds.

*3.2. Theoretical Key Ingredients in Canonical SGs*

The key ingredients in canonical SGs are considered to be randomness, frustration, and competing interactions, which are fully equipped in dilute magnetic alloys [52]. A magnetic moment interacts with other moments by means of Ruderman-Kittel-Kasuya-Yoshida (RKKY) interactions via the conductive electron. The RKKY interactions are expressed as shown in Equation (5).

$$J_{ij} \sim \frac{\cos(2k_F r_{ij})}{r_{ij}^3} \quad (5)$$

where $k_F$ denotes the Fermi energy of the hosting metal and $r_{ij}$ is the distance between two magnetic moments at $i$ and $j$ sites. Since $\cos(2k_F r_{ij})$ is an oscillating function of $r_{ij}$, $J_{ij}$ can have various plus and minus values as a function of $r_{ij}$.

The random distribution of magnetic impurities then results in random magnetic interactions between each impurity site. Under such a network of irregular interactions, some of these local sites are likely to undergo competition of interactions and be unable to settle down in a certain spin state, which satisfies the energy minimum for each interaction. Such a situation is called frustration.

The combination of randomness and frustration gives many stable states in the local energy, so that it would bring the manifold metastable states in the whole system. Accordingly, the free energy ($F$) of the SG states has many local minima (Figure 7) and this is called a multi-valley structure. This characteristic structure inherent in SGs results in various features listed in Section 3.1.

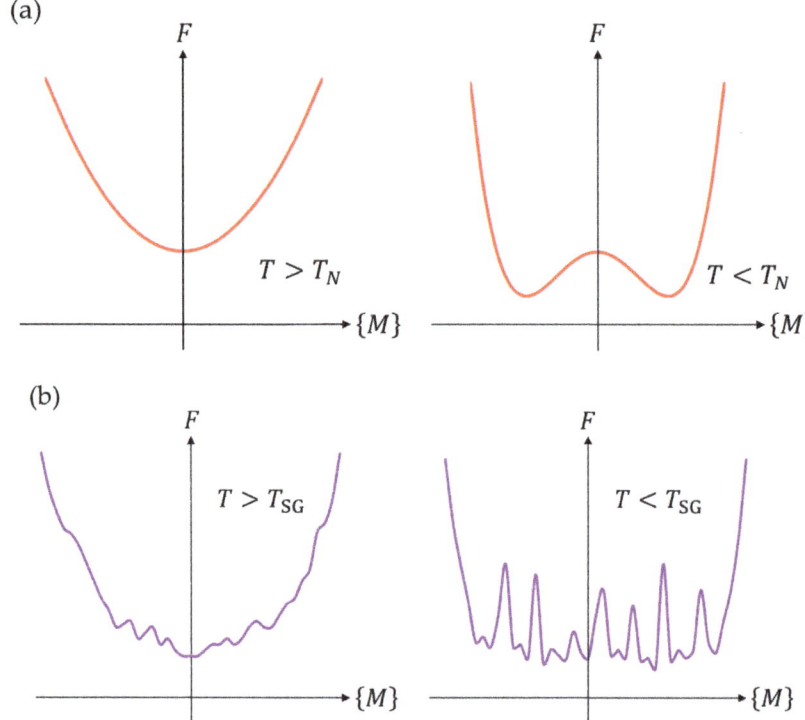

**Figure 7.** Schematic for the temperature variation of free energies ($F$) in order parameter space. The horizontal axes denote the set of magnetic order parameters {$M$} of spins. (**a**) Conventional long range second-order transition at $T = T_N$. (**b**) SG transition at $T = T_{SG}$.

## 3.3. Character of the Magnetism

### 3.3.1. Hysteresis

The hysteresis results from the multiple metastable states permitted in the material, whereas an ordinary long range ordered state has a unique ground state. As shown in Figure 6, the hysteresis in magnetic susceptibility is most noticeable when the material is cooled below the SG transition temperature ($T_{SG}$) from the paramagnetic phase. If the sample is cooled in the absence of external magnetic field [this is called 'zero field cooling' (ZFC)], the net magnetization of the sample generally does not have a large value. This is observed as a cusp near the spin freezing temperature. On the other hand, if the material is cooled in the presence of a non-zero external field [this process is called 'field cooling' (FC)], some of the spins, but not all of them, would align with the magnetic field leading to a ferromagnetic behavior. Thus, in the SG phases, the magnetization after the FC process is larger than that after the ZFC process.

### 3.3.2. Impurity Effect

In canonical SGs, the concentration ($x$) of magnetic impurities in nonmagnetic metal hosts has considerable effects on its magnetic property. In a wider range of $x$ values, each SG state takes one of various ground states. Figure 8 shows a schematic for the relationship between the $x$ values and the resulting magnetic properties. At the low limit of $x$ value, dilute magnetic impurities can be regarded as isolated spins, so that the Kondo effect would work between the conducting electron spins and the isolated spins (Figure 8a). A pair of conducting electron spin and isolated spin adopts a singlet ground state to vanish the susceptibility at low temperatures. With the increasing $x$ value, the average impurity-to-impurity distance becomes shorter enough to induce RKKY interactions, forming an SG state (Figure 8b). When the $x$ value exceeds ~10 at.%, some of the impurities exist in the nearest neighbor lattice points so that such locally high-concentrated spin clusters result in the coexistence of the SG state and ferro-magnetic or anti-ferro-magnetic domains (Figure 8c). These clusters are most likely to behave like super-para-magnets, and also exhibit a hysteresis in the temperature dependent of magnetic susceptibility, as a sort of SG behavior of the clusters. Such behavior is often called a 'cluster glass.' A further increase in the $x$ value makes the ferro-magnetic or anti-ferro-magnetic domains larger, finally, giving an almost ordered state (Figure 8d).

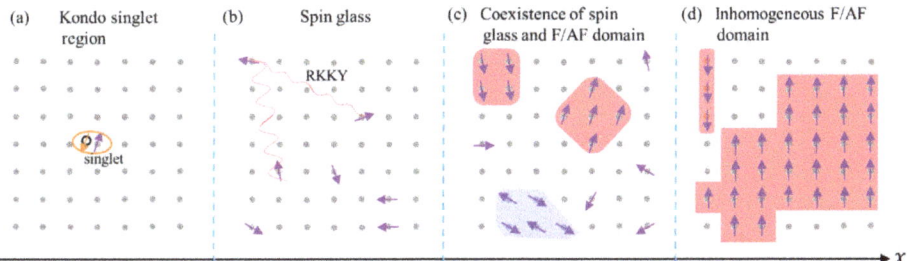

**Figure 8.** Schematic for the effect of the concentration ($x$) of magnetic impurity on the magnetic properties of alloy systems. See text for details. The grey circles represent the lattice points.

3.3.3. Long Time Scale Dynamics

Long time scale dynamics is another characteristic aspect of canonical SGs. Since there are many metastable states, hopping over the energetic barrier to other states is possible. Such metastable states are realized by a rapid change in external stimuli such as temperature or magnetic field around the SG phase transition. Rapid cooling from the temperature above $T_{SG}$ to that below $T_{SG}$ compels the system to take a certain metastable state without immediate relaxing to the lowest energy. The relaxation to the lowest energy state is called the 'aging phenomenon,' which takes a typical time of the order of $10^2$~$10^4$ s or more. Thus, the aging phenomenon can be followed by statistical DC magnetization measurements. For example, after keeping the SG state at a high temperature region in the SG phase under the ZFC conditions, the magnetization at the same temperature gradually increases with the elapse of time by applying an external magnetic field (Figure 9) [53]. After a sufficiently long time, the magnetization gets close to the value, which is observed after the FC process.

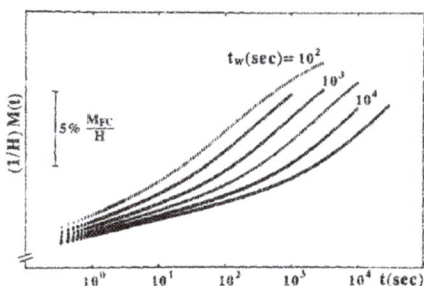

**Figure 9.** The relaxation of magnetization divided by the external field after waiting for $t_w$ at zero field. The data are obtained at the temperature of $0.91 T_{SG}$ and under the external field of 0.8 G. The horizontal axis denotes the elapse of time in the measurement. Reprint with permission [53]: Copyright 2020 AIP Publishing.

## 3.4. Non-Canonical SG-Like Behavior

Besides canonical SG alloy systems, some of the magnetic features summarized in Section 3.1 are observed in a variety of magnetic systems, which are often called 'SG-like behavior' [52]. The features can show up even if some of the key ingredients (i.e., randomness, frustration, and competing interactions) are weak or missed in the systems. In general, neither the diluted conditions of spins nor RKKY interactions are needed for the realization of SG-like behavior. The SG-like behavior is reported in many non-diluted magnets with exchange/super-exchange interactions between the nearest neighbor spins. It seems that the SG-like behavior tends to arise if only there exist some factors of randomness and/or inhomogeneity in the interaction network. They have been often found in atom-doped/substituted systems, grain boundary, systems with site-mixing or random local distortion, and more [52].

## 4. Positive Magneto-LC Effect

Here, the unique magnetism of LC nitroxide radicals, which is regarded as SG-like behavior, is described. Although the solid-state picture of usual SG-like magnets is not straightforwardly applied to the liquid crystal systems, we actually observed the SG-like features 1–3 (hysteresis, impurity effect, and long-time scale dynamics) listed in Section 3.1. In Section 4.1, our discovery of the positive magneto-LC effect, which refers to a magnetic hysteresis (i.e., thermal effect) associated with the emergence of superparamagnetic behavior in the LC state, is demonstrated. The substantial enhancement of superparamagnetic behavior by impurity effect or controlling the magnetic inhomogeneity is mentioned in Section 4.2. An example of long-time scale dynamics is presented in Section 4.1.3. In Section 4.3, the ME effect observed for the first time in the ferroelectric LC phase of the LC nitroxide radical is explained in detail.

### 4.1. Magnetic Hysteresis by a Thermal Effect

#### 4.1.1. Monoradical trans-1

In 2008, for the first time, we observed a magnetic hysteresis for nitroxide monoradical (S,S)-1 (m = n = 13) showing enantiotropic (reversible) SmC* and N* phases (Figure 3) by SQUID magnetic suscetibiliy measurement (Figure 10a) [34,35]. In the first heating process from the initial crystalline phase, the molar magnetic susceptibility ($\chi_M$) increased at the crystal–to-SmC* phase transition. An additional $\chi_M$ increase was noted at the SmC*-to-N* and N*-to-isotropic phase transitions. Interestingly, the overall 12% $\chi_M$ increase at the phase transitions in the heating process was preserved during the cooling one from the isotropic phase and even in the second heating one, and the $\chi_M$ in the first cooling and second heating processes never followed the curve drawn in the first heating process. Such magnetic hysteretic behavior, an unusual $\chi_M$ increase at the crystal–to-LC phase transition in the first heating process, was referred to as 'positive magneto-LC effect,' which was observed for all of the LC nitroxide radical compounds 1~5 with full reproducibility. Therefore, the $\chi_M$ in the first cooling process are expressed by the equation below.

$$\chi_M = \chi_{para} + \chi_{TIM} + \chi_{dia} \tag{6}$$

where $\chi_{para}$ represents the paramagnetic susceptibility of radical spins, which follows the Curie-Weiss law (Equation (7)). The $C$ and $\theta$ are Curie constant and Weiss temperature, respectively. For the mono-radical nitroxide compounds examined, the values of $C$ were close to 0.375 corresponding to the $S = 1/2$ system.

$$\chi_{para} = \frac{C}{T - \theta} \tag{7}$$

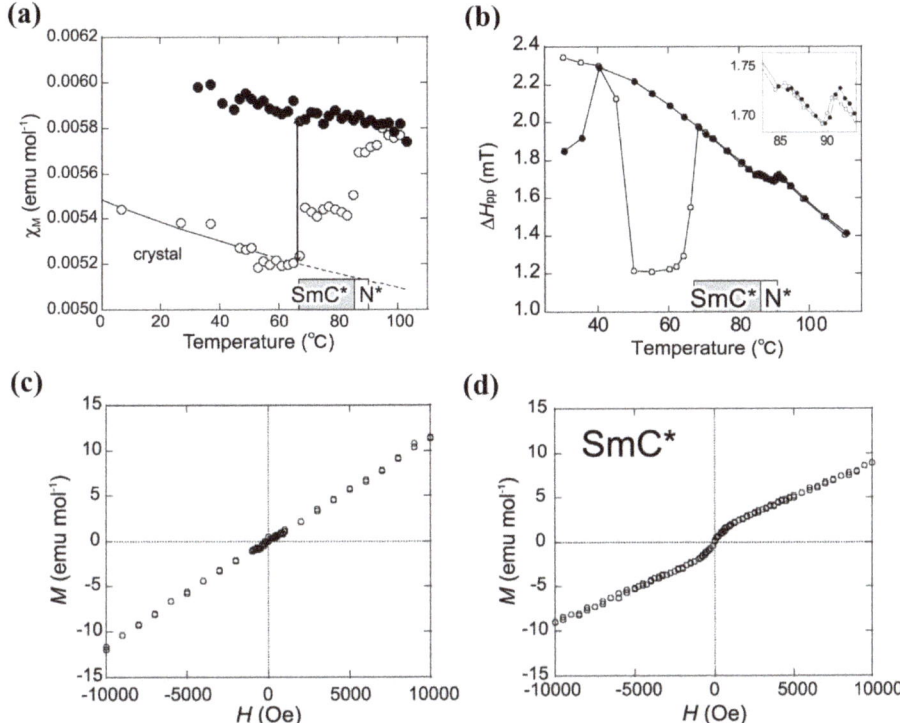

**Figure 10.** Magnetic hysteresis observed for (*S,S*)-**1** (m = n = 13) by a thermal effect. Temperature (*T*) dependence of (**a**) molar magnetic susceptibility ($\chi_M$) measured at 0.05 T and (**b**) EPR line-width ($H_{pp}$) measured at around 0.34 T in the first heating and cooling processes, and magnetic field (*H*) dependence of molar magnetization (*M*) measured at (**c**) −73 and (**d**) 77 °C in the first heating process. In panels a and b, open and filled circles represent the first heating and cooling processes, respectively. Data cited from Ref. [35].

The $\chi_{TIM}$ indicates the temperature-independent magnetic susceptibility corresponding to the $\chi_M$ increase at the crystal-to-LC phase transition, while the $\chi_{dia}$ represents the temperature-independent diamagnetic susceptibility, which arises from the orbital current and can be calculated using Pascal's constant for each atom and bond. Since the sum of $\chi_{TIM}$ and $\chi_{dia}$ can be experimentally obtained as the temperature-independent component by $\chi_M$-$T^{-1}$ plots, according to Equations (6) and (7), the $\chi_{TIM}$ value can be derived.

We observed a nonlinear relationship between the applied magnetic field (*H*) and the observed magnetization (*M*) in the chiral and achiral LC phases of *trans*-**1** (m = n = 13) by SQUID magnetometry (Figure 10d). Unlike the Brillouin function, which ordinary paramagnetic radical compounds display, a steep increase was observed at low magnetic fields (~0.1 T) and an oblique linear response was noted at higher magnetic fields. This result implied the emergence of superparamagnetic behavior in the LC phase.

Such spin glass (SG)-like hysteretic or superparamagnetic behavior was also seen for the other LC nitroxide radicals listed in Figure 3. In contrast, non-LC *trans*-**1** analogues bearing much shorter alkyl chains failed to show the magnetic hysteresis because of their common paramagnetic properties. It is worth noting that, with respect to LC *trans*-**1** and the LC derivatives, enantiomerically-enriched samples always exhibited a larger positive magneto-LC-effect (or magnetic hysteresis) at the crystal-to-LC phase transition than the corresponding racemic samples [35,36]. Thus, the positive magneto-LC effect is a unique property inherent in these LC nitroxide radicals, which suggested some relationship between the spatial symmetry of LC phases and the positive magneto-LC effect.

Here, we should stress that such a magnetic hysteresis or superparamagnetic behavior did not result from the magnetic field-induced molecular reorientation in the LC and isotropic phases. The contribution of overall magnetic anisotropy ($\Delta\chi$) is, at most, ~$10^{-5}$ emu/mol [26], while the observed $\chi_M$ increases at the crystal-to-LC phase transition points amount to much more than $10^{-4}$ emu/mol in various LC nitroxide radicals. It is also important to examine the possibility of contamination by ferromagnetically-ordered metal or metal ion spins for such nonlinear magnetization curves [54]. Consequently, we could exclude this possibility by the quantitative analysis using inductively coupled plasma–atomic emission spectroscopy (ICP-AES), verifying 0.5~1.3 ppm for the iron content and 0.3~0.6 ppm for the nickel content for LC nitroxide radicals [42]. These values are less than one-tenth of the observed $\chi_M$ increase for the individual compounds. Furthermore, these two possibilities (molecular reorientation and contamination) were ruled out by observing a large 'impurity effect' explained in Section 4.2.

The measurements of the AC magnetic susceptibilities are necessary to investigate the long time-scale spin dynamics of super-para-magnetic components in LC nitroxide radicals. However, we could not obtain these data due to the difficulties with the measurement at high temperatures. This is our future subject.

4.1.2. Biradical (S,S,S,S)-**2**

In order to investigate the relationship between the positive magnet-LC effect and the number of spins, we prepared the biradical (S,S,S,S)-**2** (>99% ee) showing an enantiotropic chiral nematic (N*) phase (Figure 3) in which the two radical spins do not interact. The $\chi_M$ increase at the crystal-to-LC phase transition in the first heating process was $0.8 \times 10^{-3}$ emu/mol (38%) (Figure 11), which was slightly larger than those of LC mono-radicals such as (S,S)-**1**. This $\chi_M$ increase was maintained during the cooling process, similarly to the case of (S,S)-**1**.

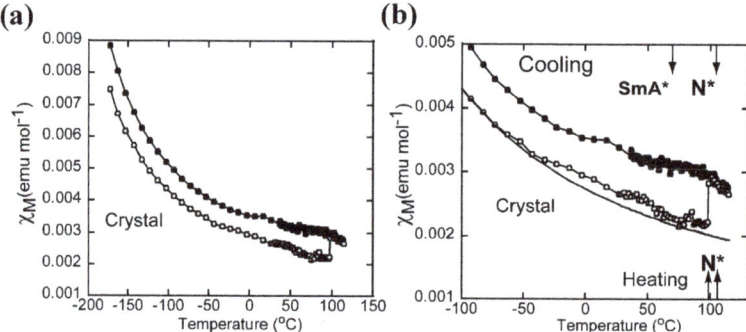

**Figure 11.** Temperature dependence of $\chi_M$ for (S,S,S,S)-**2** by SQUID magnetometry at a field of 0.05 T in the temperature range of (**a**) −170 to +120 °C and (**b**) −100 to 120 °C (magnification). Open and filled circles represent the first heating and cooling processes, respectively. The solid line indicates the Curie-Weiss fitting curve. The LC transition temperatures are shown by arrows in the lower and upper sides in panel b. (Reprint with permission [41], Copyright 2020 The Royal Society of Chemistry).

4.1.3. Diradical (R,S)-**3**

The non-π-delocalized achiral meso diradical (R,S)-**3** (Figure 3) showed an enantiotropic hexagonal columnar (Col$_h$) phase in which each disc was composed of a hydrogen-bonded molecular trimer (Figure 12) [42]. This compound displayed both thermal and impurity effects. Here, only the thermal effect is stated and the impurity effect is described in Section 4.2.

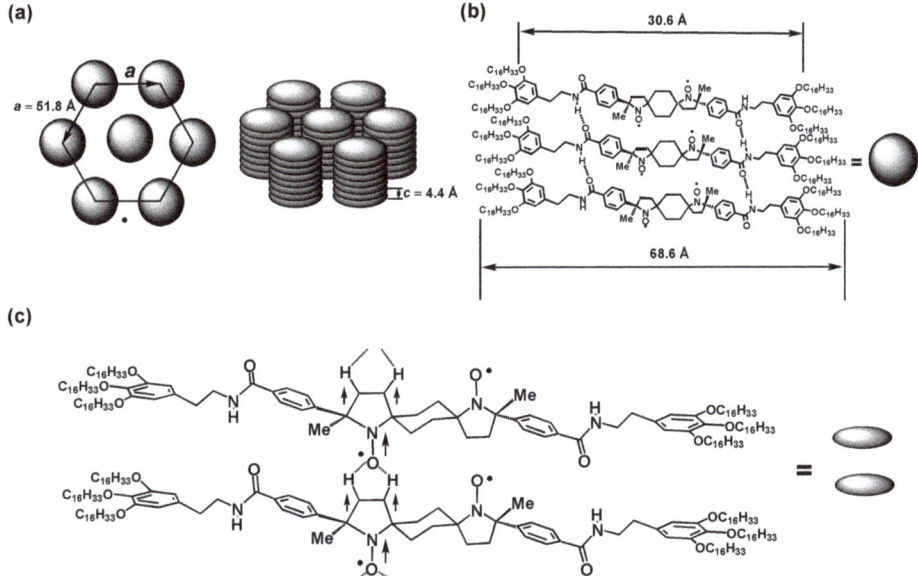

**Figure 12.** The Col$_h$ LC structure of (R,S)-3. (**a**) Hexagonal lattice and the columnar structure, (**b**) one of possible trimer structures formed by hydrogen bonds in one disc layer and (**c**) intermolecular spin polarization via CH/O interactions between two (R,S)-3 molecules in the neighboring disc layers expected in the magnetically inhomogeneous LC domains. Data cited from Ref. [42].

Measurement of the temperature dependence of $\chi_M$ for the pure (R,S)-3 indicated that (i) intramolecular anti-ferro-magnetic interactions with a singlet ground state between the two spins dominated below 100 K and (ii) the Curie-Weiss fitting could be implemented between 100 and 250 K to give the average $\theta$ value of nearly zero (Figure 13). Although the pure (R,S)-3 showed neither the $\chi_M$ increase nor the superparamagnetic M-H behavior in the first heating process at 0.05 T, distinct $\chi_M$ increase and superparamagnetic M-H behavior were observed in the first cooling from the isotropic phase and in the second heating run at 0.05 T.

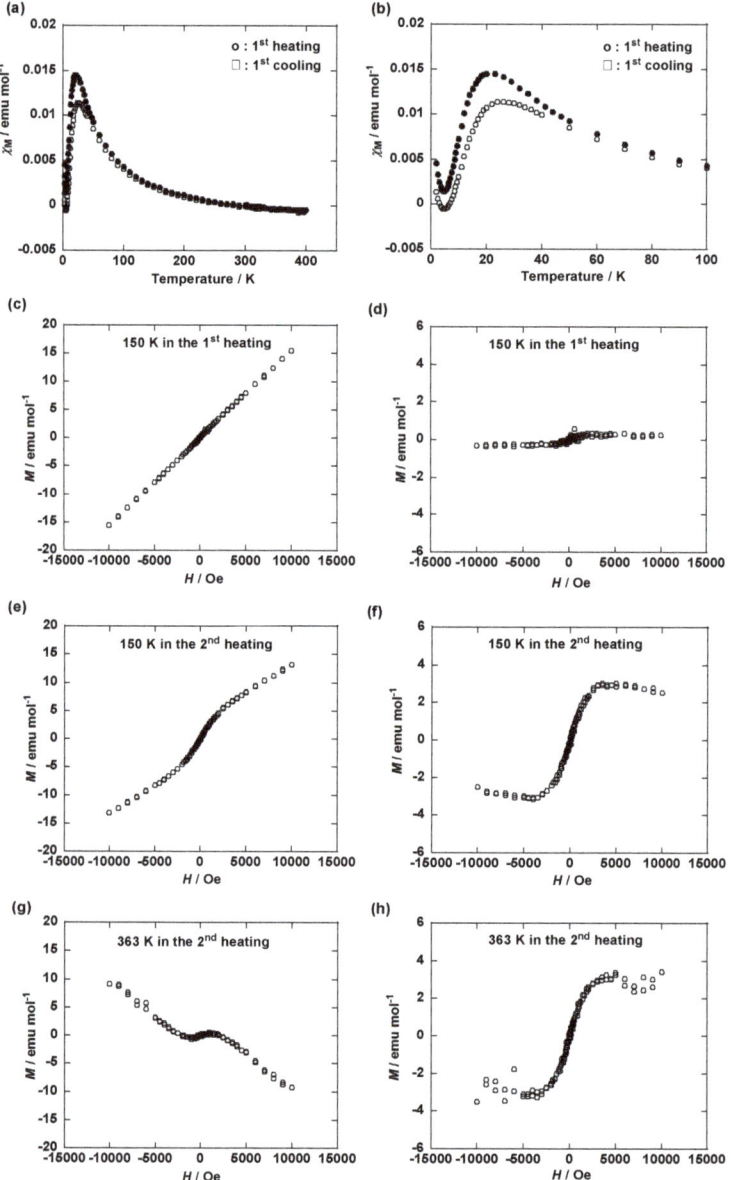

**Figure 13.** The temperature dependence of $\chi_M$ (**a**,**b**) at 0.05 T and the $H$ dependence of $M$ (**c–h**) for pure (R,S)-**3** (**a**) in the first heating and cooling processes (**b**) by magnification (between 2 and 100 K) of panel a, (**c**) in the solid at 150 K by the first heating from 2 K, (**d**) after subtraction of the oblique linear base line in panel c, (**e**) in the solid at 150 K by the second heating after the first heating up to 400 K followed by cooling to 2 K at 0.05 T, (**f**) after subtraction of the oblique linear base line in panel e, (**g**) in the Col$_h$ phase at 363 k by the second heating, and (**h**) after subtraction of the oblique linear base line in panel g. Open and filled circles in panels a and b represent the first heating and cooling processes. Data cited from Ref. [42].

Next, to learn whether a helical columnar structure can be formed in the Col$_h$ phase of achiral (R,S)-**3**, the temperature-resolved second harmonic generation (TR-SHG) microscopy was performed with an excitation wavelength of 1200 nm and SHG emission at 600 nm in the presence of a magnetic field (<0.5 T) (Figure 14). We observed a weak SHG emission from the surface of the initial crystal phase at room temperature. On the other hand, when the solid sample was heated to 90 °C and the resulting Col$_h$ phase was annealed at the same temperature, the SHG intensity increased with an elapse of time over 10 min. This observation indicates a gradual growth of domains with an inversion symmetry broken state in a magnetic field. The growth of this domain structure with broken inversion symmetry is assumed to be responsible for the generation of a positive magneto-LC effect, since this is quite analogous to the long-time relaxation in a magnetic field observed for the SG state.

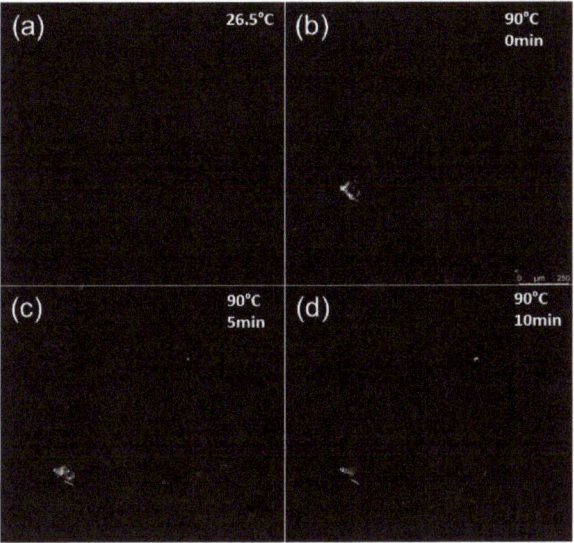

**Figure 14.** TR-SHG images for (R,S)-**1**. (**a**) No SHG signal was detected for solid I at 26.5 °C. Then, the solid I was heated at a rate of 10 °C min$^{-1}$ and the resulting LC phase was annealed at 90 °C. The SHG images were recorded at (**b**) 0, (**c**) 5, and (**d**) 10 min after the beginning of annealing. The observed SHG intensity was increased with the elapse of time. Data cited from Ref. [42].

4.1.4. Monoradical *trans*-**4**

The non-π-delocalized racemic *trans*-**4** and (R,R)-**4** (Figure 3), which formed intermolecular hydrogen bonds between the NO and OH groups in the condensed phase displayed an enantiotropic Col$_h$ phase at room temperature and formed LC glasses at lower temperatures. The SQUID magnetometry revealed that the positive magneto-LC effect was gradually increased through the LC glass-to-LC-to-Iso phase transition sequence in the first heating process at both 0.05 and 0.5 T (Figure 15) [44]. This result suggests that molecular mobility is one of the origins of the positive magneto-LC effect, which is consistent with the conclusions by means of MD simulation and DFT calculations so that the positive magneto-LC effect can originate from the conservation of the memory of spin-spin interactions between magnetic moments owing to the ceaseless molecular contacts (Figure 16) [47].

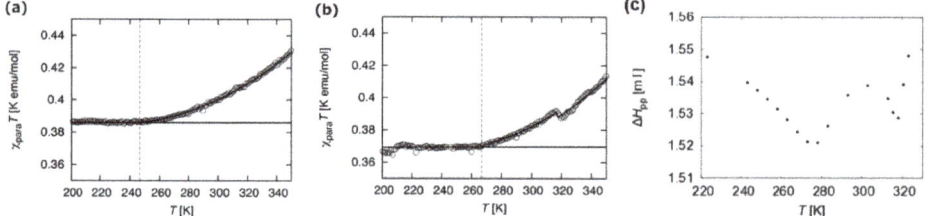

**Figure 15.** Temperature dependences of the $\chi_{para}T$ plot for (**a**) racemic *trans*-**4** and (**b**) (*R,R*)-**4** at 0.5 T by SQUID magnetometry, and of (**c**) EPR $\Delta H_{pp}$ for racemic *trans*-**4** at around 0.33 T. The circles denote the experimental data in the heating process, and the horizontal solid lines indicates the Curie-Weiss fitting curves for the LC glass state. The vertical broken lines denote the estimated glass state-to-LC phase transition temperatures, and the solid lines on the circles to the right of the broken lines represent the fitting curves for LC and isotropic phases. (Reprint with permission [44]: Copyright 2020, American Chemical Society).

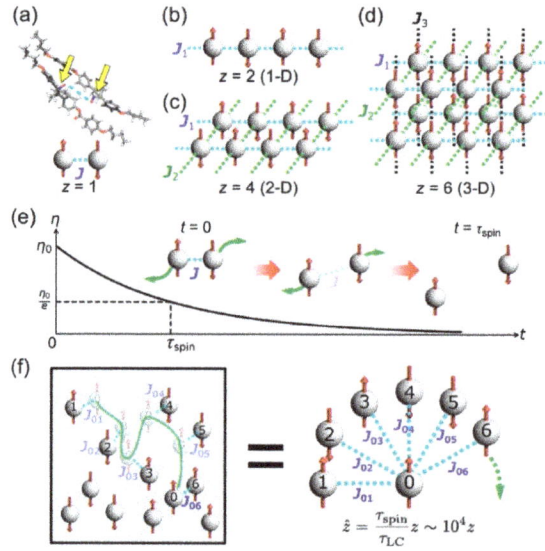

**Figure 16.** High-frequency exchange of interacting pairs detectable as an increase of the coordination number. (**a**) Molecules interact with each other in fluids. Because LC-NR molecules are bulky, each of them usually interacts with only one other molecule. In this case, the coordination number z is 1. (**b–d**) Generally speaking, z values for one, two, and three-dimensionally interacting NR crystals are 2 (**b**), 4 (**c**), and 6 (**d**), respectively. (**e**) Schematic illustration of the attenuation of incremental magnetic field ($\eta$) for a certain pair of molecules. The horizontal axis denotes the time elapsed since the interaction occurs. After a molecule interacts with another molecule, it moves and the interaction disappears. The effect of the exchange magnetic field disappears after time constant $\tau_{spin}$. (**f**) Each molecule changes the partner of the interaction many times before disappearance of the effect of the exchange magnetic field. Each molecule interacts with many molecules as if the coordination number substantially increased ($\hat{z}$). (Reprint with permission [47]: Copyright 2020, American Chemical Society).

### 4.2. Impurity Effect—Diradical (R,S)-3

It was the discovery of an impurity effect that was required to demonstrate the similarity of the positive magneto-LC effect to the emergence of superparamagnetic domains in the SG state. We implemented the $\chi_M$-T measurement for two samples with different ratios (80:20 and 95:5) of (*R,S*)-**3** and racemic cis-diastereomers (R*,R*)-**3** (Figure 3) [42]. These mixed samples could still show an Col$_h$ phase. However, pure (R*,R*)-**3** and the sample with a ratio of 50:50 displayed no LC phase. As shown in Figure 17, appreciable and substantial $\chi_M$ increases were noted during the crystal-to-Col$_h$-to-Iso phase transition sequence for the samples with the ratios of 95:5 and 80:20, respectively. The $\chi_M$ increase was $1 \times 10^{-3}$ emu/mol for the former sample and $5 \times 10^{-3}$ emu/mol for the latter one. As shown in Figure 13, the hysteresis was not observed for pure (R,S)-**3** within the experimental error limits in the first heating process. In other words, the content of superparamagnetic domains increased by adding (R*,R*)-**3** to (R,S)-**3** as an organic impurity. Furthermore, this organic impurity effect convinced us that the positive magneto-LC effect is intrinsic in LC nitroxide radicals listed in Figure 3 because the addition of diastereomers would not change the total amount of contaminating magnetic metal ions, if any.

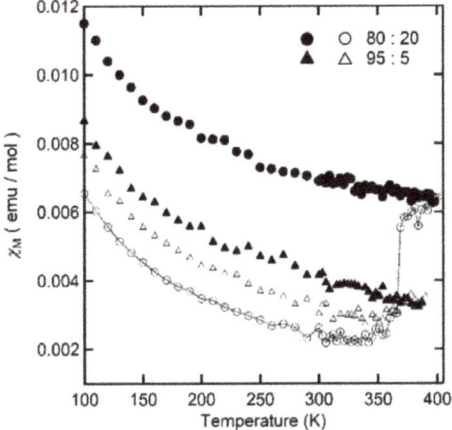

**Figure 17.** The temperature dependence of $\chi_M$ for the two samples with a different ratio (circle for 80:20 and triangle for 95:5) of (R,S)-**3** and (R*,R*)-**3** at 0.05 T in the temperature range between 2 and 400 K. Open and filled legends represent the first heating and cooling processes, respectively.

As shown in Figure 18, the H dependence of the superparamagnetic component extracted from the M-H plots for the mixed sample of (R,S)-**3** and (R*,R*)-**3** (80:20) was well fitted by Langevin function (Equation (1)), indicating the positive magneto-LC effect that originates from the generation of superparamagnetic domains in the LC phase. The amount of super-para-magnetic domains was enhanced compared to the one in the pure sample of (R,S)-**3** shown in Figure 13 h.

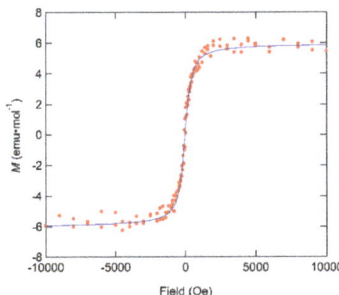

**Figure 18.** The curve fitting by Langevin function (Equation (1)) for the superparamagnetic domains extracted from the M-H plots at 363 K for the mixed sample of (R,S)-**3** and (R*,R*)-**3** (80:20).

Thus, the origin of such unique magnetic properties observed in the LC and isotropic phases of achiral (R,S)-**3** in the absence or presence of the racemic diastereomers (R*,R*)-**3** as the impurity can be rationalized mesoscopically in terms of (i) the preferential formation of SG-like inhomogeneous magnetic domains consisting of super-para-magnetic domains surrounded by para-magnetic spins in the LC phase, (ii) the gradual growth of the size and/or number of super-para-magnetic domains by thermal processing, and (iii) preservation of the overall increased super-para-magnetic domains in the solid phase by cooling until cryogenic temperatures, in weak magnetic fields (<0.1 T) (Figure 19).

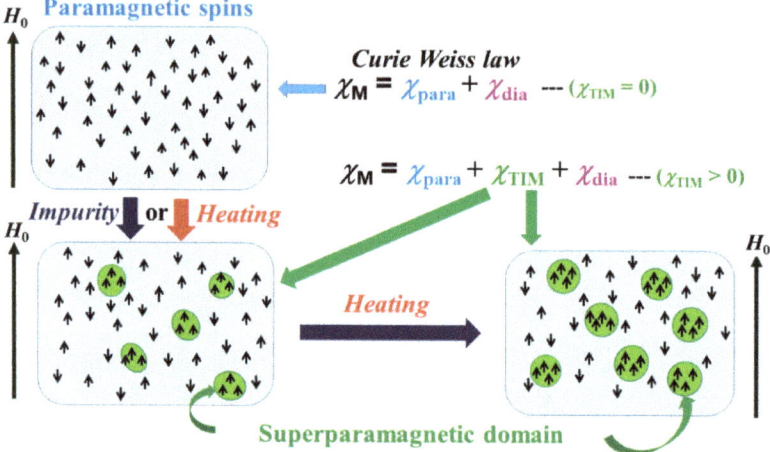

**Figure 19.** Schematic of super-para-magnetic domains in the LC and isotropic phases of (R,S)-**3**. The formation of super-para-magnetic domains surrounded by para-magnetic spins by heating or in the presence of impurity (R*,R*)-**3**, and additional growth of the size and/or number of the domains by heating.

*4.3. Magneto-Electric (ME) Effect*

The magneto-electric (ME) effect had been observed only for inorganic multiferroic materials such as $YMnO_3$, $TbMnO_3$, and so forth, which showed both ferroelectric and magnetic order (ferromagnetism or anti-ferromagnetism) at cryogenic temperatures [17–20].

In this context, it was expected that a unique ME effect might occur in the second-harmonic-generation (SHG)-active [55] and ferroelectric LC (SmC*) phase of the (S,S)-**1** (m = n = 13), which displayed both an excellent ferroelectricity [$P_S$ = 24 nC/cm$^2$, $\tau_{10\text{-}90}$ = 213 µs, $\theta$ = 29°, $\eta$ = 73.0 m Pa s at 74 °C by the triangle wave method] in a rubbed surface-stabilized liquid-crystal sandwich cell (4 µm thick) [29] and a positive magneto-LC effect (super-para-magnetic interactions) in the bulk SmC* phase [35]. Such an assumption was the case. The temperature dependences of relative paramagnetic susceptibility ($\chi_{rel}$), g-value, and $\Delta H_{pp}$ were measured by EPR spectroscopy using the thin sandwich cell into which the most appropriate sample of (S,S)-**1** (m = n = 13, 65% ee) was loaded (Figure 20).

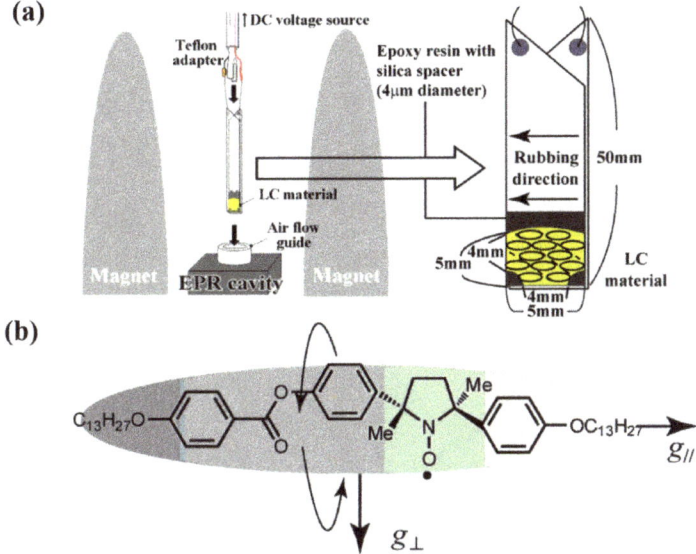

**Figure 20.** (a) Experimental setup to monitor the variable-temperature or electric field dependent EPR spectra of (*S*,*S*)-**1** (m = n = 13, 65% ee) confined in a long 4-μm thin sandwich cell. (b) Principal axes of inertia and g-values of the *trans*-**1** molecule. (Reprint with permission [37]; Copyright 2020 the Royal Society of Chemistry).

Since the magnetization measurement of the sample in the LC cell by SQUID magnetometry was technically difficult, $\chi_{para}$ was derived from the Bloch equation (Equation (8)) [56] by using the EPR parameters, such as $g$, $\Delta H_{pp}$, and maximum peak height ($I'm$ and $-I'm$) as described earlier [35,36],

$$\chi_{para} = 2\mu_B g I'_m \Delta H^2_{pp}/\left(\sqrt{3}\, h\nu H_1\right) \tag{8}$$

where $\mu_B$ is the Bohr magneton, $h$ is Planck's constant, $\nu$ is the frequency of the absorbed electromagnetic wave, and $H_1$ is the amplitude of the oscillating magnetic field. The relative paramagnetic susceptibility ($\chi_{rel,T}$) in the absence of an electric field is defined as:

$$\chi_{rel,T} = \chi_{para}/\chi_0 \tag{9}$$

where $\chi_0$ is the standard paramagnetic susceptibility at 30 °C in the heating process. First, the generation of positive magneto-LC effect and ferroelectric switching in the liquid crystal cell was confirmed. By measuring the temperature dependence of $\chi_{rel,T}$ under the conditions in which the applied magnetic field was parallel (Configuration A) and perpendicular (Configuration B) to the cell surface and the rubbing direction in the absence of an electric field, considerable $\chi_{rel,T}$ increases (ca. 40%) together with large $\Delta H_{pp}$ ones, which were noted at the crystal-to-LC phase transition in both cases (Figure 21). The ferroelectric switching at 25 V was verified by polarized optical microscopy in the absence of a magnetic field. The bright fan-shaped texture at −25 V and the dark one at +25 V were distinctly observed [37].

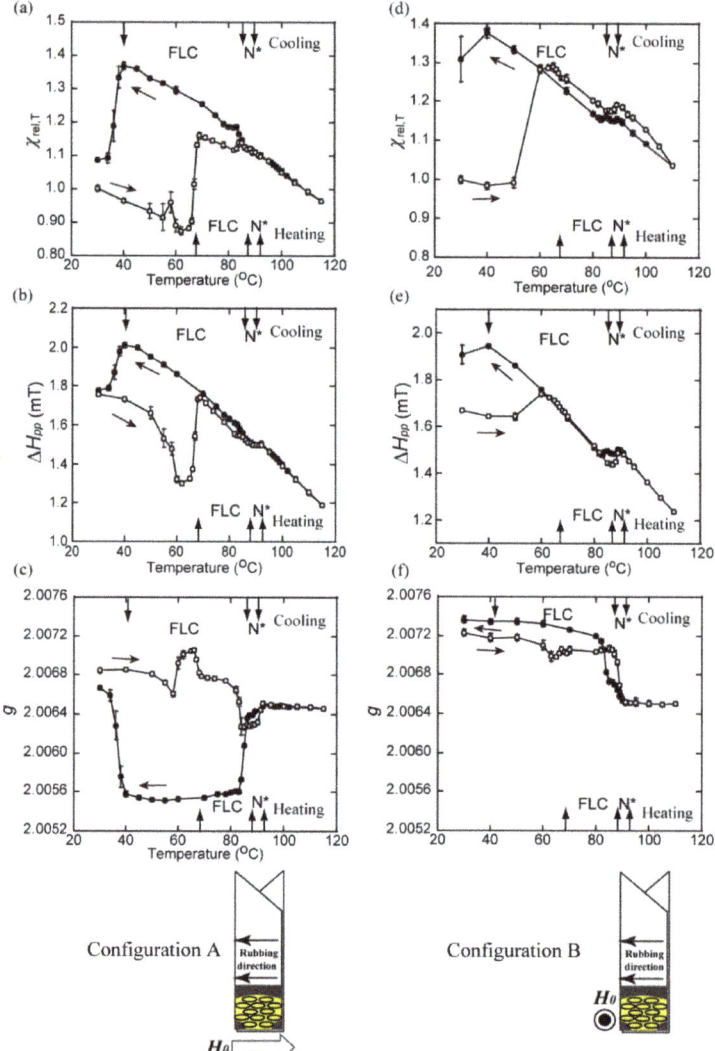

**Figure 21.** Temperature dependence of (a and d) $\chi_{rel,T}$, (b and e) $\Delta H_{pp}$ and g-value (c and f) for (S,S)-**1** (m = n = 13, 65% ee) confined in a thin rubbed sandwich cell by EPR spectroscopy at a magnetic field of around 0.33 T. The magnetic field was applied (**a–c**) parallel and (**d–f**) perpendicular to the rubbing direction. The LC transition temperatures in the heating and cooling runs are shown in the lower and upper sides of the panels, respectively. (Reprint with permission [37]; Copyright 2020 the Royal Society of Chemistry).

The difference in temperature dependence of g-values along two direction shows that the magnetic susceptibility of the (S,S)-**1** clearly has an anisotropy with respect to the molecular axis. Since the Curie constant revealed $S = 1/2$ nature of the system, the anisotropy stems from high-order perturbation term of exchange interactions with respect to spin-orbit interactions, such as Dzyaloshinkii-Moriya (DM) interactions. Moreover, the temperature dependence in the case of Configuration A (Figure 21c) shows anomalies in the vicinity of ferroelectric LC (FLC) transition. That implies some changes in magnetism in connection with ferroelectricity, while little anomaly was found in the case of configuration B (Figure 21f).

On the other hand, the relative paramagnetic susceptibility ($\chi_{rel,E}$) in the presence of an electric field is defined as:

$$\chi_{rel,E} = \chi_{para}/\chi_1 \tag{10}$$

where $\chi_1$ is the standard paramagnetic susceptibility at the initial potential of +25 V and at 75 °C. Next, the electric field dependences of $\chi_{rel,E}$, $\Delta H_{pp}$, and g-value were plotted for the (S,S)-1 (m = n = 13, 65% ee), displaying each hysteresis between +25 V and −25 V when the magnetic field was applied only perpendicularly to the electric field and parallel to the rubbing direction (Figure 22).

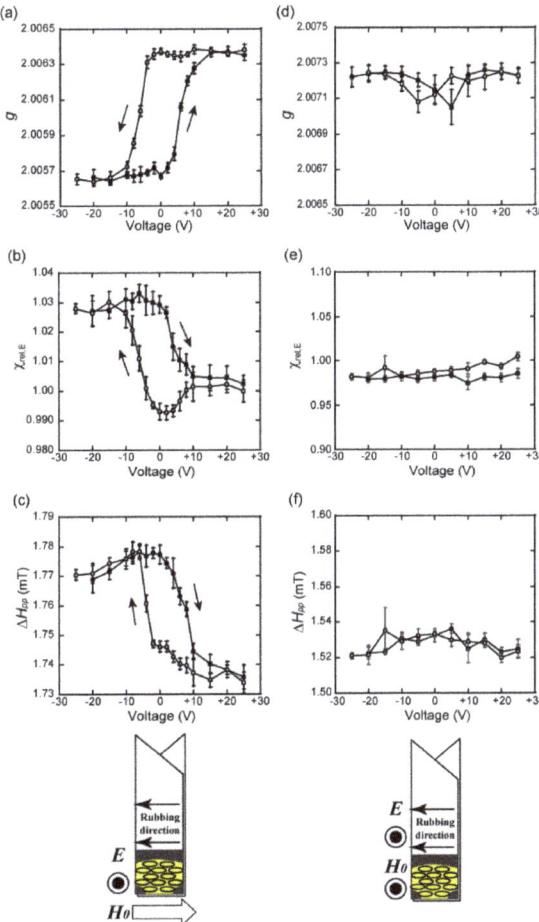

**Figure 22.** Electric field dependence of g-value, $\chi_{rel,E}$, and $\Delta H_{pp}$ for the FLC phase of (S,S)-1 (m-n = 13, 65% ee) confined in a thin rubbed sandwich cell at 75 °C by EPR spectroscopy at a magnetic field of around 0.33 T. The magnetic field was applied (**a–c**) perpendicularly to the electric field and parallel to the rubbing direction and (**d–f**) parallel to the electric field and perpendicular to the rubbing direction. Open and filled circles represent the application of electric fields from +25 V to −25 V and from −25 V to +25 V, respectively. (Reprint with permission [37]; Copyright 2020 the Royal Society of Chemistry).

It is considered that the molecules with the spontaneous electric polarization is inversed by sweeping the external electric field. Therefore, it is noteworthy that the spin inversion accompanies the inversion of the molecule, which can be regarded as a nonlinear ME effect. Since the usual paramagnetic components cannot show such a nonlinear hysteresis loop, the hysteresis indicates the spin flipping of superparamagnetic domains. This result clearly shows the predominance of the electric field over a magnetic field in controlling the super-para-magnetic behavior, since the spin flipping by the electric field occurs even under a large magnetic field of 0.33 T with a constant direction used for this EPR measurement. The direction-of-magnetic field dependence of the $g$-value, $\chi_{rel,E}$, and $\Delta H_{pp}$ indicates that the super-para-magnetic domains are directed almost perpendicularly to the spontaneous electric polarization. For both the nonlinear ME effect and the direction-of-magnetic field dependence, the high-order perturbation term of exchange interactions with respect to spin-orbit interactions is usually necessary for this $S = 1/2$ system.

In summary of Section 4, the unique magnetic properties observed for LC nitroxide radicals are characterized in connection with SG-like magnetic features (see Section 3.1) as follows:

1. The hysteresis observed by measurement of the temperature dependence of magnetic susceptibility (Figures 10a, 11 and 13) corresponds to the emergence of superparamagnetic domains in the LC and isotropic phases.
2. The superparamagnetic domains grow in the LC phases under external magnetic fields with the elapse of time in minutes (Figures 14 and 15).
3. The superparamagnetic components considerably grow with the increasing impurity content or inhomogeneity (Figure 17).

As shown in Figures 18 and 19, these results imply the formation of the super-para-magnetic domains, most likely due to the local inhomogeneity in LC phases. Therefore, this picture is analogous to the cluster glass (See Section 3.3.2), or super-para-magnetic system without magnetic interactions. A partially broken degree of freedom in LC phases is considered to affect the magnetic properties of the domains. The super-para-magnetic response to the external magnetic field is due to the domains' rotational degree of freedom. Likewise, the factors of molecular mobility and intermolecular interactions resulting in a favorable molecular correlation in LC phases are likely to contribute to the long-time scale growth of superparamagnetic domains. To comprehend the microscopic state of the superparamagnetic domains, we have to investigate the dynamics and interactions between the domains.

Elucidation of the microscopic mechanism for the formation of super-para-magnetic domains in the LC nitroxide radicals is a challenging subject. In this context, we discovered that the nonlinear ME effect occurs in the LC nitroxide radicals (Figure 22). Namely, the super-para-magnetic domains turned out to be controllable by external electric fields. This result demonstrates the ferroelectric aspects of super-para-magnetic domains. Accordingly, in Section 5, we propose and discuss the mechanism of the positive magneto-LC effect on the basis of the nonlinear ME effect.

Meanwhile, quite recently another mechanism has been suggested, in which the positive magneto-LC effect can be accounted for microscopically in terms of the molecular mobility and the resulting inhomogeneity of intermolecular interactions without assuming the formation of superparamagnetic domains by means of MD simulation and DFT calculations (Figure 16) [47]. Namely, the magnetic features seem to reflect the macroscopic inhomogeneity of the liquid crystal orientation field, even though the magnetic properties look like those resulting from the superparamagnetic domains. Since the macroscopic studies on this mechanism is under investigation, only the former mechanism based on the ME effect is discussed in Section 5.

In addition, it is fairly possible that both mechanisms operate complementarily. The latter mechanism originating from the molecular mobility is predominant in the LC and isotropic phases, while the formation of super-para-magnetic domains by the former mechanism results in the preservation of $\chi_{TIM}$ in the supercooled LC and solid phases during the cooling process.

## 5. Proposed Mechanism

The existence of superparamagnetic domains in rod-like LC nitroxide radicals means the ferromagnetic spin arrangement inside the domains. However, ferromagnetic ordering through the direct exchange interactions between radical spins at ambient or higher temperatures in metal-free organic radicals had been believed to be unrealistic. Moreover, the molecular rotation of as fast as around $10^{10}$ s$^{-1}$ in a rod-like LC phase may also make the average magnetic interactions much weaker than in a crystalline phase [25,26]. Contrary to such general believing, the positive magneto-LC effect has been observed in the rod-like LC nitroxide radicals with a specific molecular structure, but not in the homologous non-LC nitroxide radicals.

Here, we propose the parasitic ferromagnetism [57] in terms of the nonlinear ME effect, that is, the induction of ferromagnetism by the emergence of the domains with spontaneous electric polarization (Figure 23). If there is a cross-correlation term between magnetization ($M$) and electric polarization ($P$) in the free energy, it is plausible that the emergence of $M$ can lower the free energy of the system in the presence of $P$.

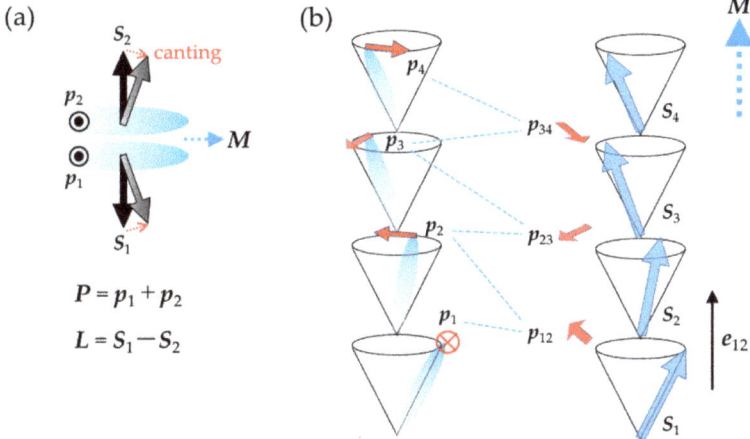

**Figure 23.** Two possible mechanisms for the emergence of ferromagnetism in LC nitroxide radicals induced via DM interactions. (**a**) The induction of weak ferromagnetism (***M***) directed perpendicularly to the electric polarization (***P***). ***L*** denotes the antiferromagnetic vector $S_1 - S_2$. (**b**) A helical magnetic structure induced by helical molecular alignment in the LC phase. The following are the explanation of individual symbols and signs in panel b—Spheroid: the rod-like LC nitroxide radical molecule, $p_i$: the electric polarization of the $i$-th molecule, $p_{ij}$: the summed electric polarization between the $i$-th and $j$-th molecules, and $e_{ij}$: translational vector directed to the screw axis.

For the operation of this mechanism, the existence of spontaneous polarization ($P_s$) is necessary, which emerges as the result of breaking the space inversion symmetry. As described in Section 4.3, the $P_s$ is already present in the LC phases composed of chiral nitroxide radical molecules. We could obtain the following three experimental results that indicated the correlation of super-para-magnetic behavior and the breaking of space inversion symmetry. Firstly, the super-para-magnetic domains grow during the crystal-to-LC-to-Iso phase transition sequence in the heating process, in which the birth and growth of domains without space inversion symmetry was revealed by the gradual increase in the SHG signal (Figure 14). Second, the result that the super-para-magnetic domains augmented with the increasing *ee* value of the LC compounds (See Section 4.1.1) is likely to demonstrate the correlation of the positive magneto-LC effect and the breaking of space inversion symmetry. Third, and more importantly, it was the external electric field that flipped the *g*-value of the EPR signal to draw a ferromagnetic-like hysteresis loop (Figure 22a). These results suggest that the positive magneto-LC effect is driven by collective alignment of LC molecules and the subsequent breaking of its space inversion symmetry.

As one of the origins of nonlinear ME effect, DM interactions, $E_{DM} \sim \sum_{ij} D_{ij} \cdot S_i \times S_j$, are known. The Dzyaloshinkii vector ($D_{ij}$) can be present when the space inversion symmetry is broken. The existence of DM interactions in LC nitroxide radicals were indicated by the anisotropic $g$-value (Figure 21c). Next, we discuss two possibilities for parasitic ferromagnetism on the basis of DM interactions.

The first possibility is the emergence of weak ferromagnetism accompanying the electric polarization (Figure 22a). Such weak ferromagnetism is suggested when the contribution of the energy term, $E_{PLM} \sim P \cdot (L \times M)$, exists in the free energy of the system [58]. This is a phenomenological term resulting from DM interactions, where $P$ denotes the electric polarization, $L$ is antiferromagnetic vector $S_1 - S_2$, and $M$ is the remanent magnetization perpendicular to $L$ (i.e., weak ferromagnetism). If $P$ is present in the system, the emergence of $M$ leads to a finite $E_{PLM}$ value. Therefore, the canting of anti-ferromagnetic pairs of $S_1$ and $S_2$ to the direction perpendicular to the electric polarization can lower the total energy of the system by $E_{PLM}$ (Figure 23a). As expressed in $L$, large anti-ferromagnetic correlations between spins by sufficient intermolecular contacts are needed for this mechanism. It is highly plausible that the large molecular mobility in the LC and isotropic phases is likely responsible for the inhomogeneous intermolecular contact (Figure 16) [47]. This picture is consistent with the results obtained from the experiment on the ME effect (Figure 22). Both the $g$-value and the line width ($\Delta H_{PP}$) are switched by the external electric field ($E$) when it is perpendicular to the external magnetic field $H_0$ (Figure 24).

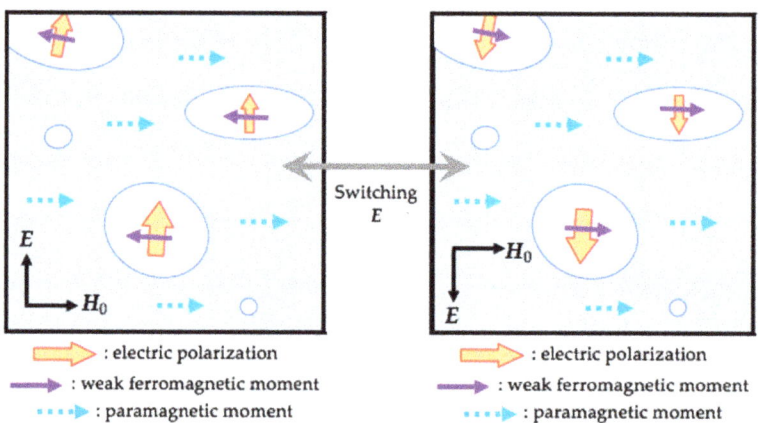

**Figure 24.** Image of domains with electric polarization and the resulting weak ferromagnetism proposed in Figure 23a. The light blue-colored background denotes the paramagnetic LC region. Response of domains to the external electric field rather than the perpendicular magnetic field in the experiment on ME effect (Figure 22). Note that the correspondence of the direction of magnetic moments with observed $g$-value depends on the internal magnetic field at resonant spin sites.

The second possibility is the emergence of helical magnetism by helical alignment of LC molecules (Figure 23b). It is known that, in the helical magnetic structures, the spins $S_i$ and $S_j$ can induce the electric polarization ($p_{ij}$) to satisfy $p_{ij} \sim e_{ij} \times (S_i \times S_j)$ due to inverse DM interactions, where $e_{ij}$ is the unit vector connecting the $i$-th and $j$-th sites [59]. In contrast, in the case of LC nitroxide radicals, the helical structure formed by local $p_{ij}$ from the $i$-th and $j$-th molecule might induce the helical structure of $S_i$ and $S_j$. This helical magnetic structure can have the net magnetization $M$ along the direction of the screw axis. To the best of our knowledge, although there is no example for the helical electric polarization-induced helical magnetization, our proposed mechanism is likely to explain the unique magnetic properties observed in chiral helical LC phases (Figures 10 and 11) or achiral LC phases containing partial chiral helical domains (Figure 14).

As for the correlation of super-para-magnetism and electric polarization, similar magnetic properties can be found in relaxors [60]. A relaxor can be regarded as an electric version of SG because it has randomly directed ferroelectric domains called Polar Nano Regions (PNR) due to the inherent randomness and frustration in the system (Figure 25). Recently, the emergence of super-para-magnetism has been observed at high temperatures with respect to some relaxors (or the system with relaxor-like behavior) containing magnetic metal ions [61,62] (Figure 26). For the mechanism of such super-para-magnetic relaxors, the presence of DM interactions are suggested, which will give weak ferromagnetism perpendicular to the polarization of PNR [61]. In addition, the MD simulation showed that the addition of spherical apolar impurity to spheroidal polar particles produces PNR to show the typical behavior of relaxors [63]. This seems analogous to the formation of ferroelectric domains in LC nitroxide radical samples by an impurity effect (See Section 4.2).

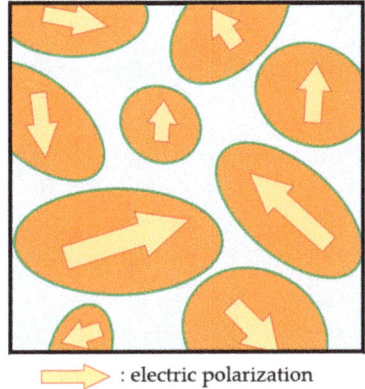

**Figure 25.** Image of PNR in relaxor systems. The electric polarization of each PNR is directed randomly.

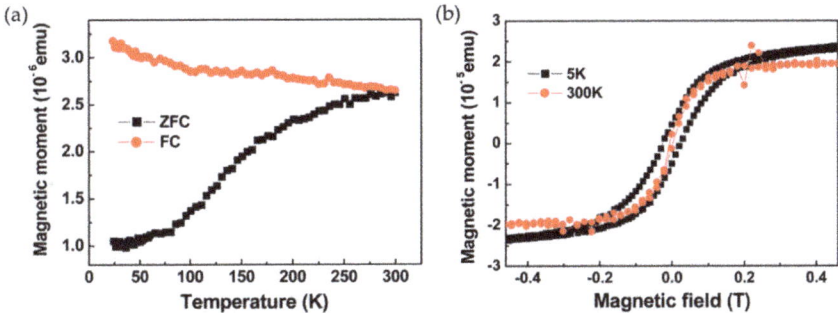

**Figure 26.** (a) The hysteretic temperature dependence of magnetic susceptibility and (b) the super-para-magnetic behavior in the magnetization curves in ZnO–Co relaxor-like nanocomposite thin films [60]. The labels (**a**,**b**) in the original graphs were moved out of the frames by the authors of this paper. (Reprint with permission [62]; Copyright 2020, American Chemical Society).

In some of these super-para-magnetic relaxors, it was revealed that the super-para-magnetic domains were identical to PNR by estimating their sizes [61]. In the case of LC nitroxide radicals, we must clarify the detailed relationship between the magnetic and electric properties in order to uncover the mechanism for the birth and grow of super-para-magnetic domains, which will lead to the elucidation of the microscopic mechanism of positive magneto-LC effect that LC nitroxide radicals exhibited.

## 6. Conclusions and Prospects

Since 2008, we have reported that a series of chiral and achiral all-organic LC nitroxide radicals having one or two PROXYL units in the core position display SG-like superparamagnetic features, such as a magnetic hysteresis (referred to as positive magneto-LC effect), and thermal and impurity effects during a heating and cooling cycle in weak magnetic fields. In general, the enantiomerically-enriched rod-like LC nitroxide radicals *trans*-**1** and their derivatives always showed a distinct positive magneto-LC effect, irrespective of the molecular positive or negative dielectric anisotropy ($\Delta\varepsilon$), whereas the corresponding racemic samples exhibited the positive or negative magneto-LC effect, depending on the negative or positive sign of molecular $\Delta\varepsilon$, respectively. Furthermore, for the first time, the magneto-electric (ME) effect was observed in the ferroelectric (SmC*) phase of S,S-enriched **1** (m = n = 13) at temperatures as high as 75 °C. Achiral meso diradical compound (R,S)-**3** became SHG-active gradually in the discotic hexagonal columnar phase by heating in the presence of a magnetic field to form a chiral helical columnar structure and eventually showed a distinct positive magneto-LC effect. It is also noticeable that a very large impurity effect was observed for this compound when 20 mol% of racemic *cis*-diastereomers (R*,R*)-**3** was added to the host (R,S)-**3** as the impurity.

By comparison of these experimental results with the well-known magnetic properties of SG materials and on the basis of the results of nonlinear ME effect, we suggest that the positive magneto-LC effect, i.e., partial formation of super-para-magnetic domains in the major paramagnetic spins in the LC phase, is most likely to originate from the emergence of weak ferromagnetism accompanying the electric polarization due to DM interactions and/or the emergence of helical magnetism by helical alignment of LC molecules.

For the practical application of the positive magneto-LC effect to organic materials science, it is essential to enlarge the ratio of superparamagnetic domains to paramagnetic spins in the LC and isotropic phases. The utilization of the impurity effect to form SG-like inhomogeneous magnetic domains is quite promising because we can select the most suitable compounds among a variety of candidate magnetic and nonmagnetic organic compounds miscible with LC phases as the impurities. The electric field control of ferromagnetic domains is another interesting choice. However, this needs further studies to understand the detailed mechanism of the positive magneto-LC effect. These studies will offer novel access to the realization and application of ferromagnetic materials composed of organic radical spins. At the same time, it will be possible to measure the AC magnetic susceptibility at low or even higher temperatures for elucidating the mechanism of positive magneto-LC effect owing to a considerable increment of the super-para-magnetic region surrounded by the para-magnetic spins.

**Author Contributions:** Conceptualization, S.S. and Y.U.; methodology, Y.U. and R.T.; validation, S.S., Y.U. and R.T.; formal analysis, S.S. and Y.U.; investigation, Y.U. and R.T.; resources, R.T.; data curation, S.S. and Y.U.; writing—original draft preparation, S.S. and R.T.; writing—review and editing, R.T. and S.S.; supervision, R.T.; administration, R.T.; funding acquisition, R.T. and Y.U. All authors have read and agreed to the published version of the manuscript.

**Funding:** This research was funded by JSPS KAENHI (Grant number 26248024).

**Conflicts of Interest:** The authors declare no conflict of interest.

## References

1. Lightman, A. *The Discoveries*; Vintage Books: New York, NY, USA, 2005.
2. Prigogine, I. *The End of Certainty*; The Free Press: New York, NY, USA, 1977.
3. Prigogine, I.; Stengers, I. *Order out of Chaos: Man's New Dialogue with Nature*; Bantam Books: New York, NY, USA, 1984.

4. Anderson, P.W. More is different. *Science* **1972**, *177*, 393–396. [CrossRef] [PubMed]
5. Waldrop, M.M. *Complexity*; Simon &Schuster Paperbacks: New York, NY, USA, 1992.
6. Kauffman, S.A. *At Home in the Universe*; Oxford University Press: New York, NY, USA, 1992.
7. Kauffman, S.A. *Investigations*; Oxford University Press: New York, NY, USA, 2000.
8. Mainzer, K. *Symmetry and Complexity: The Sprit and Beauty of Nonlinear Science*; World Scientific: Singapore, 2005.
9. Mainzer, K. *Thinking in Complexity*; Springer: Berlin, Germany, 2007.
10. Kuramoto, Y. *Chemical Oscillations, Waves, and Turbulence*; Springer: Berlin, Germany, 2003.
11. Di Bella, S.; Ratner, M.A.; Marks, T.J. Design of chromophoric molecular assemblies with large second-order optical nonlinearities. A theoretical analysis of the role of intermolecular interactions. *J. Am. Chem. Soc.* **1992**, *114*, 5842–5849. [CrossRef]
12. Bustamante, C.; Liphardt, J.; Ritort, F. The nonequilibrium thermodynamics of small systems. *Phys. Today* **2005**, *58*, 43–48. [CrossRef]
13. Kwak, H.Y.; Panton, R.L. Gas bubble formation in nonequilibrium water-gas solutions. *J. Chem. Phys.* **1983**, *7*, 5795–5799. [CrossRef]
14. Mann, S.; Heywood, B.R.; Rajam, S.; Birchall, J.D. Controlled crystallization of $CaCO_3$ under stearic acid monolayers. *Nature* **1988**, *334*, 692–695. [CrossRef]
15. Dierking, I. *Textures of Liquid Crystals*; Wiley-VCH: Weinheim, Germany, 2003.
16. Goodby, J.W.; Collings, P.J.; Kato, T.; Tschierske, C.; Gleeson, H.F.; Raynes, P. (Eds.) *Handbook of Liquid Crystals*, 2nd ed.; Wiley-VCH: Weinheim, Germany, 2014; Volume 1–8.
17. Eerenstein, W.; Mathur, N.D.; Scott, J.F. Multiferroic and magnetoelectric materials. *Nature* **2006**, *442*, 759–765. [CrossRef]
18. Rao, C.N.R.; Serrao, C.R. New routes to multiferroics. *J. Mater. Chem.* **2007**, *17*, 4931–4938. [CrossRef]
19. Felser, C.; Fecher, G.H.; Balke, B. Spintronics: A challenge for materials science and solid-state chemistry. *Angew. Chem. Int. Ed.* **2007**, *46*, 668–699. [CrossRef]
20. Seki, S. *Magnetoelectric Response in Low-Dimensional Frustrated Spin Systems*; Springer: Tokyo, Japan, 2012.
21. Rikken, G.L.J.A.; Raupch, E. Observation of magneto-chiral dichroism. *Nature* **1997**, *390*, 493–494. [CrossRef]
22. Rikken, G.L.J.A.; Raupch, E. Enantioselective magnetochiral photochemsitry. *Nature* **2000**, *405*, 932–935. [CrossRef]
23. Train, C.; Gheorghe, R.; Krisic, V.; Chamoreau, L.-M.; Ovanesyan, N.S.; Rikken, G.L.J.A.; Grussele, M.; Verdaguer, M. Strong magneto-chiral dichroism in enantiopure chiral ferromagnets. *Nat. Mater.* **2008**, *7*, 729–734. [CrossRef] [PubMed]
24. Binnemans, K.; Gröller-Walrand, C. Lanthanide-containing liquid crystals and surfactants. *Chem. Rev.* **2002**, *102*, 2302–2345. [CrossRef] [PubMed]
25. Tamura, R.; Uchida, Y.; Suzuki, K. Magnetic Properties of Organic Radical Liquid Crystals and Metallomesogens. In *Handbook of Liquid Crystals*, 2nd ed.; Goodby, J.W., Collings, P.J., Kato, T., Tschierske, C., Gleeson, H.F., Raynes, P., Eds.; Wiley-VCH: Weinheim, Germany, 2014; Volume 8, pp. 837–864.
26. Kaszyński, P. Liquid Crystalline Radicals: An Emerging Class of Organic Magnetic Materials. In *Magnetic Properties of Organic Materials*; Lahti, P.M., Ed.; Marcel Dekker: New York, NY, USA, 1999; pp. 325–344.
27. Kaszyński, P.; Kapuscinski, S.; Ciastek-Iskrzycka, S. Liquid crystalline derivtives of heterocyclic radicals. *Adv. Hetero Chem.* **2019**, *128*, 263–331.
28. Ikuma, N.; Tamura, R.; Shimono, S.; Kawame, N.; Tamada, O.; Sakai, N.; Yamauchi, J.; Yamamoto, Y. Magnetic properties of all-organic liquid crystals containing a chiral five-membered cyclic nitroxide unit within the rigid core. *Angew. Chem. Int. Ed.* **2004**, *43*, 3677–3682. [CrossRef] [PubMed]
29. Ikuma, N.; Tamura, R.; Shimono, S.; Uchida, Y.; Masaki, K.; Yamauchi, J.; Aoki, Y.; Nohira, H. Ferroelectric properties of paramagnetic, all-organic, chiral nitroxyl radical liquid crystals. *Adv. Mater.* **2006**, *18*, 477–480. [CrossRef]
30. Uchida, Y.; Tamura, R.; Ikuma, N.; Shimono, S.; Yamauchi, J.; Aoki, Y.; Nohira, H. Synthesis and characterization of novel all-organic liquid crystalline radicals. *Mol. Cryst. Liq. Cryst.* **2007**, *479*, 213–221. [CrossRef]
31. Uchida, Y.; Tamura, R.; Ikuma, N.; Yamauchi, J.; Aoki, Y.; Nohira, H. Synthesis and characterization of novel radical liquid crystals showing ferroelectricity. *Ferroelectrics* **2008**, *365*, 158–169. [CrossRef]

32. Ikuma, N.; Uchida, Y.; Tamura, R.; Suzuki, K.; Yamauchi, J.; Aoki, Y.; Nohira, H. Preparation and properties of C2-symmetric organic radical compounds showing ferroelectric liquid crystal properties. *Mol. Cryst. Liq. Cryst.* **2009**, *509*, 108–117. [CrossRef]
33. Ikuma, N.; Suzuki, K.; Uchida, Y.; Tamura, R.; Aoki, Y.; Nohira, H. Preparation and ferroelectric properties of new chiral liquid crystalline organic radical compounds. *Heterocycles* **2010**, *80*, 527–535.
34. Uchida, Y.; Tamura, R.; Ikuma, N.; Yamauchi, J.; Aoki, Y.; Nohira, H. Unusual Intermolecular Magnetic Interaction Observed in an All- Organic Radical Liquid Crystal. *J. Mater. Chem.* **2008**, *18*, 2950–2952. [CrossRef]
35. Uchida, Y.; Suzuki, K.; Tamura, R.; Ikuma, N.; Shimono, S.; Noda, Y.; Yamauchi, J. Anisotropic and inhomogeneous magnetic interactions observed in all-organic nitroxide radical liquid crystals. *J. Am. Chem. Soc.* **2010**, *132*, 9746–9752. [CrossRef] [PubMed]
36. Suzuki, K.; Uchida, Y.; Tamura, R.; Shimono, S.; Yamauchi, J. Observation of positive and negative magneto-LC effects in all-organic nitroxide radical liquid crystals by EPR spectroscopy. *J. Mater. Chem.* **2012**, *22*, 6799–6806. [CrossRef]
37. Suzuki, K.; Uchida, Y.; Tamura, R.; Noda, Y.; Ikuma, N.; Shimono, S.; Yamauchi, J. Influence of applied electric fields on the positive magneto-LC effects observed in the ferroelectric liquid crystalline phase of a chiral nitroxide radical compound. *Soft Matter* **2013**, *9*, 4687–4692. [CrossRef]
38. Suzuki, K.; Uchida, Y.; Tamura, R.; Noda, Y.; Ikuma, N.; Shimono, S.; Yamauchi, J. Electric field dependence of molecular orientation and anisotropic magnetic interactions in the ferroelectric liquid crystalline phase of an organic radical compound by EPR spectroscopy. *Adv. Sci. Tech.* **2013**, *82*, 50–54. [CrossRef]
39. Tamura, R.; Uchida, Y.; Suzuki, K. *Advances in Organic Crystal Chemistry: Comprehensive Reviews 2015*; Tamura, R., Miyata, M., Eds.; Springer: Tokyo, Japan, 2015; pp. 689–706.
40. Takemoto, Y.; Uchida, Y.; Shimono, S.; Yamauchi, J.; Tamura, R. Preparation and magnetic properties of nitroxide radical liquid crystalline physical gels. *Mol. Cryst. Liq. Cryst.* **2017**, *647*, 279–289. [CrossRef]
41. Suzuki, K.; Takemoto, Y.; Takaoka, S.; Taguchi, K.; Uchida, Y.; Mazhukin, D.G.; Grigor'ev, I.A.; Tamura, R. Chiral all-organic nitroxide biradical liquid crystals showing remarkably large positive magneto-LC effects. *Chem. Commun.* **2016**, *52*, 3935–3938. [CrossRef]
42. Takemoto, Y.; Zaytseva, E.; Suzuki, K.; Yoshioka, N.; Takanishi, Y.; Funahashi, M.; Uchida, Y.; Akita, T.; Park, J.; Sato, S.; et al. Unique superparamagnetic-like behavior observed in non-π-delocalized nitroxide diradical compounds showing discotic liquid crystalline phase. *Chem. Eur. J.* **2018**, *24*, 17293–17302. [CrossRef]
43. Uchida, Y.; Suzuki, K.; Rui, T. Magneto-LC effects in hydrogen-bonded all-organic radical liquid crystal. *J. Phys. Chem. B* **2012**, *116*, 9791–9795. [CrossRef]
44. Nakagami, S.; Akita, T.; Kiyohara, D.; Uchida, Y.; Tamura, R.; Nishiyama, N. Molecular mobility effect on magnetic interactions in all-organic paramagnetic liquid crystal with nitroxide radical as a hydrogen bonding acceptor. *J. Phys. Chem. B* **2018**, *122*, 7409–7415. [CrossRef]
45. Uchida, Y.; Tamura, R.; Ikuma, N.; Shimono, S.; Yamauchi, J.; Shimbo, Y.; Takezoe, H.; Aoki, Y.; Nohira, H. Magnetic-field-induced molecular alignment in an achiral liquid crystal spin-labeled by a nitroxyl group in the mesogen core. *J. Mater. Chem.* **2009**, *19*, 415–418. [CrossRef]
46. Vorobiev, A.K.; Chumakova, N.A.; Pomogailo, D.A.; Uchida, Y.; Suzuki, K.; Noda, Y.; Tamura, R. Determination of structure characterizaton of all-organic radical liquid crystals based on analysis of the dipole-dipole broadened EPR spectra. *J. Phys. Chem. B* **2014**, *118*, 1932–1942. [CrossRef] [PubMed]
47. Uchida, Y.; Watanabe, G.; Akita, T.; Nishiyama, N. Thermal molecular motion can amplify intermolecular magnetic interactions. *J. Phys. Chem. B* **2020**, *124*, 6175–6180. [CrossRef]
48. Bean, C.P.; Livingston, J.D. Superparamagnetism. *J. Appl. Phys.* **1959**, *30*, S120–S129. [CrossRef]
49. Fischer, K.H.; Hertz, J.A. *Spin Glasses*; Cambridge Univeristy Press: Cambridge, UK, 1991.
50. Cannella, V.; Mydosh, J.A. Magnetic ordering in gold-iron alloys. *Phys. Rev. B* **1972**, *6*, 4220–4237. [CrossRef]
51. Nagata, S.; Keesom, P.H.; Harrison, H.R. Low-dc-field susceptibility of Cu Mn spin glass. *Phys. Rev. B* **1979**, *19*, 1633–1638. [CrossRef]
52. Mydosh, J.A. Spin glasses: Redux: An updated experimental/materials survey. *Rep. Prog. Phys.* **2015**, *78*, 052501. [CrossRef]
53. Sandlund, L.; Svedlindh, P.; Granberg, P.; Nordblad, P.; Lundgren, L. Experimental evidence for the existence of an overlap length in spin glasses. *J. Appl. Phys.* **1988**, *64*, 5616–5618. [CrossRef]

54. Jain, R.; Kabir, K.; Gilroy, J.B.; Mitchell, K.A.R.; Wong, K.C.; Hicks, R.G. High-temperature metal-organic magnets. *Nature* **2007**, *445*, 291–294. [CrossRef]
55. Kogo, R.; Araoka, F.; Uchida, Y.; Tamura, R.; Ishikawa, K.; Takezoe, H. Second harmonic generation in a paramagnetic all-organic chiral smectic liquid crystal. *Appl. Phys. Express* **2010**, *3*, 041701. [CrossRef]
56. Bloch, F. Nuclear induction. *Phys. Rev.* **1946**, *70*, 460–474. [CrossRef]
57. Néel, L. Some New Results on Antiferromagnetism and Ferromagnetism. *Rev. Mod. Phys.* **1953**, *25*, 58–61. [CrossRef]
58. Fennie, C.J. Ferroelectrically Induced Weak Ferromagnetism by Design. *Phys. Rev. Lett.* **2008**, *100*, 1–4. [CrossRef]
59. Katsura, H.; Nagaosa, N.; Balatsky, A.V. Spin Current and Magnetoelectric Effect in Noncollinear Magnets. *Phys. Rev. Lett.* **2005**, *95*, 1–4. [CrossRef] [PubMed]
60. Cross, L.E. Relaxor ferroelectrics: An overview. *Ferroelectrics* **1994**, *151*, 305–320. [CrossRef]
61. Soda, M.; Matsuura, M.; Wakabayashi, Y.; Hirota, K. Superparamagnetism Induced by Polar Nanoregions in Relaxor Ferroelectric (1-x)BiFeO$_3$-xBaTiO$_3$. *J. Phys. Soc. Jpn.* **2011**, *80*, 2–5. [CrossRef]
62. Li, D.Y.; Zeng, Y.J.; Batuk, D.; Pereira, L.M.C.; Ye, Z.Z.; Fleischmann, C.; Menghini, M.; Nikitenko, S.; Hadermann, J.; Temst, K.; et al. Relaxor ferroelectricity and magnetoelectric coupling in ZnO-Co nanocomposite thin films: Beyond multiferroic composites. *ACS Appl. Mater. Interfaces* **2014**, *6*, 4737–4742. [CrossRef] [PubMed]
63. Takae, K.; Onuki, A. Ferroelectric glass of spheroidal dipoles with impurities: Polar nanoregions, response to applied electric field, and ergodicity breakdown. *J. Phys. Condens. Matter* **2017**, *29*, 4–6. [CrossRef]

**Publisher's Note:** MDPI stays neutral with regard to jurisdictional claims in published maps and institutional affiliations.

© 2020 by the authors. Licensee MDPI, Basel, Switzerland. This article is an open access article distributed under the terms and conditions of the Creative Commons Attribution (CC BY) license (http://creativecommons.org/licenses/by/4.0/).

*Review*

# Chiral Symmetry Breaking in Liquid Crystals: Appearance of Ferroelectricity and Antiferroelectricity

Yoichi Takanishi

Department of Physics, Graduate School of Scince, Kyoto University, Kitashirakawaoiwake, Sakyo, Kyoto 606-8502, Japan; ytakanis@scphys.kyoto-u.ac.jp

Received: 30 September 2020; Accepted: 17 November 2020; Published: 19 November 2020

**Abstract:** The study of chiral symmetry breaking in liquid crystals and the consequent emergence of ferroelectric and antiferroelectric phases is described. Furthermore, we show that the frustration between two phases induces a variety of structural phases called subphases and that resonant X-ray scattering is a powerful tool for the structural analysis of these complicated subphases. Finally, we discuss the future prospects for clarifying the origin of such successive phase transition.

**Keywords:** chirality; symmetry breaking; ferroelectricity; antiferrolelectricity; subphases; resonant X-ray scattering

## 1. Introduction: Liquid Crystals

Liquid Crystals (LCs) are stationary state meso-phases appearing between crystals and isotropic liquids, and have intermediate properties between them, i.e., fluidity and complete disorder as liquids and three-dimensional long-rage anisotropic ordering as crystals. Spontaneous meso-scale organization is also characteristic of LCs, and comparatively easy fabrication of fine structures can be done. Usually rod-like or disk-like shaped molecules with a hard-core part at their center and flexible end chains tend to exhibit liquid crystal phases. Here, we mainly treat the LCs of rod-like molecules, in which the phase transition occurs by a thermal process (called "thermotropic liquid crystals"). By the degree of ordering, LCs are classified into many types. The most disordered type of LC is called the nematic phase, which has no positional order but the orientational order of a molecular long axis. The nematic phase appears just below the isotropic liquid phase, and, between the nematic phase and crystal phase, another LC phase frequently appears. This LC phase has a one-dimensional positional order (layer order) in addition to the molecular orientational order, called smectic phase. Smectic phases are additionally classified in many types by molecular orientation along the layer normal or in a layer. Two common smectic phases are SmA and SmC. In SmA, the molecular long axis is parallel to the layer normal, and, in SmC, it is tilted, as shown in Figure 1.

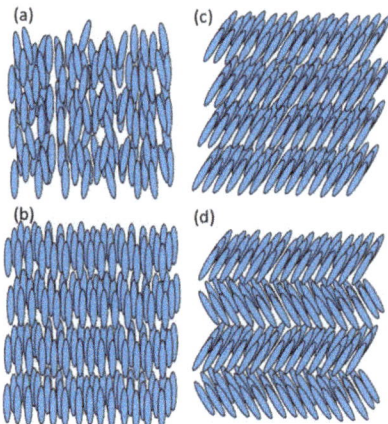

**Figure 1.** Molecular arrangement of liquid crystal phases: (**a**) nematic, (**b**) SmA, (**c**) SmC(*) and (**d**) SmCA(*).

## 2. Introduction of Chiral Symmetry Breaking to Liquid Crystals and the Appearance of Ferroelectricity in Liquid Crystals

Chirality has an important factor for many scientific fields, such as biology, chemistry, and physics. For example, in drag development, one of the enantiomers is active and good for health but the other becomes poison. In physics, chirality strongly influences the optical properties, such as optical rotation and circular dichroism. Circular optical polarization properties are the signature of the sample dissymmetry leading to parity breaking; this is what is defined as chemical chirality. Parity breaking is a consequence of the absence of any Sn (improper rotation) symmetry element in the sample symmetry. The disappearance of inversion symmetry due to the chirality induces several physical properties such as pyroelectricity, piezoelectiricity and ferroelectricity. Of the 32 different space groups, there are 21 that do not have inversion symmetry and they exhibit piezoelectricity. Among them, 10 space groups show pyroelectricity with spontaneous polarization. However not all materials exhibiting pyroelectricity exhibit ferroelectricity [1]. In the case that the spontaneous polarization can be reversed in the application of an external reverse electric field, it is called ferroelectricty. Even in LCs with fluidity, the ferroelectricity may appear. Most of rod-like molecules exhibiting LC phases have permanent electric dipole moments parallel and/or perpendicular to the molecular long axes. If their dipole moments are aligned properly, it is possible to accomplish the occurrence of ferroelectricity in liquid crystals. However, some factors prevent the occurrence of ferroelectricity: (1) thermal fluctuation is large, so that dipole-dipole interaction is weak because thermal kinetic energy is larger than dipole-dipole interaction energy. (2) Due to the high symmetry, free rotation around the molecular long axis and head and tail equivalence disturbs ferroelectric property. (3) In the case that the dipole moment is too strong, fluidity induces dimmer organization, and the antiparallel arrangement of the dipole moment occurs and total polarization is cancelled out.

Since usually the symmetry of the SmC phase is $C_{2h}$, it is impossible to possess spontaneous polarization because of the existence of inversion symmetry. However, by introducing the chirality in the system, its symmetry is broken to $C_2$ without inversion symmetry, so it is possible to produce the spontaneous polarization along the $C_2$ axis. In SmC(*), molecules still rotate around the molecular long axis, but the rotation is biased and then the biaxiality appears. Hence, if molecules have a transverse dipole moment, and the average biasing direction of the transverse dipole moment tends to align normal to the tiling plane, macroscopic polarization is able to appear in the smectic layer because the $C_2$ axis is normal to the molecular tilting plane. In such deep consideration for the symmetry of the LCs, Meyer et al. discovered ferroelectricity in LCs [2,3]. In the SmC* phase,

$$P_S = P_0 \frac{z \times n}{|z \times n|} \tag{1}$$

Here, $n$ is the director (average direction of molecular long axes), and $z$ is a unit vector normal to the smectic layer. Tilt angle $\theta$, which is the angle between $z$ and $n$, is defined to zero when $z$ is coincident with $n$.

Strictly speaking, due to the chirality, molecular tilting direction (azimuthal angle in Figure 2) is slightly shifted from layer to layer, so that the helical structure whose axis is parallel to the layer normal is formed in the bulk state. Hence, net polarization of the thick sample may become zero, because the direction of the polarization in a smectic layer is normal to the tilting direction. The helical pitch is typically much longer than the smectic layer spacing, which has be of the order of thousands times of the layer spacing, and the deviation of azimuthal angle between molecules in neighboring layers is small (less than 1 deg.). Therefore, SmC* is not a real ferroelectric phase in the case of the bulk state. However, by confining this LC into thin sandwich cells, which are composed of two clean glass plates and have a gap less than the helical pitch, and aligning the helical axis parallel to the substrate surface (called "*planar alignment*") as shown in Figure 3, helix is suppressed (unwound) by the surface effect. As a result, in such a condition, the molecular tilting plane is parallel to the surface, so that the spontaneous polarization exists along the surface normal direction. In such a thin cell whose substrate is coated with optically transparent electrodes such as ITO (Indium-Thin-Oxide), molecules are tilted in two states (A or B) by molecular tilt angle ± θ with respect to the layer normal. If an electric field with proper magnitude whose direction is parallel to the smectic layer is applied to the ferroelectric SmC* phase in this condition, the polarization is forced to align along the field direction. Since the molecular tilting (and also director) is normal to the polarization, it is also normal to the field direction. Therefore, the director is tilted to the z-axis of Figure 3 and parallel to the substrate (xy) plane (A or B). If the sense of an applied field is inverted, polarization and its corresponding molecular tilting direction are also inverted correspondingly, so that the director is still parallel to the substrates but the tilting sense is opposite. As shown in Figure 4, apparent tilt angle and corresponding polarization shows a single hysteresis behavior with respect to the applied field, which is characteristic for the ferroelectric property. Hence, the indirect ferroelectricity appears due to the surface effect, and this state is called surface-stabilized ferroelectric liquid crystals (SSFLCs) [4].

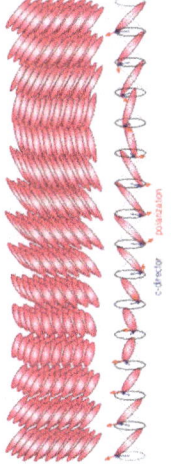

**Figure 2.** Bulk structure of the SmC* phase. C-director (blue arrow) means the projection of directors (molecular long axis) on the smectic layer plane, and the polarization (red arrow) is parallel to the smectic plane but normal to the C-director.

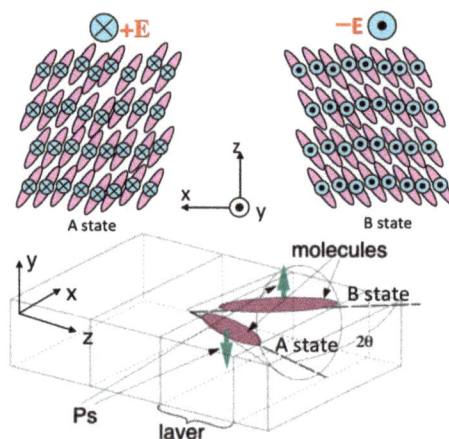

**Figure 3.** Molecular arrangement of SmC* in the thin planar alignment. Due to the surface effect, molecules are located at A or B states, and the corresponding polarization is oriented up or down to the substrate normal. By applying the electric field normal to the substrates, molecules switch to A and B states as the sign of transverse polarization is consistent with the sign of electric field.

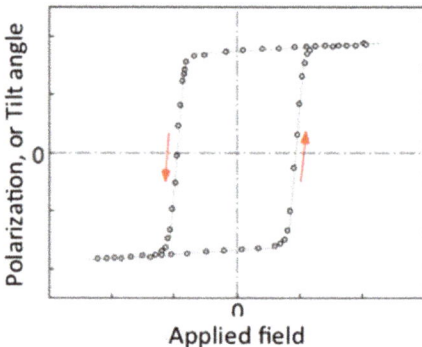

**Figure 4.** Apparent tilt angle of SmC* as a function of the electric field, and the corresponding molecular arrangement under the electric field.

## 3. Discovery of Antiferroelectricity in Liquid Crystals

As mentioned later, the ferroelectric SmC* phase shows a fast electro-optic response, whose speed is in the order of microseconds, much faster than that in the nematic phase using flat panel displays (msec order). This is because of the direct coupling between the applied electric field and the spontaneous polarization, which is different from the dielectric coupling in the nematic phase. Hence, FLC has been expected to utilize next-generation fast flat panel displays, and more than thousands of FLC compounds were synthesized and extensively studied in order to commercialize it as a new fast response display [5]. In this situation for the study on ferroelectric liquid crystals, an antiferroelectric phase was discovered in 1989 [6]. A first compound in which antiferroelectric liquid crystal (AFLC) was confirmed was 4-(1-methylheptyloxycarbonyl) phenyl 4′-octyloxybiphenyl- 4-carboxylate (MHPOBC), whose chemical structure is shown in Figure 5. At first, it was reported that this compound shows so-called tristable switching in the lower-temperature smectic phase by the Tokyo Tech group [7,8]; the switching shows two states, which is the same as ferroelectric bistable states mentioned above and another state without any electric field, but was not clarified why such a switching occurs. Later, considering previous results and the result of obliquely incident transmission spectra, Chandani et al. concluded that molecules are tilted in the opposite sense in neighboring layers and the corresponding polarization was cancelled out in two layers without any electric field, as shown in Figure 6b [6]. When the proper electric field is applied, electric-field induced phase transition to ferroelectric states occurs. Apparent tilt angle and corresponding polarization shows a double hysteresis behavior with respect to the applied field, which is characteristic for antiferroelectric property, as shown in Figure 6a. In the bulk state, a helical structure was also formed like SmC* due to the chirality, in such a molecular arrangement with helix, obliquely incident transmitted spectra are clearly different from that in the ferroelectric SmC* phase. Details are explained as follows.

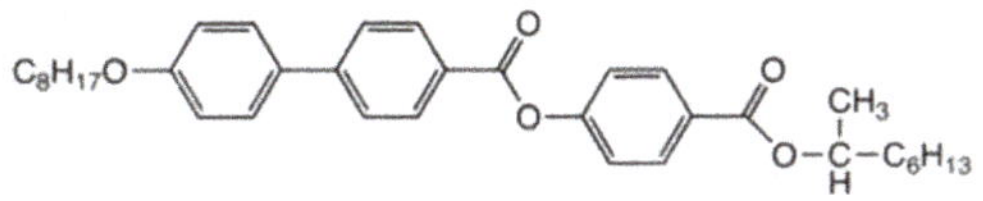

**Figure 5.** Chemical structure of MHPOBC, in which antiferroelectric SmCA* was discovered.

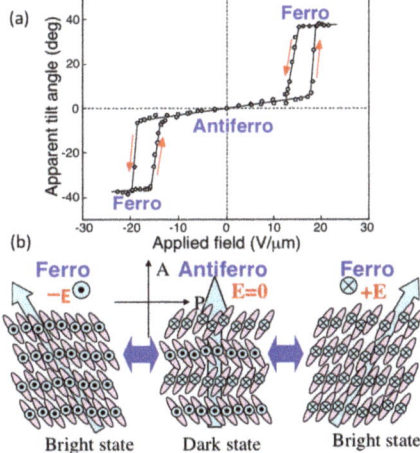

**Figure 6.** Apparent tilt angle of SmCA* as a function of the electric field (**a**), and the corresponding molecular arrangement under the electric field (**b**).

The important difference between ferroelectric SmC* and antiferroelectric SmCA* is symmetry; SmC* and SmCA* have $C_2$ and $D_2$ symmetries, respectively. Reflecting this symmetry difference, the physical properties between them are distinguished. The most remarkable difference of the optical transmission spectra is as mentioned previously [6]. Bulk states in both phases have helical structures induced by twisting power due to the chirality. If the helical pitch (strictly speaking, optical pitch, which is physical pitch multiplied by refractive index) is consistent with the visible wavelength, the circular polarized light with the same wavelength and the same handedness is selectively reflected, called selective reflection. Hence, when the transmission spectrum is measured, a clear dip is observed at this wavelength [6]. When the incident light is parallel to the helical axis, the periodicity of the optical birefringence corresponds to half the pitch in both SmC* and SmCA*. However, when the incident light is oblique to the helical axis, the situation changes; in antiferroelectric SmCA*, the periodicity of the optical birefringence corresponds to half the pitch, but in ferroelectric SmC*, it corresponds to the full pitch. Hence, from the oblique incident transmission spectra, both phases are clearly distinguished as shown in Figure 7 [6], which comes from the difference from the molecular arrangement.

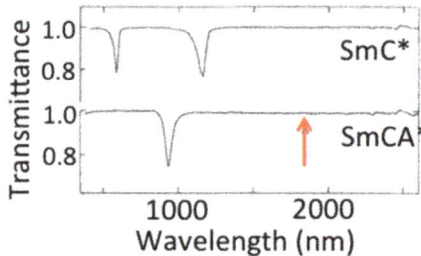

**Figure 7.** Transmitted spectra at obliquely incident light in SmC* and SmCA*. In SmCA*, the periodicity of optical birefringence is half the helical pitch, so that the dip of transmission due to the selective reflection of the full pitch band is not observed (at the red arrow position) [6].

The structure in SmC* showing the ferroelectric property and that in SmCA* showing the antiferroelectric property are very similar. However, due to the difference of system symmetry, a defect structure is also drastically different, and this difference can judge the anticlinic structure in SmCA*. Now we consider the helical pitch in SmC* and SmCA* to be very long or infinite. When we make homeotropic alignment (in which the layer plane is parallel to the substrate), typical defect texture called "Schlieren" is frequently observed. In the nematic phase, similar schlieren textures are also observed in the planar alignment, and it is caused by the spatially deviation of director $n$ orientation, which is the average direction of molecular long axis around the visible wavelength region. Schlieren texture in the homeotropic alignment of tilted smectic phase, on the other hand, is caused by the deviation of the direction of the projection of the director to the smectic layer ($c$-director). Due to the symmetry, $n$ and $-n$ are equivalent, but $c$ and $-c$ are not, because they indicate alternately anticlinic tilting orientation of director. Hence, only four blush schlieren texture, indicating disclination defect with the strength s = ±1, is observed in the ferroelectric SmC* as shown in Figure 8a. However, in antiferroelectric SmCA*, two blush schlieren texture, indicating disclination defect with the strength s = ±1/2, is also observed in addition to the four-blush one [9,10]. Considering $c$ is not equal to $-c$ like SmC*, it is difficult to attain this defect structure, because it is necessary to provide discontinuous plane along the solid line as shown in Figure 8b for the appearance of the s = ±1/2 defects and the defect energy becomes extremely high. However, in the SmCA*, it has an anticilinic structure, so that $c$-director in adjacent layers are antiparallel. Therefore, by introducing a screw dislocation with a Burgers vector of the same magnitude as the single layer thickness together with wedge disclination with s = ±1/2 at the same position, discontinuous boundary plane can disappear. This defect is called dispiration [10], which is a combined defect of dislocation and disclination. As far as I know, this is the first obvious experimental proof of the existence of a dispiration. Additionally, conversely speaking, the appearance of dispirations provides experimental evidence of the anticlinic structure in the antiferroelectric liquid crystalline phase [11]. Results of the transmission spectra measurement and texture observation seem to be indirect experimental proofs of the structure of AFLCs. However, this is enough to explain the anticlinic structure in AFLCs and electro-optic property. The final direct decision of the structure of AFLCs was performed by the resonant X-ray scattering technique; the results are shown in Section 5.

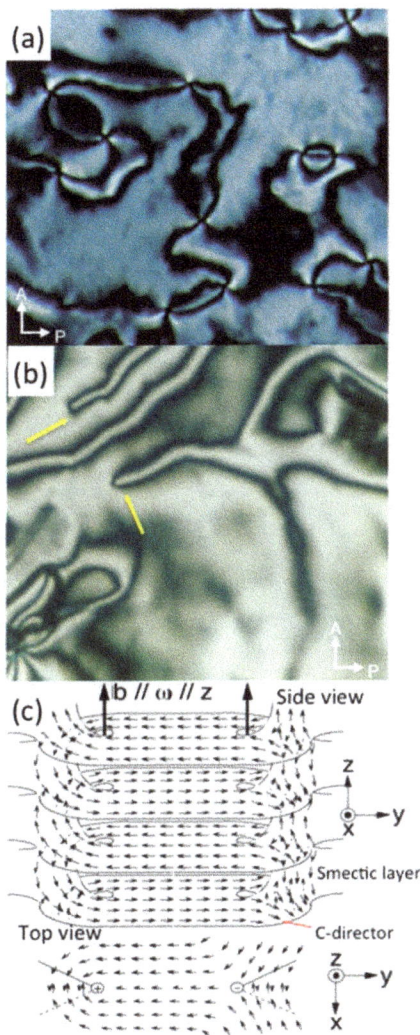

**Figure 8.** Texture under the crossed polarized observation in the homeotropic alignment in SmC* (**a**) and SmCA* (**b**). In SmCA*, two brushed texture by wedge disclination with s = ±1/2 was observed, which proves the existence of wedge-screw dispiration. (**c**) The structure of wedge-screw dispiration in SmCA*. The small arrows indicate c-directors, which is the projection of directors on the smectic layer plane. Wedge disclination lines with Frank vector (0, 0, π) as well as the screw dislocation lines with Burgers vector (0, 0, d) are indicated by bold arrows, which parallel the layer normal (z axis). Here, d is the layer thickness. This defect can appear by the anticlinic structure in SmCA* [11].

## 4. Frustration of Ferroelectric and Antiferroelectric Phases; Successive Phase Transition to Several Subphases between Two Phases

When the antiferroelectric SmCA* phase was discovered in 4-(1-methylheptyloxycarbonyl) phenyl 4′-octyloxybiphenyl-4-carboxylate (MHPOBC), three phases with a narrow temperature range called "*subphase*" or "SmC* variants" were already observed between SmA and SmCA*. Fukui et al. first noticed the existence of subphases by differential scanning calorimetry (DSC) [12]. At almost the same time, Takezoe et al. [13] and Chandani et al. [14] reported three subphases between SmA and SmCA* in MHPOBC, and these subphases were tentatively designated as SmCα*, SmCβ* and SmCγ* in order of temperature at the beginning, and later found that SmCβ* was normal ferroelectric SmC*. The SmCγ* phase shows four-state switching in the planar cells [15], and, by conoscope observation [16], this phase shows an averagely three-layer periodic structure in which two molecules are tilted to the one side and one molecule is tilted to another side with respect to the layer normal, as shown in Figure 9b. After further study, some other phases are frequently observed between ferroelectric SmC* and antiferroelectric SmCA* phases. These phases exist in a very narrow temperature range (several K to 0.2 K) but are clearly distinguished by texture observation [17], conoscope observation [18–20], dielectric [21,22] and thermal measurements [23,24], etc. Later complicated successive phase transition to many subphases is extensively being studied. In particular, subphases are frequently observed in the binary mixture of FLCs and AFLCs [18–20,25].

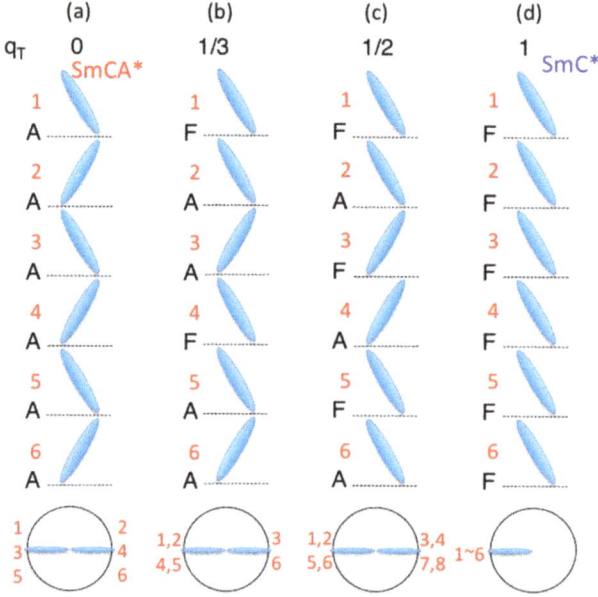

**Figure 9.** Molecular arrangement of subphase between SmC* and SmCA*, and the corresponding qT number. (**a**) antiferroelectric SmCA*, (**b**) correponds to SmCγ* phase, (**c**) is the subphase with four-layer periodicity, and (**d**) ferroelectric SmC*.

First, discovered subphase (SmCγ*) forms a three-layered periodic structure mentioned above, and four-layered periodic subphases were next discovered in 4-(1-methylheptyloxycarbonyl) phenyl 4′-octylbiphenyl- 4-carboxylate (MHPBC) [26]. Typical molecular arrangements are shown in Figure 9c. In the four-layer periodic phase, two molecules in the nearest layer tilt with the same sense, but the molecules in the next two layers tilt with opposite sense, and such a molecular arrangement is formed repeatedly. Here, molecular ordering in adjacent layers is defined as follows: when the molecules in adjacent layers tilt with opposite sense, the molecular ordering is "A", which means antiferroelectric (or anticlinic) ordering. Additionally, when they tilt with the same sense, the molecular ordering is "F", which means ferroelectric (or synclinic) ordering. Here, any subphase expected to appear is specified by an irreducible rational number, qT [25],

$$qT = [F]/([A] + [F])$$

where [F] and [A] are the number of synclinic ferroelectric and anticlinic antiferroelectric orderings, respectively. In the three-layered periodic subphase (SmC$\gamma$*), qT = 1/3, while qT = 1/2 in the four-layered periodic subphase. In antiferroelectric SmCA* and ferroelectric SmC*, qT = 0 and 1, respectively, so that the qT number of the subphase step by step and gradually increases on heating.

At first, at least five subphases are proposed in the electro-optic and conoscope measurements [25], but because of the experimental difficulty such as the temperature accuracy being less than ±0.1 °C due to the narrow temperature range of each subphase and indirect information about the structure obtained by such an experiment, only two subphases got the public's attention. For SmCA*(qT = 1/3) and SmCA*(qT = 1/2), the structures are relatively recognized because these subphases are observed more frequently in many liquid crystal compounds and mixtures, and there are many experimental data. Finally, the three-layer or four-layer periodic structures have been decisively recognized world wide by the resonant X-ray scattering technique about 15 years ago [27].

## 5. Resonant X-ray Scattering and the Determination of Molecular Arrangement of Subphases

The conventional X-ray scattering comes from the electron density distribution of the materials. Because the spherical symmetry of electron density was supposed in each atom, the X-ray susceptibility of the system is usually regarded as a scholar. In this technique, density periodicity can be detected by the Bragg condition. The smectic phase has a one-dimensional positional order with a two-dimensional liquid order, and the density has a periodicity along the layer normal. Since the X-ray susceptibility is treated as a scholar, the densities of the smectic layer in which tilt of the molecule is left and right are not distinguished. Hence, in synclinic SmC* and anticlinic SmCA* (and other suphases), X-ray diffraction profiles reflecting the density distribution along the layer normal are the same.

On the other hand, the resonant X-ray scattering (RXS) technique treats the X-ray susceptibility as a tensor but not a scholar at the absorption edge energy of a specific element as surfa, selenium and bromine, etc. Thus, RXS intensity reflects the system symmetry, and, as a result, the prohibited Bragg diffraction in the conventional method can appear as satellite peaks, which reflects the corresponding system symmetry. Hence, the orientational order (local layer structure) of different subphases can be clarified by measuring and analyzing resonant scattering satellite peaks. Using this technique, Mach et al. firstly directly clarified the two-, three-, and four-layer periodic structures in SmCA*(qT = 0), qT = 1/3 and qT = 1/2 [27]. RXS satellite peaks obtained from the subphase at the resonant condition appear at

$$Q/Q_0 = l + m/\nu \pm \varepsilon$$

where $Q$ is the scattering vector and $Q_0 = 2\pi/d$, d is the smectic layer spacing, $l = 1, 2, 3 \ldots$ is the diffraction order due to the smectic layer spacing, $\nu$ is the number of layers in the unit cell of a subphase, $m = \pm 1, \pm 2, \ldots, \pm(\nu - 1)$ defines RXS peak positions due to the super-lattice periodicity. $\varepsilon = d/P$, and $P$ is the pitch of the macroscopic helix. Later, using this technique, some groups have extensively studied subphases with 3- and 4-layer periodic structures [28–31]. Related to the RXS experiment, Levelut and Pansu calculated the tensorial X-ray structure factor in the smectici phase with the various structure models [32].

The author and his coworkers also performed a new experiment using microbeam resonant X-ray scattering to determine the structure of the novel subpahse except qT = 1/3 and qT = 1/2. Here, we synthesized Br-containing chiral molecule, (*S,S*)-*bis*-[4'-(1-methylheptyloxycarbonyl)-4-biphenyl] 2-bromo-terephthalate (compound **1**) [33], and measured RXS in the mixture of compound **1** and (*S,S*)-$\alpha$, $\omega$ -bis (4-{[4'-(1-methylheptyloxycarbonyl)biphenyl-4-yl] oxycarbonyl}phenoxy)hexane [34]. The results are shown in Figure 10. The most notable result was Figure 10d; 1 ± 0.17 satellite peaks are observed, which clearly indicates a six-layer periodic superstructure just above the SmCA*(qT = 1/2) phase. Based on the paper written by Osipov and Gorkunov [35], we calculated the satellite peak intensity of six-layered structures. Considering the calculated results and finite dielectric constants caused by a non-zero net spontaneous polarization are observed for the observed six-layer subphase

($qT = 2/3$), we conclude that the ferrielectric structure of Figure 10e is considered to be more suitable to explain our experimental results. This structure was theoretically predicted by Emelyanenko and Ishikawa [36]. The smectic phase with six-layer periodicity has been already found between the SmC* and SmCα* phases by Wang et al. [37], but this is the first evidence of the subphase except $qT = 1/3$ and $1/2$ between SmC* and SmCA*.

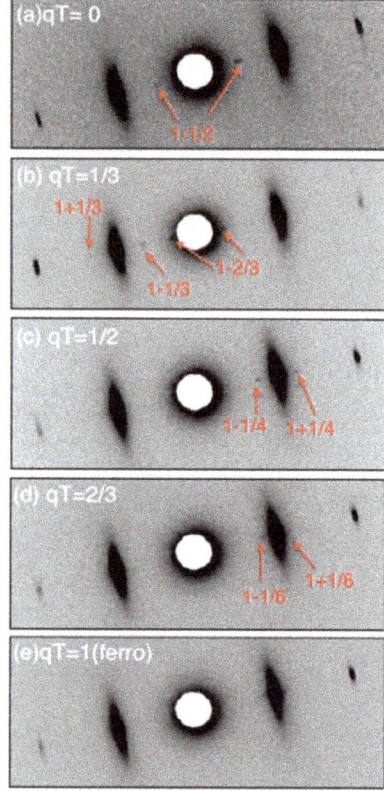

**Figure 10.** Two-dimensional microbeam resonant X-ray scattering profiles at various temperatures: (**a**) SmCA*($qT = 0$), (**b**) $qT = 1/3$, (**c**) $qT = 1/2$, (**d**) $qT = 2/3$, and (**e**) SmC*($qT = 1$) [34]. The red arrows indicate the satellite peak positions.

Further experiments were conducted and three new subphases were found by Feng et al. [38]. The results are shown in Figure 11, obtained from the mixture of Se-containing chiral liquid crystalline molecules. Between SmCA*($qT = 0$) and SmCA*($qT = 1/3$), subphases showing 3/8 and 5/8 resonant scattering are observed. Phase transition to this subphase was also observed by in situ texture observation by the polarized microscope equipped in the RXS system. From the calculation of relative intensities of RXS peaks based on Osipov and Gorkunov's paper [35], we could conclude that this subphase is SmCA*($qT = 1/4$) with an eight-layer periodic structure, as shown in Figure 11d. This summary is also consistent with to the simplest Farey sequence number, and the results of electric-field induced optical birefringence measurement. Furthermore, this subphase was universally observed in the different mixture of two Se-containing chiral molecules.

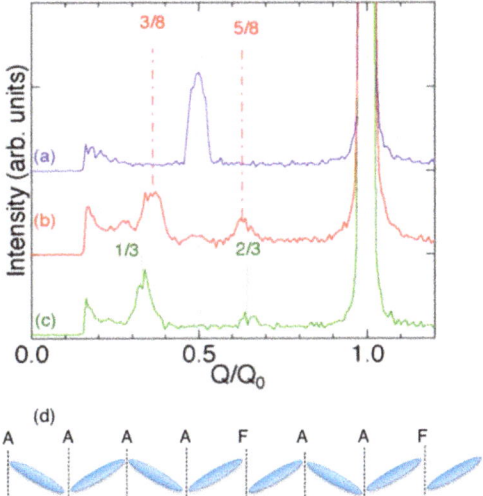

**Figure 11.** Microbeam resonant X-ray scattering intensity profiles in a binary mixture including a Se-containing chiral molecule for two-layer SmCA* (**a**), eight-layer qT = 1/4 (**b**), and three-layer qT = 1/3 (**c**) [38]. (**d**) Molecular arrangement of qT = 1/4 with an eight-layer periodic structure obtained from the calculation using the Osipov–Gorkunov formula [35].

Between SmCA*(qT = 1/3) and SmCA*(qT = 1/2), other different subphases are confirmed [38]. Figure 12 shows one-dimensional RXS intensity profiles along the layer normal of two Se-containing chiral molecules at various temperatures. Just above SmCA*(qT = 1/3), RXS satellite peaks were observed at $Q/Q_0$ = 0.3 and 0.7, which suggest that the subphase has a ten-layer periodic structure. Calculating RXS relative peak intensities for the respect to ten-layer periodic structures using the Osipov–Gorkunov formula and comparing electric-field induced optical birefringence measurement, we could summarize this subphase is SmCA*(qT = 2/5) and its molecular arrangement in the ten-layer periodic structure shown in Figure 12b. Furthermore, another RXS satellite peak at about $Q/Q_0$ = 0.286 (~2/7) was observed at the temperature between SmCA*(qT = 2/5) and SmCA*(qT = 1/2), suggesting the subphase SmCA*(qT = 3/7) with ferrielectric structure detecting by electric-field induced optical birefringence measurement. However, this satellite peak sometimes overlaps the RXS satellite peak at $Q/Q_0$ = 0.3 due to the coexistence of two subphases, and it is difficult to determine the structure decisively. We would like to perform more delicate measurements in the future.

In this way, the existence of suphases, which have not been recognized before, was clarified in addition to the three- and four-layered structures (SmCA*(qT = 1/3) and SmCA*(qT = 1/2)). Similar subphases are theoretically predicted by Emelynaneko and Osipov [39]; they considered the frustration between ferroelectric SmC* and antiferroelectric SmCA* and resulting degeneracy lifting by long-range interlayer interaction due to the discrete flexoelectric effect, they predicted subphases that could emerge, and, in these subphases, qT = 1/4 is contained between SmCA*(qT = 0) and SmCA*(qT = 1/3). They also predicted the emergence of qT = 3/7 (seven-layer periodic structure) but not the emergence of qT = 2/5 (ten-layer periodic structure), because their numerical calculations were limited to nine smectic layers in the unit cell. When the calculations are expanded to ten layers, qT = 2/5 with ten-layer periodicity appears [40].

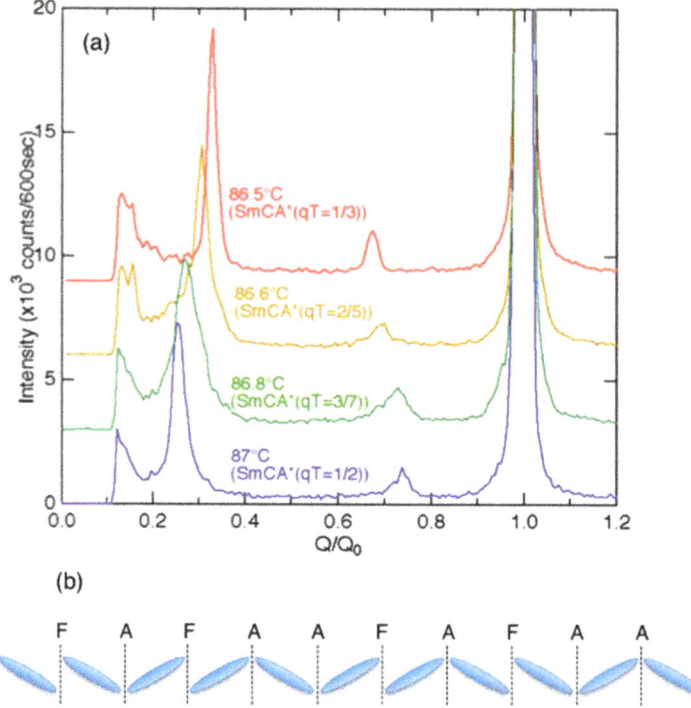

**Figure 12.** (**a**) Microbeam resonant X-ray scattering intensity profiles along the layer normal as a function of the normalized scattering vector ($Q/Q_0$, where $Q_0 = 2\pi/d$ and $d$ is a layer spacing) in of the mixture of AS657 (80 wt. %) and AS620 (20 wt. %). Between three-layer (qT = 1/3) and four-layer (qT = 1/2) subphases, ten-layer (qT = 2/5) and seven-layer (qT = 3/7) subphases are observed [38]. (**b**) Molecular arrangement of qT = 2/5 with a ten-layer periodic structure obtained from the calculation using the Osipov–Gorkunov formula [35].

## 6. Brief Summary: My Future Plan and Expectations for Applications

Finally, in the future, it is necessary to find additional suphases in different compounds to demonstrate their universality. For this purpose, not only resonant hard X-ray scattering, which requires specific elements, but also resonant soft X-ray scattering (RSoXS) using the absorption K-edge of carbon atom would be a powerful tool [41–43], and we have just started the study using RSoXS. As for applications, large flat panels may be difficult, but applications to small, high-definition display devices [44], optical modulators [45], and microwave and millimeter-wave control devices [46] can still be expected with the advantage of a fast response time (μsec order).

**Funding:** This research received no external funding.

**Acknowledgments:** This work is strongly supported by many collaborators outside, our staff and former and students. I acknowledge all these collaborators. In particular, special thanks are due to A. Fukuda, H. Takezoe, A. Iida, and J. K. Vij who led me to the interesting field and experiments, and made significant discussion. Some of the works was carried out under the approval of the Photon Factory Advisory committee (Proposal No. 2009G586, 2011G581, 2012G105 and No. 2014G154), and the second hutch of SPring-8 BL03XU constructed by the Consortium of Advanced Softmaterial Beamline (FSBL), with Proposal No. 2015A7255.

**Conflicts of Interest:** The author declares no conflict of interest.

## References

1. Känzig, W. Ferroelectrics and antiferroeletrics. *Solid State Phys.* **1957**, *4*, 1–197.
2. Meyer, R.B.; Liebert, L.; Strzelecki, L.; Keller, P. Ferroelectric liquid crystals. *J. Phys.* **1975**, *36*, L69–L71. [CrossRef]
3. Meyer, R.B. Ferroelectric liquid crystals; a review. *Mol. Cryst. Liq. Cryst.* **1977**, *40*, 33–48. [CrossRef]
4. Clark, N.A.; Lagerwall, S.T. Submicrosecond bistable electro-optic switching in liquid crystals. *Appl. Phys. Lett.* **1980**, *36*, 899–901. [CrossRef]
5. *FLC Displays with the Size of 15inch were Commercialized by Canon in 1993*; Canon: Tokyo, Japan, 1993.
6. Chandani, A.D.L.; Gorecka, E.; Ouchi, Y.; Takezoe, H.; Fukuda, A. Antiferroelectric chiral smectic phases responsible for the tristable switching in MHPOBC. *Jpn. J. Appl. Phys.* **1989**, *28*, L1265–L1268. [CrossRef]
7. Hiji, N.; Chandani, A.D.L.; Nishiyama, S.; Ouchi, Y.; Takezoe, H.; Fukuda, A. Layer structure and electro-optic properties in surface stabilized ferroelectric liquid crystal cells. *Ferroelectrics* **1988**, *85*, 99. [CrossRef]
8. Chandani, A.D.L.; Hagiwara, T.; Suzuki, Y.; Ouchi, Y.; Takezoe, H.; Fukuda, A. Tristable switching in surface stabilized ferroelectric liquid crystals with a large spntaneous polarization. *Jpn. J. Appl. Phys.* **1988**, *27*, L729. [CrossRef]
9. Takanishi, Y.; Takezoe, H.; Fukuda, A.; Komura, H.; Watanabe, J. Simple method for confirming the antiferroelectric structure of smectic liquid crystals. *J. Mater. Chem.* **1992**, *2*, 71–73. [CrossRef]
10. Harris, W.F. The dispiration: A distinct new crystal defect of the Weingarten-Volterra type. *Philos. Mag.* **1970**, *22*, 949. [CrossRef]
11. Takanishi, Y.; Takezoe, H.; Fukuda, A.; Watanabe, J. Visual observation of dispirations in liquid crystals. *Phys. Rev. B* **1992**, *45*, 7684–7689. [CrossRef]
12. Fukui, M.; Orihara, H.; Yamada, Y.; Yamamoto, N.; Ishibashi, Y. New phases in the ferroelectric liquid crystal MHPOBC studied by differential scanning calorimetry. *Jpn. J. Appl. Phys.* **1989**, *28*, L849. [CrossRef]
13. Takezoe, H.; Lee, J.; Chandani, A.D.L.; Gorecka, E.; Ouchi, Y.; Fukuda, A.; Terashima, K.; Furukawa, K. Antiferroelectric phase and tristable-switching in MHPOBC. *Ferroelectrics* **1991**, *114*, 187–197. [CrossRef]
14. Chandani, A.D.L.; Ouchi, Y.; Takezoe, H.; Fukuda, A.; Terashima, K.; Furukawa, K.; Kishi, A. Novel phases exhibiting tristable switching. *Jpn. J. Appl. Phys.* **1989**, *28*, L1261–L1264. [CrossRef]
15. Lee, J.; Chandani, A.D.L.; Itoh, K.; Takezoe, Y.O.H.; Fukuda, A. Frequency-dependent switching behavior under triangular waves in antiferroelectric and ferroelectric chiral smectic phases. *Jpn. J. Appl. Phys.* **1990**, *29*, 1122. [CrossRef]
16. Gorecka, E.; Chandani, A.D.L.; Ouchi, Y.; Takezoe, H.; Fukuda, A. Molecular orientational structures in ferroelectric, ferrielectric and antiferroelectric smectic liquid crystal phases as studied by conoscope observation. *Jpn. J. Appl. Phys.* **1990**, *29*, 131. [CrossRef]
17. Miyachi, K.; Kabe, M.; Ishikawa, K.; Takezoe, H.; Eukuda, A. Fluctuations in the ferrielectric smectic-C$\gamma^*$ phase as observed by laser beam diffraction and photon correlation spectroscopy. *Ferroelectrics* **1993**, *147*, 147. [CrossRef]
18. Isozaki, T.; Fujikawa, T.; Takezoe, H.; Fukuda, A.; Hagiwara, T.; Suzuki, Y.; Kawamura, I. Competition between ferroelectric and antiferroelectric interactions stabilizing varieties of phases in binary mixtures of smectic liquid crystals. Japanese journal of applied physics. *Jpn. J. Appl. Phys.* **1992**, *31*, L1435. [CrossRef]
19. Isozaki, T.; Fujikawa, T.; Takezoe, H.; Fukuda, A.; Hagiwara, T.; Suzuki, Y.; Kawamura, I. Devil's staircase formed by competing interactions stabilizing the ferroelectric smectic-C* phase and the antiferroelectric smectic-C A* phase in liquid crystalline binary mixtures. *Phys. Rev. B* **1993**, *48*, 13439. [CrossRef]
20. Isozaki, T.; Takezoe, H.; Fukuda, A.; Suzuki, Y.; Kawamura, I.J. Devil's staircase and racemization in antiferroelectric liquid crystals. *Muter. Chem.* **1994**, *4*, 237–343. [CrossRef]
21. Hiraoka, K.; Taguchi, A.; Ouchi, Y.; Takezoe, H.; Fukuda, A. Observation of three subphases in smectic C* of MHPOBC by dielectric measurements. *Jpn. J. Appl. Phys.* **1990**, *29*, L103. [CrossRef]
22. Isozaki, T.; Suzuki, Y.; Kawamura, I.; Mori, K.; Yamamoto, N.; Yamada, Y.; Orihara, H.; Ishibashi, Y. Successive phase transitions in antiferroelectric liquid crystal 4-(1-methylheptyloxycarbonyl) phenyl 4'-octylcarbonyloxybiphenyl-4-carboxylate (MHPOCBC). *Jpn. J. Appl. Phys.* **1991**, *30*, L1573. [CrossRef]
23. Ema, K.; Yao, H.; Kawamura, I.; Chan, T.; Garland, C.W. High-resolution calorimetric study of the antiferroelectric liquid crystals methylheptyloxycarbonylphenyl octyloxybiphenyl carboxylate and its octylcarbonylbiphenyl analog. *Phys. Rev. E* **1993**, *47*, 1203. [CrossRef] [PubMed]

24. Asahina, S.; Sorai, M.; Fukuda, A.; Takezoe, H.; Furukawa, K.; Terashima, K.; Suzuki, Y.; Kawamura, I. Heat capacities and phase transitions of the antiferroelectric liquid crystals MHPOBC and MHPOCBC. *Liq. Cryst.* **1997**, *23*, 339–348. [CrossRef]
25. Fukuda, A.; Takanishi, Y.; Isozaki, T.; Ishikawa, K.; Takezoe, H. Antiferroelectric chiral smectic liquid crystals. *J. Mater. Chem.* **1994**, *4*, 997–1016. [CrossRef]
26. Okabe, N.; Suzuki, Y.; Kawamura, I.; Isozaki, T.; Takezoe, H.; Fukuda, A. Reentrant antiferroelectric phase in 4-(1-methylheptyloxycarbnyl) phenyl 4′-octylbiphenyl-4-carboxylate. *Jpn. J. Appl. Phys.* **1992**, *31*, L793–L796. [CrossRef]
27. Mach, P.; Pindak, R.; Levelut, A.-M.; Barois, P.; Nguyen, H.T.; Huang, C.C.; Furenlid, L. Structural characterization of various chiral smectic-C phases by resonant X-ray scattering. *Phys. Rev. Lett.* **1998**, *81*, 1015–1018. [CrossRef]
28. Hirst, L.S.; Watson, S.T.; Gleeson, H.F.; Cluzeau, P.; Barois, P.; Pindak, R.; Pitney, J.; Cady, A.; Johnson, P.M.; Huang, C.C. Interlayer structure of the chiral smectic liquid crystal phases revealed by resonant x-ray scattering. *Phys. Rev. E* **2002**, *65*, 041705. [CrossRef]
29. Matkin, L.S.; Watson, S.T.; Gleeson, H.F.; Pindak, R.; Pitney, J.; Johnson, P.M.; Huang, C.C.; Barois, P.; Levelut, A.M.; Srajer, G. Resonant X-ray scattering study of the antiferroelectric and ferroelectric phases in liquid crystal devices. *Phys. Rev. E* **2001**, *64*, 021705. [CrossRef]
30. Brimicombe, P.D.; Roberts, N.W.; Jaradat, S.; Southern, C.; Wang, S.-T.; Huang, C.-C.; DiMasi, E.; Pindak, R.; Gleeson, H.F. Deduction of the temperature-dependent structure of the four-layer intermediate smectic phase using resonant X-ray scattering. *Eur. Phys. J. E* **2007**, *23*, 281. [CrossRef]
31. Roberts, N.W.; Jaradat, S.; Hirst, L.S.; Thurlow, M.S.; Wang, Y.; Wang, S.T.; Liu, Z.Q.; Huang, C.C.; Bai, J.; Pindak, R. Biaxiality and temperature dependence of 3- and 4-layer intermediate smectic–phase structures as revealed by resonant X-ray scattering. *Europhys. Lett.* **2005**, *76*, 976. [CrossRef]
32. Levelut, A.M.; Pansu, B. Tensorial X-ray structure factor in smectic liquid crystals. *Phys. Rev. E* **1999**, *60*, 6803–6815. [CrossRef] [PubMed]
33. Takanishi, Y.; Nishiyama, I.; Yamamoto, J.; Ohtsuka, Y.; Iida, A. Remarkable effect of a lateral substituent on the molecular ordering of chiral liquid crystal phases: A novel bromo-containing dichiral compound showing SmC* variants. *J. Mater. Chem.* **2011**, *21*, 4465–4469. [CrossRef]
34. Takanishi, Y.; Nishiyama, I.; Yamamoto, J.; Ohtsuka, Y.; Iida, A. Smectic-C* liquid crystals with six-layer periodicity appearing between the ferroelectric and antiferroelectric chiral smectic phases. *Phys. Rev. E* **2013**, *87*, 050503. [CrossRef] [PubMed]
35. Osipov, M.A.; Gorkunov, M.V. Model-independent structure and resonant X-ray spectra of intermediate smectic phases. *Liq. Cryst.* **2006**, *33*, 1133–1141. [CrossRef]
36. Emelyanenko, A.V.; Ishikawa, K. Smooth transitions between biaxial intermediate smectic phases. *Soft Matter* **2013**, *9*, 3497–3508. [CrossRef]
37. Wang, S.; Pan, L.D.; Pindak, R.; Liu, Z.Q.; Nguyen, H.T.; Huang, C.C. Discovery of a novel smectic-C* liquid-crystal phase with six-layer periodicity. *Phys. Rev. Lett.* **2010**, *104*, 027801. [CrossRef]
38. Feng, Z.; Perera, A.D.L.C.; Fukuda, A.; Vij, J.K.; Iida, K.I.A.; Takanishi, Y. Definite existence of suphases with eight- and ten-layer unit cells as studied by complementary methods, electric-field-induced birefringence and microbeam resonant X-ray scattering. *Phys. Rev. E* **2017**, *96*, 012701. [CrossRef]
39. Emelyanenko, A.V.; Osipov, M.A. Theoretical model for the discrete flexoelectric effect and a description for the sequence of intermediate smectic phases with increasing periodicity. *Phys. Rev. E* **2003**, *68*, 051703. [CrossRef]
40. Takanishi, Y.; Iida, A.; Yadav, N.; Perera, A.D.L.C.; Fukuda, A.; Osipov, M.A.; Vij, J.K. Unexpected electric-field-induced antiferroelectric liquid crystal phase in the SmCα* temperature range and the discrete flexoelectric effect. *Phys. Rev. E* **2019**, *100*, 010701. [CrossRef]
41. Virgili, J.M.; Tao, Y.; Kortright, J.B.; Balsara, N.P.; Segalman, R.A. Analysis of order formation in block copolymer thin films using resonant soft X-ray scattering. *Macromolecules* **2007**, *40*, 2092–2099. [CrossRef]
42. Zhu, C.; Wang, C.; Young, A.; Liu, F.; Gunkel, I.; Chen, D.; Walba, D.; Maclennan, J.; Clark, N.; Hexemer, A. Probing and controlling liquid crystal helical nanofilaments. *Nano Lett.* **2015**, *15*, 3420–3424. [CrossRef] [PubMed]

43. Salamonczyk, M.; Vaupotič, N.; Pociecha, D.; Wang, C.; Zhu, C.; Gorecka, E. Structure of nanoscale-pitch helical phases: Blue phase and twist-bend nematic phase resolved by resonant soft X-ray scattering. *Soft Matter* **2017**, *13*, 6694. [CrossRef] [PubMed]
44. Iio, K.; Kondoh, S. Bistable FLC panels with film substrates using a novel adhesive patterned spacer technology. *Ferroelectrics* **2006**, *344*, 197. [CrossRef]
45. Martinez, A.; Beaudoin, N.; Moreno, I.; Lopez, M.S.; Velasquez, P. Optimization of the contrast ratio of a ferroelectric liquid crystal optical modulator. *J. Opt. A* **2006**, *8*, 1013. [CrossRef]
46. Moritake, H.; Morita, S.; Ozaki, R.; Kamei, T.; Utsumi, Y. Fast-switching microwave phase shifter of coplanar waveguide using ferroelectric liquid crystal. *Jpn. J. Appl. Phys.* **2007**, *46*, L519. [CrossRef]

**Publisher's Note:** MDPI stays neutral with regard to jurisdictional claims in published maps and institutional affiliations.

© 2020 by the author. Licensee MDPI, Basel, Switzerland. This article is an open access article distributed under the terms and conditions of the Creative Commons Attribution (CC BY) license (http://creativecommons.org/licenses/by/4.0/).

*Review*

# Spontaneous and Controlled Macroscopic Chiral Symmetry Breaking by Means of Crystallization

Gérard Coquerel * and Marine Hoquante

SMS EA3233, Place Emile Blondel, University of Rouen Normandy, CEDEX, F-76821 Mont-Saint-Aignan, France; marine.hoquante@univ-rouen.fr
* Correspondence: gerard.coquerel@univ-rouen.fr

Received: 21 September 2020; Accepted: 26 October 2020; Published: 30 October 2020

**Abstract:** In this paper, macroscopic chiral symmetry breaking refers to as the process in which a mixture of enantiomers departs from 50–50 symmetry to favor one chirality, resulting in either a scalemic mixture or a pure enantiomer. In this domain, crystallization offers various possibilities, from the classical Viedma ripening or Temperature Cycle-Induced Deracemization to the famous Kondepudi experiment and then to so-called Preferential Enrichment. These processes, together with some variants, will be depicted in terms of thermodynamic pathways, departure from equilibrium and operating conditions. Influential parameters on the final state will be reviewed as well as the impact of kinetics of the R ⇔ S equilibrium in solution on chiral symmetry breaking. How one can control the outcome of symmetry breaking is examined. Several open questions are detailed and different interpretations are discussed.

**Keywords:** chirality; deracemization; preferential enrichment; thermodynamics; phase diagrams; kinetics

## 1. Context, Introduction

*1.1. Chiral Discrimination between Pairs of Enantiomers in the Solid State*

Two behaviors of the R–S system regarding racemization need to be distinguished. In the first case, the two enantiomers do not interconvert under the operating condition or in the time scale of the experiment. Therefore, the system is a symmetrical binary system where the two components have exactly the same thermodynamic properties (temperature, enthalpy and entropy of fusion, density, Cp versus T, etc.). This behavior is represented in Figure 1A–D. In the second case, the two enantiomers racemize rapidly in the liquid state. There is a relationship of interdependence between the two components, so the system is actually a degenerated binary system, as depicted in Figure 1E.

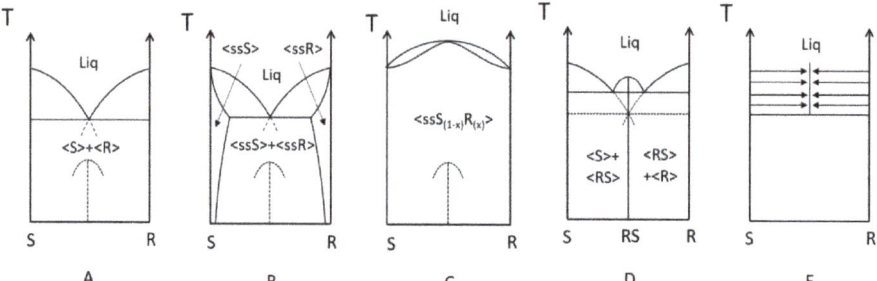

**Figure 1.** Binary system between a pair of enantiomers showing different types of chiral discriminations in the solid state: (**A–D**) in the absence of racemization in the liquid state; (**E**) with racemization in the liquid state. In cases (**A–C**), the vertical dashed lines represent metastable racemic compounds. System E represents a conglomerate with fast racemization in the liquid state.

There are several degrees of chiral discrimination in the solid state [1]. The top chiral discrimination occurs when every single crystal contains a single enantiomer, which is called conglomerate. This is often defined as "spontaneous resolution" (Figure 1A). The same case can exist with a partial solid solution (Figure 1B). By extension, there is also the possibility of a complete solid solution with a maximum or minimum point where, in the full composition range, both enantiomers are randomly distributed over the crystallographic sites. (Figure 1C). In the latter case, if there is no miscibility gap in the solid state, the chiral discrimination in this single solid phase is very poor. Then, there is the important heterochiral recognition corresponding to the formation of a stable <1-1> stoichiometric intermediate solid phase, called the "racemic compound" (Figure 1D). The latter case is by far the most popular: it accounts for 90–95% of all derivatives of a given couple of enantiomers. Crystallographic surveys show that most of these racemic compounds crystallize in a centro-symmetric space group ($P2_1/c$, P-1, Pbca, C2/c have the greatest occurrence). Nevertheless, there are several hundreds of kryptoracemic compounds (KRCs) in CSD version 2020. These KRCs account for ca. 1% of the racemic compounds. In those structures corresponding to chiral space groups (also named Sohncke space groups), the two enantiomers are in equal amount in the unit cell, but as independent molecules [2–5]. Therefore, Z' is an even number. When Z' > 2, other possibilities can arise such as anomalous conglomerates [6–8].

When changing the temperature, the chiral discrimination can be slightly or even completely altered. Indeed, from a racemic compound at low temperature, a stable conglomerate can be obtained at higher temperature through a three-phase peritectoid invariant [1]. The opposite situation is also well known. The three-phase invariant is then a eutectoid [9]. The switch from a racemic compound to a conglomerate-forming system can also appear with the addition of a particular solvent or co-crystal former or both [10]. The addition of crystal co-formers (isolated or in a mixture of solvents, counter-ions, co-crystal former, etc.) greatly increases the number of possibilities to explore in order to spot at least one conglomerate [11,12]. This increases the order of the system, which is no longer binary, but ternary, quaternary, quinary, etc. [13]. High-throughput techniques are of great help to alleviate the amount of work due to the overwhelming number of tests [14,15].

A complete solid solution at high temperature does not prevent the existence of a large miscibility gap in the solid state at low temperature [16]. The chiral discrimination in the solid state increases continuously as the two symmetrical solvus curves become more apart and towards a low temperature.

We will see that (paragraph on Preferential Enrichment: PE hereafter), for some special cases, and for systems initially very far from equilibrium reproducible chiral symmetry breaking can be observed. Conversely, departure from equilibrium can be detrimental to the chiral discrimination in the solid state. Fast cooling can lead to solid solutions without PE effects and even more severe cooling (i.e., quenching) can lead to amorphous material without any kind of macroscopic chiral recognition [17].

Partial solid solutions between enantiomers are also well known [5,18]. Those cases are intermediate between conglomerates without solid solution and with complete solid solution. In some systems, the crystal growths of the pure enantiomers can lead to peculiar microstructures named lamellar conglomerates, which should not be confused with racemic compounds [19].

*1.2. Equilibrium in Solution*

In solution, we can consider two different extreme situations: no racemization and fast racemization [10]. In the former, the chirality of the chemical entity is blocked in the solid state as well as in the liquid state. The latter encompasses: (i) the loss of chirality in a solution such as sodium chlorate ("racemization" is instantaneous here); (ii) some atropisomers with a low energetic barrier between the enantiomers in solution; (iii) enantiomers that can rapidly interconvert by the action of a catalyst (a base, an acid, an enzyme, etc.) (Figure 2).

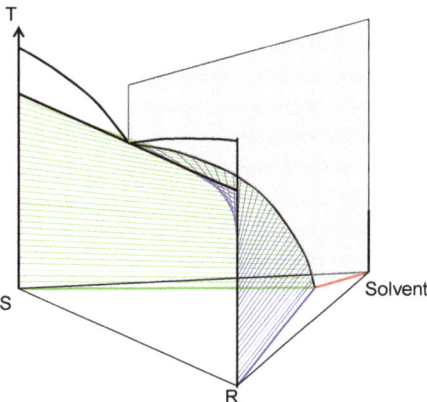

**Figure 2.** Degenerated conglomerate-forming ternary system with fast racemization in the liquid phase. The vertical plane contains the racemic liquor from saturated solution at a different temperature to the vertical line of the pure solvent. All other compositions of the liquid phase are simply not accessible. The green and blue tie lines, respectively, connect the S enantiomer and the R enantiomer to the saturated racemic solution at a given temperature.

## 2. Macroscopic Spontaneous Chiral Symmetry Breaking Induced by Crystallization

*Deracemization Induced by a Flux of Energy Crossing the Suspension (DIFECS)*

When a fast racemization takes place in solution under the same condition as the crystallization of a conglomerate, it is possible to observe a spontaneous macroscopic break in symmetry. The corresponding phase diagrams are no longer those displayed in Figure 1A–D, as detailed in [20,21]. Indeed, the phase diagrams are degenerated because the liquid phase can only contain an equimolar amount of enantiomers (i.e., a racemic composition see Figures 1E and 2). Under those circumstances, any energy flux passing through the suspension for long enough will lead to the disappearance of one population of homochiral crystals. A constant mechanical stress such as grinding (known as Viedma ripening) [22], numerous temperature cycles [23], long exposure to ultrasound [24], pressure stress [25] or microwaves [26] are general methods enabling the evolution of the initial dual population of particles to a single population of crystals containing a single enantiomer only, i.e., deracemization. Those methods operate rather close to thermodynamic equilibrium.

In Figure 3, the isotherm shows the main feature of the process. V stands for the solvent, while S and R are the two enantiomorphous chemical entities. Due to the fast racemization in solution or simply the absence of chirality in solution, the attainable states of the system are inside the triangle S-R- $L_{SAT}$ and along the racemic line V- $L_{SAT}$. $L_{SAT}$ is the point representative of the saturated racemic solution at that temperature. Under strong enough continuous attrition, the initial suspension represented by point I evolves towards $F_R$ or $F_S$ (Figure 3A,B). Simultaneously, the racemic mixture of solids represented by point M evolves towards S or R, i.e., the pure enantiomers. The tie lines connecting the constant saturated liquid $L_{SAT}$, the point representative of the overall synthetic mixture and the point representative of the solid composition move from $L_{SAT}$–I–M to $L_{SAT}$–$F_S$–S or $L_{SAT}$–$F_R$–R. If the system does not contain any chiral impurity and the initial two populations of crystals are symmetrical in terms of Crystal Size Distribution (CSD) and Growth Rate Dispersion (GRD), the last stage of evolution, that is, crystals of pure S or else pure R in equilibrium with $L_{SAT}$, is purely stochastic. Usually, the kinetics of this spontaneous evolution are of the first order. In other words, logarithm of enantiomeric excess (e.e.) versus time is linear. The more the overall composition departs from 50–50 composition, the faster the evolution towards homochirality is. Thus, it is an auto-catalytic process. However, if the system contains chiral impurities, the growth and dissolution rates of the two enantiomers become different, promoting one enantiomer over the other. The final evolution of Viedma ripening can be directed by using a chiral impurity and the e.e. versus time can also evolve linearly [27].

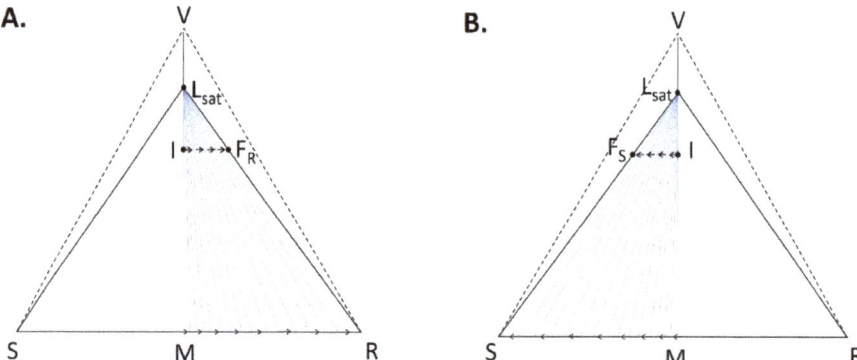

**Figure 3.** Spontaneous deracemization by Viedma ripening, ultrasound, microwaves towards R enantiomer (part (**A**)) and towards S enantiomer (part (**B**)). Starting from a suspension represented by point I, composed of a doubly saturated solution $L_{SAT}$ and an equimolar mixture of pure chiral solids, the system will spontaneously break its symmetry by an evolution towards $L_{SAT}$–$F_R$–R (represented) or conversely to a symmetrical system: $L_{SAT}$–$F_S$–S (not represented). $F_R$ and $F_S$ represent the two possible final compositions of the overall synthetic mixture after symmetry breaking.

Viedma ripening can be implemented directly during the synthesis; this method is called asymmetric synthesis, involving dynamic enantioselective crystallization [28].

For example, isoindolinones, a class of compounds used as core structures for pharmaceutical applications, were resolved successfully using Viedma ripening by the group of Vlieg [29]. Indeed, isoindolinones 1–3 (Figure 4) crystallize in a conglomerate-forming system and racemize quickly in solution without a catalyst.

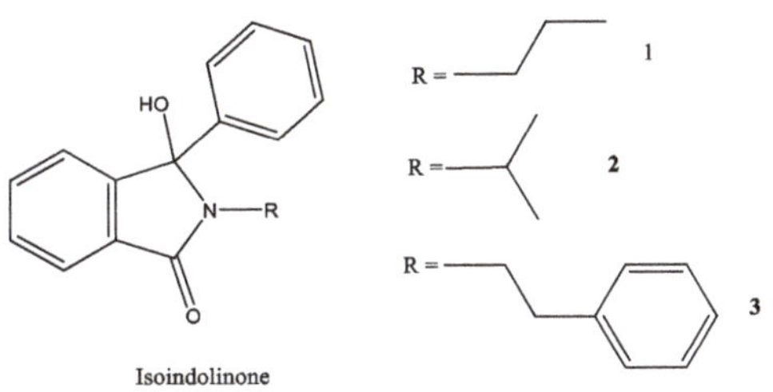

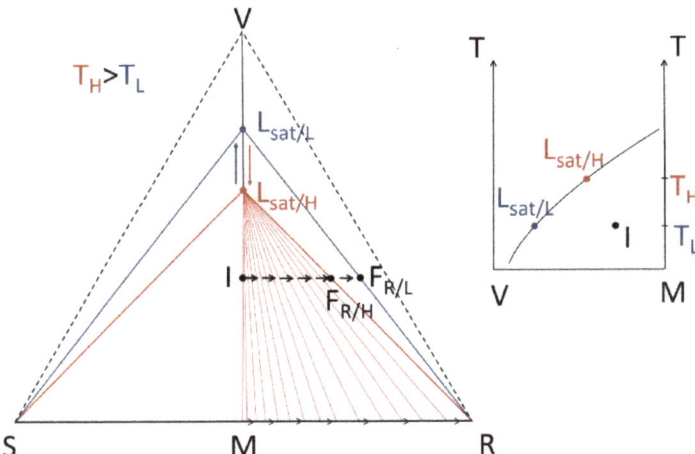

**Figure 4.** Isoindolinones that could be deracemized by Viedma ripening.

Repetitive oscillations of temperature are also able to induce deracemization in the system without limiting scale. This spontaneous process is known as Temperature Cycle-Induced Deracemization (known as TCID) [23]. The temperature gradient could be in space [30] or in time [31], or even both [32]. Figure 5 shows the corresponding phase diagram, which illustrates the evolution of the system when the temperature of the whole system fluctuates between $T_L$ and $T_H$, the low and high temperatures, respectively. When the e.e. of the solid phase strongly departs from zero, it is beneficial to decrease the amplitude of the temperature oscillation. The damped effect avoids wasting time and energy by preventing the dissolution of the enantiomer in excess [33].

**Figure 5.** TCID experiment: the inset shows the racemic vertical section versus temperature M–V–T, starting from a suspension of composition I, composed of a racemic mixture M of pure enantiomers in equilibrium with their saturated solution at $T_L$. Repetitive thermal oscillations of the suspension from $T_L$ to $T_H$ back and forth lead to complete deracemization (i.e., chiral symmetry breaking). Here, only the evolution to R is represented, but, in the absence of any chiral bias, the system can evolve to S as well. The amplitude and kinetics of the temperature oscillation impact the kinetics of deracemization.

Figure 5 schematizes the principle of deracemization by TCID; only one spontaneous evolution is shown: towards the R enantiomer. Two isotherms are represented: one at $T_L$ and the other one at $T_H$. When the initial suspension I is put back and forth from $T_L$ to $T_H$, it undergoes partial dissolution and recrystallization cycles. Starting from a suspension represented by point I, the overall synthetic mixture evolves towards $F_{R/H}$ at $T_H$ and $F_{R/L}$ at $T_L$. Simultaneously, the composition of the solid phase evolves towards the pure enantiomer R (or S; the latter case is symmetrical to the one represented in Figure 5). It is worth noting that there is an initial period without a noticeable evolution in the e.e. of the solid phase. However, the CSD and GRD and probably other solid-phase attributes, change during that period. This first step ends with what is called the "take-off", a colloquial expression for the significant macroscopic evolution of the system. Kinetics can have a rather odd aspect, e.g., the system can remain seated on the "racemic fence" for several days before the take off. This phenomenon illustrates the stochastic aspect of spontaneous symmetry breaking. After this first period, the system shows the classical sigmoid evolution of the enantiomeric excess (e.e.) of the solid phase vs. time, which means that it possesses first-order kinetics.

The temperature versus time profile must be tuned for the achievement of the deracemization and for its productivity, together with the minimization of the chemical degradation, if there is any. The variation in solubility versus temperature is, of course, an important factor, but the cooling rate is also important for the generation of small nuclei via secondary nucleation. This phenomenon participates to the turnover of the particles of the two populations of crystals [34].

Deracemization using temperature fluctuations was demonstrated on a precursor of paclobutrazol, a molecule of interest because of its role as a plant growth inhibitor (Figure 6) [31]. In this case, the temperature fluctuations range between 20 °C and 25 °C or 30 °C and the racemization is induced by sodium hydroxide.

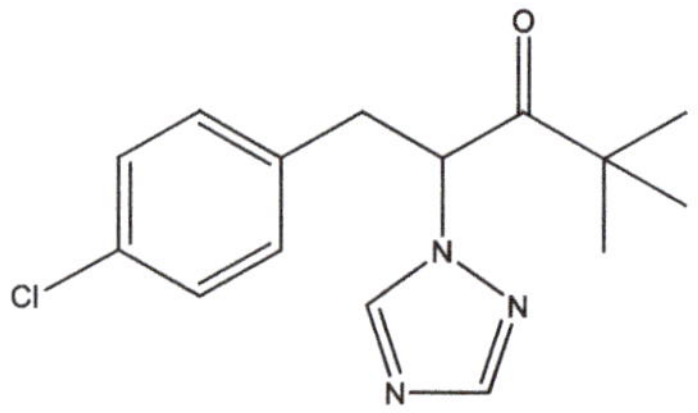

**Figure 6.** 1-(4-chlorophenyl)-4,4-dimethyl-2-(1H-1,2,4-triazol-1-yl)pentan-3-one, a precursor of paclobutrazol, which served as a model compound for TCID.

In addition to mechanical or thermal energy fluxes (see above), other energetic fluxes passing through the suspension lead to complete deracemization (Deracemization Induced by a Flux of Energy Crossing the Suspension (DIFECS). For instance, periodic variations in pressure [25] or pressure and temperature [35], microwaves [26], and photons for light-sensitive molecules [11] have proved to induce complete chiral symmetry breaking. This is not a limitative list. On top of this, these stimuli have agonist effects and thus can be cumulated to speed-up the macroscopic chiral symmetry breaking [36]. The common features of these processes are that they are operated somewhat close to thermodynamic equilibrium with a stochastic character regarding the final stages, R or S. They show first-order kinetics, that is, an autocatalytic global behavior. This does not mean that the predominant mechanisms are all the same. For instance, the application of ultrasound could be faster than attrition to induce complete deracemization; nevertheless, the final crystals are bigger [24]. The agonist effects of those various fluxes of energy seem more consistent with several—concomitant—possible pathways. Several mathematical models have been proposed that fit pretty well with the sigmoid shape of e.e. variation in the solid versus time [37–41]. DIFECS has been proven to be suitable for general application, provided the two following conditions are fulfilled: a conglomerate is in equilibrium with a doubly saturated solution and the chemical entities undergo rapid racemization in solution if not instantaneous—e.g., in the case of a loss of asymmetry such as $NaClO_3$ and $NaBrO_3$.

For instance, glutamic acid could be deracemized by microwave-assisted temperature cycling with a much shorter process time compared to conventional temperature cycles [26].

Kondepudi's experiment [42] is another illustration of spontaneous chiral macroscopic symmetry breaking by using crystallization in a conglomerate-forming system. Figure 7 illustrates this experiment. Typically, a racemic solution is cooled down with given kinetics in a stirred medium. If the system is able to generate a single nucleus only (the "Eve" crystal) for a sufficient period of time and if the stirring rate and stirring mode [43] are adequately tuned, numerous offspring crystals will be created by collision with the stirrer or the wall of the reactor and thus a fast secondary nucleation originates from this "Eve" crystal. This phenomenon drops the supersaturation of the medium. At the end of the process, a homochiral population of crystals is generated, which are all descendants of the one which nucleated first. Ideally, by repeating numerous experiments, the results should be a random series of (−) and (+) populations of crystals that are statistically close to 50–50%.

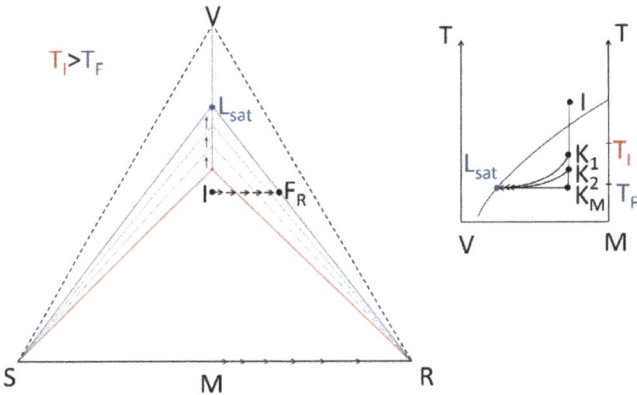

**Figure 7.** Kondepudi's experiment: in the inset, the temperature versus composition pathway of the mother liquor is represented. This is the racemic section as the liquid phase cannot deviate from enantiomeric excess (e.e.) = 0. In this experiment, it is possible that I-K1-$L_{SAT}$ and I-K2-$L_{SAT}$ pathways lead to the homochirality of the solid phase. By contrast, $K_M$ might correspond to an excessive undercooling which leads to several quasi simultaneous primary nucleation events and thus a heterochiral final population of crystals.

This process was first described by Kondepudi, who found that spontaneous symmetry breaking upon the crystallization of sodium chlorate occurs in stirred solutions, whereas static conditions give an equal amount of L and D crystals [44]. Kondepudi's experiment is also applicable to the molten state, as a supercooled melt of 1,1'-binaphthyl could crystallize with a large enantiomeric excess when vigorously stirred [45].

Preferential Enrichment (PE hereafter) also appears as a spontaneous macroscopic symmetry breaking phenomenon [46]. It was discovered and has been developed for 25 years by Professor Rui Tamura at Kyoto University. Initially confined to a series of non-racemizable organic salts (first generation), it has since been extended to other families of compounds such as amino acids and chiral pharmaceutical drugs (second generation compounds, hereafter considered non-racemizable chemical entities). It is, in a way, the opposite of preferential crystallization. Indeed, the binary system corresponds to an intermediate state between Figure 1B,C (see below for a discussion on the nature of the solid phase close to racemic composition), it is run with a considerable supersaturation and the phase that is very much enriched (>90% e.e.) is the mother liquor. By contrast, the solid phase exhibits a poor and opposite deviation (ca 4–5% e.e.). In opposition, preferential crystallization [47] is run rather close to equilibrium with a conglomerate-forming system (Figure 1A) and the very much enriched phase is the solid. The e.e. of the mother liquor is opposite to that of the solid and usually remains lower than 20% (this value is merely for the best cases). Several conditions have been pointed out in relation to the success of PE:

(i) A large solubility difference between the racemic composition (poorly soluble) and the pure enantiomer (much more soluble) is necessary. The alpha molar ratio, $\alpha = s(\pm)/s(+) = s(\pm)/s(-)$, is thus very small (this constitutes another contrast with preferential crystallization for which $\alpha$

is usually comprised between: $\sqrt{2}$ and 2). Various analyses of racemic solutions in different solvents led to the surmised existence of solvated homochiral assemblies.

(ii) For a globally racemic composition, the crystal structure permits a certain degree of disorder between homochiral chains and/or planes, even if single crystals obtained from poorly supersaturated racemic solution (e.g., point $\Omega_{1.2}$ in Figure 8 representing a supersaturation $C/C_{SAT} = 1.2$) could have their structures resolved by X-ray diffraction in centrosymmetric space groups such as P-1 (the most frequent for PE) or $P2_1/c$. However, crystals obtained under high supersaturation (i.e., from a clear solution represented by point $\Omega_8$ in Figure 8) clearly reveal, by Second Harmonic Generation (SHG), homochiral domains. If this effect cannot be detected, PE experiments fail [48].

(iii) The first generation of compounds showing PE effect have all exhibited solid–solid transitions between various disordered phases. For the second generation of compounds showing PE effect in some cases, no such solid–solid transition could be detected. A solid–solid transition during PE does not appear anymore as a mandatory condition for its success.

(iv) From the first solid crystallized, which can have a high degree of stacking faults, a selective dissolution of domains containing the same enantiomer as that in excess in the solution occurs. A unique, detailed study [49] has shown that this dissolution is actually concomitant to the re-incorporation of the opposition enantiomer. It is thus the exchange of opposite enantiomers that is likely to be a concerted process. This results in a clear enrichment of the mother liquor and, simultaneously, a slight enrichment of the solid phase in the opposite enantiomer At the end of PE, the solid phase appears to be composed of Heterogeneous Nearly-Racemic Crystals (HNRC). This is different from a genuine solid solution (i.e., mixed crystals) where a random distribution of the two enantiomeric molecules is observed over the crystallographic sites. In HNRC, there are some homochiral domains of sub-micron to micron sizes.

(v) The HNRC could remain kinetically stable for months without a return to stable equilibrium if the system remains unstirred in a quiescent state.

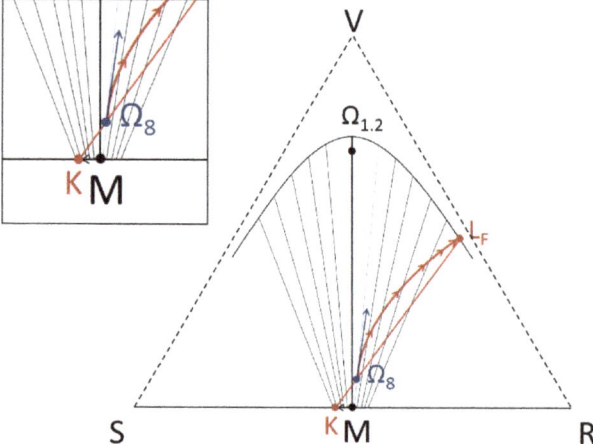

**Figure 8.** Second generation of compound presenting Preferential Enrichment (PE) effect (without polymorphism), when left very far from equilibrium, e.g., solution $\Omega_{8+}$. From the eightfold supersaturation, crystallization takes place and the solution point moves along a trajectory schematized by the red curve. The final evolution of the system is represented by the solution $L_F$ and solid K—with opposite chirality—both aligned with point $\Omega_8$, representing the overall synthetic mixture.

It is important to mention that PE has to be run at very high initial supersaturation and the system should not undergo any mechanical stress. For instance, smooth stirring introduced by the slow motion of a rocking plate is enough to derail PE and to return to normal thermodynamics without macroscopic symmetry breaking. Moreover, if the initial supersaturation is not high enough ($\beta >$ four, six, eight or more (!)) PE is not observed. Thus, PE is clearly a far from equilibrium process. In Figure 8, the red tie line connecting the composition of the mother liquor and the solid at the end of the PE intersects the racemic line M–L$_{RAC}$, which is a clear violation of the thermodynamics of equilibrium. The thermodynamics of equilibrium are represented by the black tie lines and, within their context, it is impossible to have a break in symmetry, i.e., to have a solution and a solid of opposite chirality for a long period of time. When the macroscopic evolution of the system is over—this can take several days—the composition of the mother liquor (e.g., +95% e.e.) and that of the solid (e.g., ca. −4–5% e.e.) can remain unchanged for months, maybe more, in a stagnant medium.

An efficient Preferential Enrichment phenomenon could be observed for the (DL)-phenylalanine and fumaric acid co-crystal (Figure 9) [50].

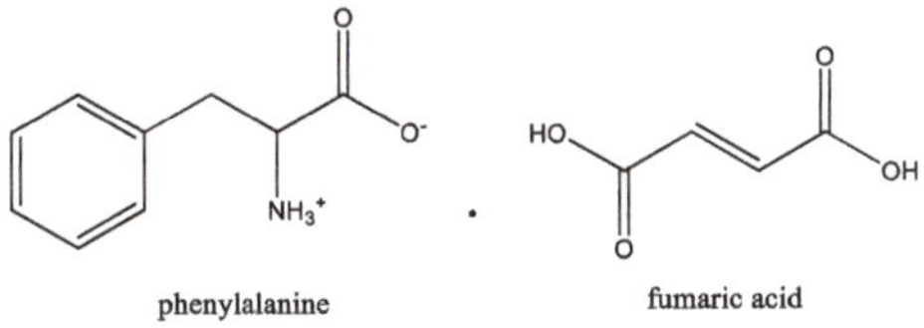

**Figure 9.** Cocrystal of (DL)-phenylalanine and fumaric acid, a system that exhibits Preferential Enrichment.

There is an interesting analogy between Lamellar conglomerates [19] and fluctuations in enantiomeric composition around the racemic composition observed in PE. On the one hand, lamellar conglomerates correspond to the stacking of homochiral domains [51]. A particle, crystallized from a racemic solution, could look very much like a nice, single crystal, but could actually be constituted by the alternation of homochiral domains. At the interface of opposite domains, a 2D racemic compound is formed, but, for unclear reasons, this packing does not expand in the third direction. Quite often, the Flack parameter [52] reveals trouble in the absolute configuration assignment of the molecule and the non-linear optics of the powder show an enhanced SHG effect compared to that of a single enantiomer with the same CSD. The global composition is thus quasi-racemic, but this is not actually a single crystal. On the other hand, when operating PE with large supersaturation and under stagnant conditions (and therefore far from equilibrium thermodynamics), there are also fluctuations in the composition of solid particles. This could also be revealed by the SHG effect. Thus, the PE effect is linked to the formation of Heterogeneous Nearly-Racemic Crystals (HNRC) or, in other words, racemic compounds with local fluctuations in their enantiomeric composition. Racemic crystals that do not display the possibility of local enantiomeric fluctuation do not exhibit any PE effects.

Stirring the crystallizing suspension curbs the fluctuations in its composition and totally inhibits PE. Local deviations in the composition of the stagnant mother liquor in the vicinity of the growing surfaces are the driving forces of those phenomena. For lamellar conglomerates, when an S crystal is growing, the R enantiomer is overrepresented in the neighboring solution. The heteronucleation of R on top of the S crystal is more likely as preferential crystallization proceeds towards the end of a run (see Figure 10 and caption). The resulting epitaxy is a sort of regulatory phenomenon that diminishes the entrainment effect, i.e., drops the magnitude of the transient symmetry breaking. One study has shown that these local fluctuations could be amplified at a macroscopic scale in the mother liquor [53]. In the case of PE, the opposite effect occurs: the incorporation of the minor enantiomer in the mother liquor around the crystal and the liberation of the minor enantiomer in the solid (slightly in default in the solid) constitute an amplification of local dissymmetry (see Figure 10 and caption).

One can perceive this phenomenon as another type of ripening resulting from an initial, far from equilibrium crystallization.

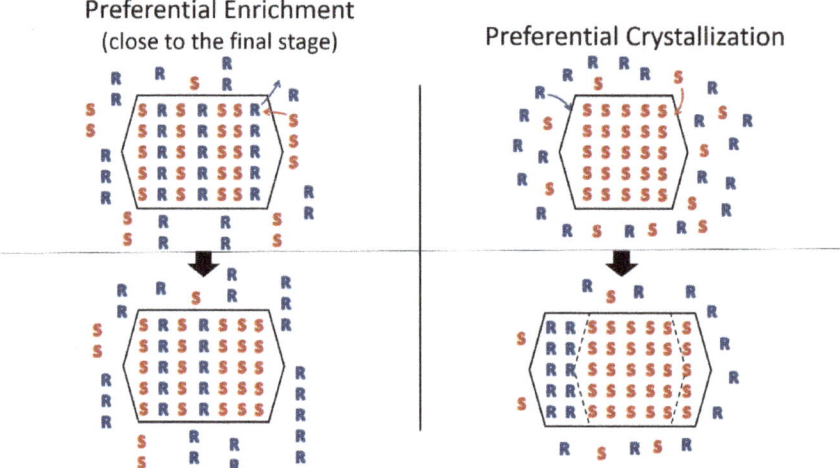

**Figure 10.** Schematic representation of the mechanisms of formation of heterochiral domains in single particles in PE and Preferential Crystallization (PC). Left: as explained in conditions for the success of PE, solvated homochiral associations are likely to exist, and are symbolized as vertical redundant letters—S or R. During the process, even when the supersaturation becomes fairly weak, the minor enantiomer in solution can substitute for the minor enantiomer in the defective solid state. Thus, there is an amplification of the e.e. in the solid and in the mother liquor. Right: focus of the crystal growth of the S enantiomer during PC, the mother liquor is depleted in the S enantiomer in the vicinity of the single crystal. As R is supersaturated, it is possible for this enantiomer to heteronucleate on top of its antipode by means of an epitaxy. In both cases PE and PC, the resulting particles are constituted by homochiral domains: sub-micron to several microns size in PE and up to hundreds of microns in PC.

One of the deep-seated reasons for those effects is the existence of an exact match between the crystal lattices of the enantiomorphous components (i.e., the Friedel and Royer conditions for the junction of the two crystal lattices are perfectly fulfilled) [54].

## 3. Control of Macroscopic Chiral Symmetry Breaking by Means of Crystallization

It is possible to orientate the symmetry breaking towards a desired enantiomer by using several robust methods. For instance, Deracemization Induced by a Flux of Energy Crossing the Suspension (DIFECS) could lead to the eutomer (the desire enantiomer) by adding a small investment prior to the beginning of the process [55]. For example, only a small percentage of e.e. (+) in the solid state is sufficient to conduct the complete deracemization by using Viedma ripening, TCID, ultrasound, microwaves, etc., towards the (+) enantiomer. This statement is valid if there is no chiral impurity in the medium that could overcompensate the initial bias introduced purposely [56], that is, without this imbalance, the system would have a stochastic behavior. This is illustrated by the Viedma ripening of sodium chlorate (used as received; see Figure 11).

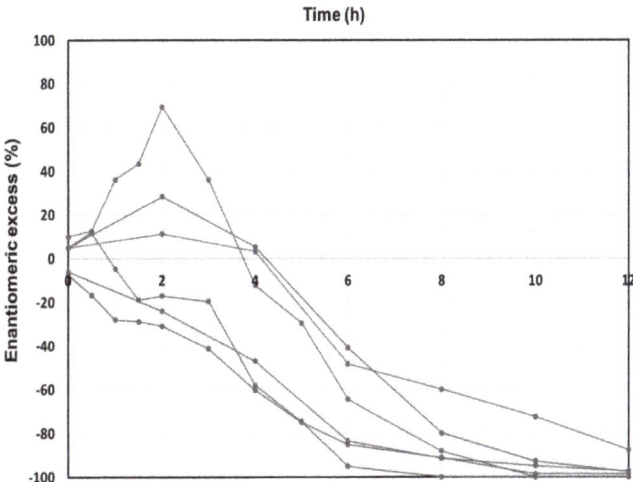

**Figure 11.** Impurity effect on the handedness of the attrition-enhanced deracemization in a non-recrystallized batch of sodium chlorate [57]. After a simple recrystallization in water, this effect disappeared.

The evolution is systematically towards the same chirality even if an initial investment in the counter enantiomer is performed. One experiment shows an evolution of up to 70% e.e. before a turn back. A simple recrystallization of the initial racemic mixture is enough to remove that effect. It is likely that a small (maybe minute) amount of a chiral impurity "pushes" the system towards the same enantiomers. This phenomenon has been observed with organic compounds [58]. Dissymmetry in crystal size distributions is also able to systematically direct the symmetry breaking towards the population of bigger crystals ("big is beautiful"). A study has shown that there is actually a balance between the initial e.e. and the dissymmetry of crystal size distributions [59]. Bigger crystals of (+) can, for instance, compensate a slight initial excess in small (−) crystals [60].

Kondepudi's experiment, seeded with very pure enantiomer prior to any primary nucleation of either enantiomer, is equivalent to preferential crystallization. A detailed analysis of the process is given elsewhere with different protocols for the inoculation of seeds and temperature profiles [47]. The symmetry of the system is purposely broken by the seeding: if the solid is the (+) enantiomer, the mother liquor evolves towards an excess of (−) in the absence of racemization. This induced symmetry breaking lasts for some minutes to some hours. The fine enantiopure inoculated crystals lead to stereoselective growth and secondary nucleation; during this period, the counter enantiomer remains in the supersaturated solution. If the system is left for too long, the second enantiomer starts to spontaneously crystallize so that the ultimate evolution of the system is a mixture of crystallized enantiomers in equilibrium with a doubly saturated racemic solution. If fast racemization takes place in the system (or in the absence of chirality in solution), the mother liquor cannot deviate from 0% e.e, as illustrated by Figure 2). In that case, the preferential crystallization receives another name: Second-Order Asymmetric Transformation (SOAT; represented in Figure 12) [61]. This elegant process could be two orders of magnitude more productive than any variant of DIFECS [62]. Supersaturation has to be kept within reasonable limits so that crystal growth and secondary nucleation remain stereoselective.

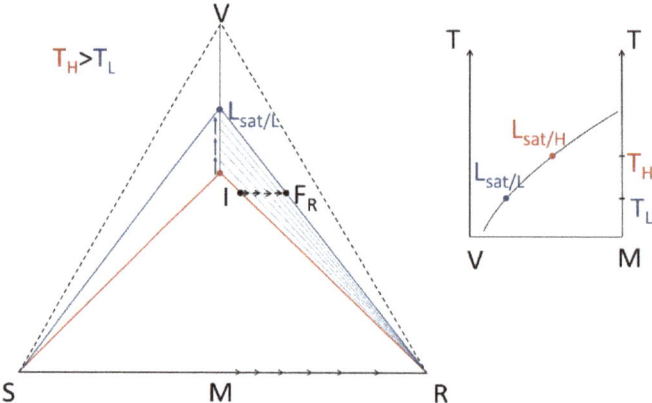

**Figure 12.** Second-Order Asymmetric Transformation (SOAT) experiment: the inset shows the temperature versus composition pathway of the mother liquor. Point I does not belong to the isopleth section represented in the inset, as the seed crystals are enantiopure.

In 1979, SOAT was used to resolve (DL)-α-amino-ε-caprolactam, a lysine precursor, with the racemization being induced by potassium hydroxide in ethanol at 80 °C (Figure 13) [63]. Likewise, the enantiomers of the precursors of paclobutrazol, which served as an example for TCID, were also obtained in high enantiomeric purity with SOAT by Black et al. in 1989 [64].

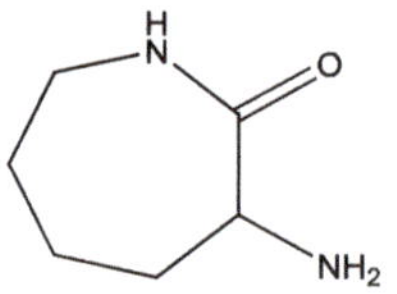

**Figure 13.** (DL)-α-amino-ε-caprolactam can be resolved by SOAT to produce enantiopure lysine after hydrolysis.

Lamellar epitaxies between crystals of opposite handedness could be a serious problem for Preferential Crystallization and Kondepudi's experiments [19]. The remedy is to rely on deracemization, which has been proven to achieved 100% e.e. even when the crystals of the two enantiomers can produce repeated epitaxies [65].

Kondepudi's experiments could also be controlled by the addition of specific chiral impurities. In this process, to a racemic supersaturated solution of a conglomerate-forming system, a small amount of R* additive is added to the medium. If R* has a sufficient degree of analogy with R, the nucleation and growth of R and S become significantly different and precipitation leads to an excess of S in the solid. This effect is known as the rule of reversal [66]. Of course, there is a reversibility of the rule of reversal, which means that if S* and R* also crystallize as a stable conglomerate, R could stereoselectively delay the nucleation of R*, giving way to the nucleation and growth of S*. It is also possible to induce stereoselective nucleation by using a polarized laser beam in a supersaturated solution. This process, known as Non-Photochemical Laser Induced Nucleation (NPLIN), has received some attention [67].

In the case of PE, the final state could also be controlled by an initial minor investment. As the initial step is a total dissolution of non-racemizable enantiomers (at least in the context of the crystallization), the CSD of the solid has no influence on the outcome of the process. Successive applications of PE lead to the alternation of (−), (+), (−), (+), etc., a slightly enriched solid and the opposite series for the liquid phase, strongly enriched in (+), (−), (+), (−), (+), etc. These alternations are clearly reproducible.

## 4. Conclusions and Perspectives

Crystallization offers a variety of methods for spontaneous macroscopic chiral symmetry breaking. The corresponding processes are run close to equilibrium for Deracemization Induced by a Flux of Energy Crossing the Suspension (DIFECS) (Viedma ripening, TCID and other deracemization variants where a single flux or several fluxes of energy pass through a suspension) or with a significant departure from equilibrium for the Kondepudi experiment with preferential secondary nucleation, or even very far from equilibrium in the case of Preferential Enrichment (PE). Those processes include an important amplification of local fluctuations and their mechanisms are currently the focus of many studies. Interestingly, as the system departs more from equilibrium, the mechanical stress imposed on the system has to be softened in order to observe the spontaneous chiral symmetry breaking (see Table 1). Indeed, in Viedma ripening performed close to equilibrium, all sorts of abrasions, breakages, defects induced by shear forces, etc., are beneficial to the advancement of chiral symmetry breaking. In the Kondepudi experiment, the mechanical stress must be softer to avoid primary and secondary heteronucleation in the system. In the case of Preferential Enrichment (PE), the very large departure from equilibrium has to be associated with almost stagnant conditions. Indeed, for the latter case a simple magnetic stirrer is sufficient to return the solution to normal conditions of crystallization without macroscopic chiral symmetry breaking (see Table 1 below).

**Table 1.** Spontaneous macroscopic break of symmetry by means of crystallization. In theory, the final state of the system is not predictable. There is a stochastic evolution during the first step and then amplification. In Deracemization Induced by a Flux of Energy Crossing the Suspension (DIFECS) processes, there are agonist effects between ultrasound, microwaves, attrition, TCID and damped TCID (and photons for light-sensitive molecules).

| Towards greater departure from equilibrium ↑ | Preferential Enrichment | Heterogenous Nearly-Racemic Crystals (HNRC) with possibility of alternating homochiral domains. Largest domains corresponding to the minor enantiomer | 0: Stagnant conditions to stay away from stable equilibrium | ↓ |
|---|---|---|---|---|
| ↑ | Kondepudi's Experiment | Conglomerate forming system Fast racemization in solution or non-chiral entity in solution | Moderate, collision but avoid too strong shearing effects | ↓ |
| ↑ | Deracemization Induced by a Flux of Energy Crossing the Suspension (DIFECS): -Ultrasounds -Microwaves -TCID -Viedma Ripening (close to 0) | Conglomerate forming system Fast racemization in solution or non-chiral entity in solution A lamellar conglomerate does not hinder deracemization | From soft to medium for TCID Strong with shearing effect for attrition enhanced i.e., Viedma Ripening | ↓ Towards greater intensity of the mechanical stress |

It is also possible to control a macroscopic chiral symmetry breaking such as Viedma ripening, TCID and its variants, or SOAT or preferential crystallization. Some of these elegant processes could be made productive enough for industrial applications. Thus, there are fundamental and applied interests in these crystallization processes associated with macroscopic symmetry breaking.

**Author Contributions:** Both authors contributed to this publication. All authors have read and agreed to the published version of the manuscript.

**Funding:** This review is part of a PhD funded by the University of Rouen Normandy.

**Conflicts of Interest:** The authors declare no conflict of interest.

**List of Symbols and Abbreviations**

| | |
|---|---|
| CSD: | Crystal Size Distribution |
| DIFECS: | Deracemization Induced by a Flux of Energy Crossing the Suspension |
| e.e.: | Enantiomeric excess = (R − S)/(R + S) |
| GRD: | Growth Rate Dispersion |
| HNRC: | Heterogeneous Nearly-Racemic Crystals |
| NPLIN: | Non-Photochemical Laser Induced Nucleation |
| PC: | Preferential Crystallization |
| PE: | Preferential Enrichment |
| SOAT: | Second Order Asymmetric Transformation |
| TCID: | Temperature Cycle-Induced Deracemization |
| US: | Ultrasound |

**References**

1. Coquerel, G. Review on the Heterogeneous Equilibria between Condensed Phases in Binary Systems of Enantiomers. *Enantiomer* **2000**, *5*, 481–498. [PubMed]
2. Clevers, S.; Coquerel, G. Kryptoracemic Compound Hunting and Frequency in the Cambridge Structural Database. *CrystEngComm* **2020**. [CrossRef]
3. Fábián, L.; Brock, C.P. A List of Organic Kryptoracemates. *Acta Crystallogr. B* **2010**, *66*, 94–103. [CrossRef] [PubMed]
4. Tiekink, E.R.T. Kryptoracemates. In *Advances in Organic Crystal Chemistry, Comprehensive Reviews 2020*; Springer: Cham, Switzerland, 2020; Chapter 19; pp. 381–404.
5. Rekis, T. Crystallization of Chiral Molecular Compounds: What Can Be Learned from the Cambridge Structural Database? *Acta Crystallogr. Sect. B Struct. Sci. Cryst. Eng. Mater.* **2020**, *76*, 307–315. [CrossRef]
6. Bishop, R.; Scudder, M.L. Multiple Molecules in the Asymmetric Unit (Z′ > 1) and the Formation of False Conglomerate Crystal Structures. *Cryst. Growth Des.* **2009**, *9*, 2890–2894. [CrossRef]
7. Dryzun, C.; Avnir, D. On the Abundance of Chiral Crystals. *Chem. Commun.* **2012**, *48*, 5874–5876. [CrossRef]
8. Gubaidullin, A.T.; Samigullina, A.I.; Bredikhina, Z.A.; Bredikhin, A.A. Crystal Structure of Chiral Ortho-Alkyl Phenyl Ethers of Glycerol: True Racemic Compound, Normal, False and Anomalous Conglomerates within the Single Five-Membered Family. *CrystEngComm* **2014**, *16*, 6716–6729. [CrossRef]
9. Druot, S.; Petit, N.; Petit, S.; Coquerel, G.; Chanh, N. Experimental Data and Modelling of the Interactions in Solid State and in Solution between (R) and (S) N-Acetyl-α-Methylbenzylamine. Influence on Resolution by Preferential Crystallization. *Mol. Cryst. Liq. Cryst. Sci. Technol. Sect. Mol. Cryst. Liq. Cryst.* **1996**, *275*, 271–291. [CrossRef]
10. Harfouche, L.C.; Brandel, C.; Cartigny, Y.; Ter Horst, J.H.; Coquerel, G.; Petit, S. Enabling Direct Preferential Crystallization in a Stable Racemic Compound System. *Mol. Pharm.* **2019**. [CrossRef]
11. Belletti, G.; Tortora, C.; Mellema, I.D.; Tinnemans, P.; Meekes, H.; Rutjes, F.P.J.T.; Tsogoeva, S.B.; Vlieg, E. Photoracemization-Based Viedma Ripening of a BINOL Derivative. *Chem. Eur. J.* **2020**, *26*, 839–844. [CrossRef]
12. Wacharine-Antar, S.; Levilain, G.; Dupray, V.; Coquerel, G. Resolution of (±)-Imeglimin-2,4-Dichlorophenylacetate Methanol Solvate by Preferential Crystallization. *Org. Process Res. Dev.* **2010**, *14*, 1358–1363. [CrossRef]
13. Coquerel, G. Solubility of Chiral Species as Function of the Enantiomeric Excess. *J. Pharm. Pharmacol.* **2015**, *67*, 869–878. [CrossRef] [PubMed]
14. Galland, A.; Dupray, V.; Berton, B.; Morin-Grognet, S.; Sanselme, M.; Atmani, H.; Coquerel, G. Spotting Conglomerates by Second Harmonic Generation. *Cryst. Growth Des.* **2009**, *9*, 2713–2718. [CrossRef]
15. Simon, F.; Clevers, S.; Dupray, V.; Coquerel, G. Relevance of the Second Harmonic Generation to Characterize Crystalline Samples. *Chem. Eng. Technol.* **2015**, *38*, 971–983. [CrossRef]
16. Gendron, F.-X.; Mahieux, J.; Sanselme, M.; Coquerel, G. Resolution of Baclofenium Hydrogenomaleate by Using Preferential Crystallization. A First Case of Complete Solid Solution at High Temperature and a Large Miscibility Gap in the Solid State. *Cryst. Growth Des.* **2019**, *19*, 4793–4801. [CrossRef]
17. Coquerel, G.; Tamura, R. "Enantiomeric disorder" pharmaceutically oriented. In *Disordered Pharmaceutical Materials*; John Wiley & Sons, Ltd.: Hoboken, NJ, USA, 2016; pp. 135–160. [CrossRef]

18. Mbodji, A.; Gbabode, G.; Sanselme, M.; Cartigny, Y.; Couvrat, N.; Leeman, M.; Dupray, V.; Kellogg, R.M.; Coquerel, G. Evidence of Conglomerate with Partial Solid Solutions in Ethylammonium Chlocyphos. *Cryst. Growth Des.* **2020**, *20*, 2562–2569. [CrossRef]
19. Zlokazov, M.V.; Pivnitsky, K.K. Lamellar Conglomerates. *Mendeleev Commun.* **2020**, *30*, 1–6. [CrossRef]
20. Coquerel, G. Chiral discrimination in the solid state: Applications to resolution and deracemization. In *Advances in Organic Crystal Chemistry: Comprehensive Reviews 2015*; Tamura, R., Miyata, M., Eds.; Springer: Tokyo, Japan, 2015; pp. 393–420. [CrossRef]
21. Oketani, R.; Marin, F.; Tinnemans, P.; Hoquante, M.; Laurent, A.; Brandel, C.; Cardinael, P.; Meekes, H.; Vlieg, E.; Geerts, Y.; et al. Deracemization in a Complex Quaternary System with a Second-Order Asymmetric Transformation by Using Phase Diagram Studies. *Chem. Eur. J.* **2019**, *25*, 1–10. [CrossRef]
22. Viedma, C. Chiral Symmetry Breaking during Crystallization: Complete Chiral Purity Induced by Nonlinear Autocatalysis and Recycling. *Phys. Rev. Lett.* **2005**, *94*, 065504. [CrossRef]
23. Intaraboonrod, K.; Lerdwiriyanupap, T.; Hoquante, M.; Coquerel, G.; Flood, A.E. Temperature Cycle Induced Deracemization. *Mendeleev Commun.* **2020**, *30*, 395–405. [CrossRef]
24. Rougeot, C.; Guillen, F.; Plaquevent, J.-C.; Coquerel, G. Ultrasound-Enhanced Deracemization: Toward the Existence of Agonist Effects in the Interpretation of Spontaneous Symmetry Breaking. *Cryst. Growth Des.* **2015**, *15*, 2151–2155. [CrossRef]
25. Iggland, M.; Fernández-Ronco, M.P.; Senn, R.; Kluge, J.; Mazzotti, M. Complete Solid State Deracemization by High Pressure Homogenization. *Chem. Eng. Sci.* **2014**, *111*, 106–111. [CrossRef]
26. Cameli, F.; Xiouras, C.; Stefanidis, G.D. Intensified Deracemization via Rapid Microwave-Assisted Temperature Cycling. *CrystEngComm* **2018**, *20*, 2897–2901. [CrossRef]
27. Steendam, R.R.E.; Dickhout, J.; van Enckevort, W.J.P.; Meekes, H.; Raap, J.; Rutjes, F.P.J.T.; Vlieg, E. Linear Deracemization Kinetics during Viedma Ripening: Autocatalysis Overruled by Chiral Additives. *Cryst. Growth Des.* **2015**, *15*, 1975–1982. [CrossRef]
28. Sakamoto, M. Asymmetric synthesis involving dynamic enantioselective crystallization. In *Advances in Organic Crystal Chemistry: Comprehensive Reviews 2020*; Sakamoto, M., Uekusa, H., Eds.; Springer: Singapore, 2020; pp. 433–456. [CrossRef]
29. Steendam, R.R.E.; Brouwer, M.C.T.; Huijs, E.M.E.; Kulka, M.W.; Meekes, H.; van Enckevort, W.J.P.; Raap, J.; Rutjes, F.P.J.T.; Vlieg, E. Enantiopure Isoindolinones through Viedma Ripening. *Chem. Eur. J.* **2014**, *20*, 13527–13530. [CrossRef]
30. Viedma, C.; Cintas, P. Homochirality beyond Grinding: Deracemizing Chiral Crystals by Temperature Gradient under Boiling. *Chem. Commun.* **2011**, *47*, 12786–12788. [CrossRef]
31. Suwannasang, K.; Flood, A.E.; Rougeot, C.; Coquerel, G. Using Programmed Heating–Cooling Cycles with Racemization in Solution for Complete Symmetry Breaking of a Conglomerate Forming System. *Cryst. Growth Des.* **2013**, *13*, 3498–3504. [CrossRef]
32. Suwannasang, K.; Flood, A.E.; Coquerel, G. A Novel Design Approach to Scale Up the Temperature Cycle Enhanced Deracemization Process: Coupled Mixed-Suspension Vessels. *Cryst. Growth Des.* **2016**, *16*, 6461–6467. [CrossRef]
33. Suwannasang, K.; Flood, A.E.; Rougeot, C.; Coquerel, G. Use of Programmed Damped Temperature Cycles for the Deracemization of a Racemic Suspension of a Conglomerate Forming System. *Org. Process Res. Dev.* **2017**, *21*, 623–630. [CrossRef]
34. Schindler, M.; Brandel, C.; Kim, W.-S.; Coquerel, G. Temperature Cycling Induced Deracemization of $NaClO_3$ under the Influence of $Na_2S_2O_6$. *Cryst. Growth Des.* **2020**, *20*, 414–421. [CrossRef]
35. Maggioni, G.M.; Fernández-Ronco, M.P.; Meijden, M.W.v.d.; Kellogg, R.M.; Mazzotti, M. Solid State Deracemisation of Two Imine-Derivatives of Phenylglycine Derivatives via High-Pressure Homogenisation and Temperature Cycles. *CrystEngComm* **2018**, *20*, 3828–3838. [CrossRef]
36. Xiouras, C.; Fytopoulos, A.A.; Ter Horst, J.H.; Boudouvis, A.G.; Van Gerven, T.; Stefanidis, G.D. Particle Breakage Kinetics and Mechanisms in Attrition-Enhanced Deracemization. *Cryst. Growth Des.* **2018**, *18*, 3051–3061. [CrossRef]
37. Uchin, R.; Suwannasang, K.; Flood, A.E. Model of Temperature Cycle-Induced Deracemization via Differences in Crystal Growth Rate Dispersion. *Chem. Eng. Technol.* **2017**, *40*, 1252–1260. [CrossRef]
38. Bodák, B.; Maggioni, G.M.; Mazzotti, M. Population-Based Mathematical Model of Solid-State Deracemization via Temperature Cycles. *Cryst. Growth Des.* **2018**, *18*, 7122–7131. [CrossRef]

39. Uwaha, M. A Model for Complete Chiral Crystallization. *J. Phys. Soc. Jpn.* **2004**, *73*, 2601–2603. [CrossRef]
40. Katsuno, H.; Uwaha, M. Monte Carlo Simulation of a Cluster Model for the Chirality Conversion of Crystals with Grinding. *J. Cryst. Growth* **2009**, *311*, 4265–4269. [CrossRef]
41. Katsuno, H.; Uwaha, M. Mechanism of Chirality Conversion by Periodic Change of Temperature: Role of Chiral Clusters. *Phys. Rev. E* **2016**, *93*, 013002. [CrossRef]
42. Kondepudi, D.K.; Bullock, K.L.; Digits, J.A.; Hall, J.K.; Miller, J.M. Kinetics of Chiral Symmetry Breaking in Crystallization. *J. Am. Chem. Soc.* **1993**, *115*, 10211–10216. [CrossRef]
43. Ni, X.; Shepherd, R.; Whitehead, J.; Liu, T. Chiral Symmetry Breaking Due to Impeller Size in Cooling Crystallization of Sodium Chlorate. *CrystEngComm* **2018**, *20*, 6894–6899. [CrossRef]
44. Kondepudi, D.K.; Kaufman, R.J.; Singh, N. Chiral Symmetry Breaking in Sodium Chlorate Crystallizaton. *Science* **1990**, *250*, 975–976. [CrossRef]
45. Kondepudi, D.K.; Laudadio, J.; Asakura, K. Chiral Symmetry Breaking in Stirred Crystallization of 1,1′-Binaphthyl Melt. *J. Am. Chem. Soc.* **1999**, *121*, 1448–1451. [CrossRef]
46. Tamura, R.; Takahashi, H.; Coquerel, G. Twenty-Five years' history, mechanism, and generality of preferential enrichment as a complexity phenomenon. In *Advances in Organic Crystal Chemistry, Comprehensive Reviews 2020*; Springer: Cham, Switzerland, 2020; Chapter 20; pp. 405–432.
47. Coquerel, G. Preferential Crystallization. *Top. Curr. Chem.* **2007**, *269*, 1–51. [CrossRef] [PubMed]
48. De Saint Jores, C.; Brandel, C.; Gharbi, N.; Sanselme, M.; Cardinael, P.; Coquerel, G. Limitations of Preferential Enrichment: A Case Study on Tryptophan Ethyl Ester Hydrochloride. *Chem. Eng. Technol.* **2019**, *42*, 1500–1504. [CrossRef]
49. De Saint Jores, C. Towards a Deeper Understanding of Preferential Enrichment. A Case Study: DL Arginine Fumarate in Ethanol-Water 50-50 Mixture. Ph.D. Thesis, University of Rouen, Mont-Saint-Aignan, France, 2019.
50. Gonnade, R.G.; Iwama, S.; Mori, Y.; Takahashi, H.; Tsue, H.; Tamura, R. Observation of Efficient Preferential Enrichment Phenomenon for a Cocrystal of (Dl)-Phenylalanine and Fumaric Acid under Nonequilibrium Crystallization Conditions. *Cryst. Growth Des.* **2011**, *11*, 607–615. [CrossRef]
51. Gervais, C.; Beilles, S.; Cardinaël, P.; Petit, S.; Coquerel, G. Oscillating Crystallization in Solution between (+)- and (−)-5-Ethyl-5-Methylhydantoin under the Influence of Stirring. *J. Phys. Chem. B* **2002**, *106*, 646–652. [CrossRef]
52. Mbodji, A.; Gbabode, G.; Sanselme, M.; Couvrat, N.; Leeman, M.; Dupray, V.; Kellogg, R.M.; Coquerel, G. Family of Conglomerate-Forming Systems Composed of Chlocyphos and Alkyl-Amine. Assessment of Their Resolution Performances by Using Various Modes of Preferential Crystallization. *Cryst. Growth Des.* **2019**, *19*, 5173–5183. [CrossRef]
53. Potter, G.A.; Garcia, C.; McCague, R.; Adger, B.; Collet, A. Oscillating Crystallization of (+) and (−) Enantiomers during Resolution by Entrainment of 2-Azabicyclo[2.2.1]Hept-5-En-3-One. *Angew. Chem. Int. Ed. Engl.* **1996**, *35*, 1666–1668. [CrossRef]
54. Kern, R. Fundamentals of epitaxy. In *Crystal Growth in Science and Technology*; Arend, H., Hulliger, J., Eds.; NATO ASI Series; Springer US: Boston, MA, USA, 1989; pp. 143–165. [CrossRef]
55. Noorduin, W.L.; Izumi, T.; Millemaggi, A.; Leeman, M.; Meekes, H.; Van Enckevort, W.J.P.; Kellogg, R.M.; Kaptein, B.; Vlieg, E.; Blackmond, D.G. Emergence of a Single Solid Chiral State from a Nearly Racemic Amino Acid Derivative. *J. Am. Chem. Soc.* **2008**, *130*, 1158–1159. [CrossRef]
56. Steendam, R.R.E.; Harmsen, B.; Meekes, H.; van Enckevort, W.J.P.; Kaptein, B.; Kellogg, R.M.; Raap, J.; Rutjes, F.P.J.T.; Vlieg, E. Controlling the Effect of Chiral Impurities on Viedma Ripening. *Cryst. Growth Des.* **2013**, *13*, 4776–4780. [CrossRef]
57. Schindler, M. Deracemization of Sodium Chlorate with or without the Influence of Sodium Dithionate. Ph.D. Thesis, University of Rouen, Mont-Saint-Aignan, France, 2020.
58. Belletti, G.; Meekes, H.; Rutjes, F.P.J.T.; Vlieg, E. Role of Additives during Deracemization Using Temperature Cycling. *Cryst. Growth Des.* **2018**, *18*, 6617–6620. [CrossRef]
59. Hawbaker, N.A.; Blackmond, D.G. Energy Threshold for Chiral Symmetry Breaking in Molecular Self-Replication. *Nat. Chem.* **2019**, *11*, 957–962. [CrossRef]
60. Hein, J.E.; Huynh Cao, B.; Viedma, C.; Kellogg, R.M.; Blackmond, D.G. Pasteur's Tweezers Revisited: On the Mechanism of Attrition-Enhanced Deracemization and Resolution of Chiral Conglomerate Solids. *J. Am. Chem. Soc.* **2012**, *134*, 12629–12636. [CrossRef]

61. Sheldon, R.A. *Chirotechnology: Industrial Synthesis of Optically Active Compounds*; Marcel Dekker—CRC Press: New York, NY, USA, 1993.
62. Oketani, R.; Hoquante, M.; Brandel, C.; Cardinael, P.; Coquerel, G. Resolution of an Atropisomeric Naphthamide by Second-Order Asymmetric Transformation: A Highly Productive Technique. *Org. Process Res. Dev.* **2019**, *23*, 1197–1203. [CrossRef]
63. Boyle, W.J.; Sifniades, S.; Van Peppen, J.F. Asymmetric Transformation of.Alpha.-Amino-.Epsilon.-Caprolactam, a Lysine Precursor. *J. Org. Chem.* **1979**, *44*, 4841–4847. [CrossRef]
64. Black, S.N.; Williams, L.J.; Davey, R.J.; Moffatt, F.; Jones, R.V.H.; McEwan, D.M.; Sadler, D.E. The Preparation of Enantiomers of Paclobutrazol: A Crystal Chemistry Approach. *Tetrahedron* **1989**, *45*, 2677–2682. [CrossRef]
65. Kaptein, B.; Noorduin, W.L.; Meekes, H.; van Enckevort, W.J.P.; Kellogg, R.M.; Vlieg, E. Attrition-Enhanced Deracemization of an Amino Acid Derivative That Forms an Epitaxial Racemic Conglomerate. *Angew. Chem. Int. Ed.* **2008**, *47*, 7226–7229. [CrossRef]
66. Addadi, L.; Van Mil, J.; Lahav, M. Useful Impurities for Optical Resolutions. 2. Generality and Mechanism of the Rule of Reversal. *J. Am. Chem. Soc.* **1981**, *103*, 1249–1251. [CrossRef]
67. Niinomi, H.; Sugiyama, T.; Tagawa, M.; Murayama, K.; Harada, S.; Ujihara, T. Enantioselective Amplification on Circularly Polarized Laser-Induced Chiral Nucleation from a NaClO$_3$ Solution Containing Ag Nanoparticles. *CrystEngComm* **2016**, *18*, 7441–7448. [CrossRef]

**Publisher's Note:** MDPI stays neutral with regard to jurisdictional claims in published maps and institutional affiliations.

© 2020 by the authors. Licensee MDPI, Basel, Switzerland. This article is an open access article distributed under the terms and conditions of the Creative Commons Attribution (CC BY) license (http://creativecommons.org/licenses/by/4.0/).

Article

# Absolute Asymmetric Synthesis Involving Chiral Symmetry Breaking in Diels–Alder Reaction

Naohiro Uemura [1], Seiya Toyoda [1], Waku Shimizu [1], Yasushi Yoshida [1,2], Takashi Mino [1,2] and Masami Sakamoto [1,2,*]

1. Department of Applied Chemistry and Biotechnology, Graduate School of Engineering, Chiba University, Yayoi-Cho, Inage-Ku, Chiba 263-8522, Japan; aapa2584@chiba-u.jp (N.U.); ahxa5079@chiba-u.jp (S.T.); aeya2450@chiba-u.jp (W.S.); yoshiday@chiba-u.jp (Y.Y.); tmino@faculty.chiba-u.jp (T.M.)
2. Molecular Chirality Research Center, Chiba University, Yayoi-Cho, Inage-Ku, Chiba 263-8522, Japan
* Correspondence: sakamotom@faculty.chiba-u.jp

Received: 27 April 2020; Accepted: 18 May 2020; Published: 1 June 2020

**Abstract:** Efficient generation and amplification of chirality from prochiral substrates in the Diels–Alder reaction (DA reaction) followed by dynamic crystallization were achieved without using an external chiral source. Since the DA reaction of 2-methylfuran and various maleimides proceeds reversibly, an *exo*-adduct was obtained as the main product as the reaction proceeded. From single crystal X-ray structure analysis, it was found that five of ten *exo*-adducts gave conglomerates. When 2-methylfuran and various maleimides with a catalytic amount of TFA were reacted in a sealed tube, the *exo*-DA adducts were precipitated from the solution, while the reaction mixtures were continuously ground and stirred using glass beads. Deracemization occurred and chiral amplification was observed for four of the substrates. Each final enantiomeric purity was influenced by the crystal structure, and when enantiomers were included in the disorder, they reached an enantiomeric purity reflecting the ratio of the disorder. The final ee value of the 3,5-dimethylphenyl derivative after chiral amplification was 98% ee.

**Keywords:** amplification of chirality; dynamic crystallization; Diels–Alder reaction; absolute asymmetric synthesis; conglomerate; racemization; attrition-enhanced deracemization; Viedma ripening; reversible reaction; enantiomorphic crystal; polymorphism

## 1. Introduction

The Diels–Alder (DA) reaction is one of the most important and fundamental organic synthesis reactions, achieving the concerted [4 + 2] cycloaddition of a diene and an alkene [1,2]. Because it can form two C-C single bonds in one step, it is used to create many cyclic compounds including polycyclic compounds (Figure 1) [3–10]. In addition, a number of asymmetric reactions have been reported, since these can theoretically construct four asymmetric centers at once. Excellent catalytic asymmetric synthesis [11–20] and diastereoselective reactions of chiral substrates have also been reported [21–24]. Each of these reactions is an asymmetric induction method using an enantiomerically active catalyst or substrate and constructs a diastereoselective environment in the transition state of the reaction.

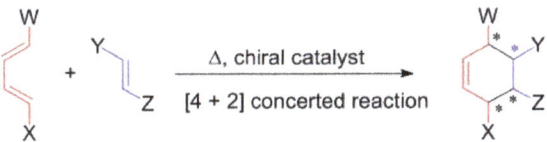

**Figure 1.** Valuable Diels–Alder reaction constructing numerous cyclic compounds.

In contrast to reactions using chiral sources as starting materials and catalysts, asymmetric synthesis using chirality that occurs naturally when organic compounds crystallize has been reported in recent years (Figure 2) [25]. In this method, a compound having a chiral center is generated from a prochiral substrate, and a dynamic preferential crystallization accompanied by the racemization of the generated chiral center is performed continuously without using any external asymmetric source. It is possible to obtain a crystal of the product having a high enantiomeric purity. The racemization process of the product includes a reaction that regenerates a prochiral starting material via a reverse reaction or a process via a direct racemization reaction of an asymmetric center.

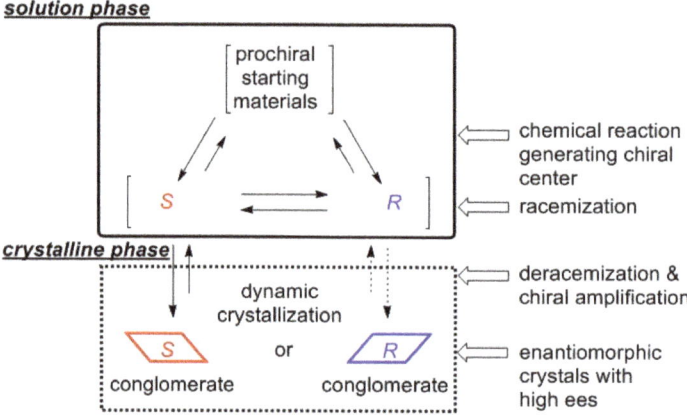

**Figure 2.** Absolute asymmetric synthesis involving dynamic crystallization from prochiral starting materials under achiral conditions.

These synthetic reactions are absolute asymmetric syntheses that provide enantiomerically active compounds from prochiral substrates without using an external chiral source; this is a phenomenon that is of wide interest to researchers in many academic fields [26–30]. Successful absolute asymmetric synthesis by fusion of the reaction that forms the chiral center from prochiral starting materials and the subsequent dynamic crystallization process have been reported in several reaction systems. For example, the Mannich-type reaction [31], aldol reaction [32], stereoisomerization of succinimide [25], synthesis of isoindolinone [33], aza-Michael addition [34–36], Strecker reaction [37,38], and a photochemical reaction [39,40] have all been achieved under achiral conditions. In these limited examples, crystals with high ee were obtained from prochiral materials, and this method is expected to be applied to many reaction systems.

We recently reported the asymmetric Diels–Alder reaction of prochiral starting materials leading to a conglomerate crystal of the adduct in enantiomerically active form [41]. When 2-methylfuran and N-phenylmaleimide were reacted with a small amount of solvent and a catalytic amount of trifluoroacetic acid in a sealed tube at 80 °C, the racemic *exo*-adduct quickly precipitated. Subsequently, the continuous suspension of the reaction mixture with glass beads promoted the chiral amplification to 90% ee by attrition-enhanced deracemization (Figure 3). In this phenomenon, a racemic product having a chiral center is first formed by a DA reaction from a prochiral substance, followed by preferential crystallization of conglomerate crystals. In the mother liquor, by returning to the prochiral substrate by the reverse reaction (*retro*-DA reaction), the racemic condition is always maintained, preventing the excessive formation of the enantiomer.

In this asymmetric amplification by dynamic crystallization, attrition-enhanced deracemization was quite effective in promoting Viedam ripening by continuous grinding of the crystals using glass beads, and finally, the enantiomer crystals converged to high enantiomeric purity [42]. This technique was followed in the deracemization reaction from a racemic mixture of $NaClO_3$ and has recently been applied to deracemization of conglomerate crystals of organic compounds such as amino acids and pharmaceutical and agricultural chemical intermediates [43–54]. A key feature of this technique is that it can be applied to deracemization in a system with a relatively low racemization rate, as compared with the method of promoting crystallization using a solvent evaporation method or a temperature gradient [55–58].

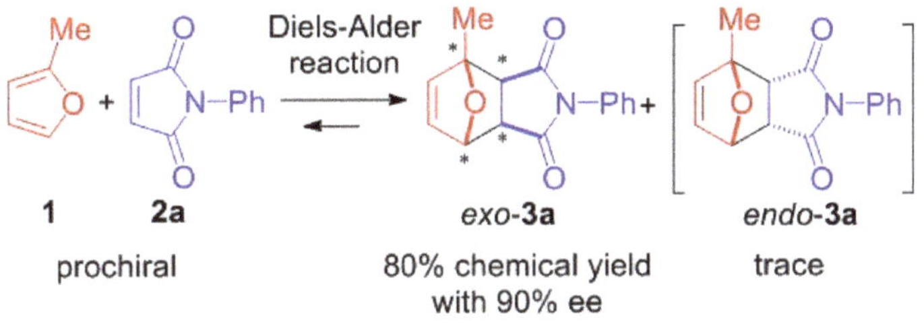

**Figure 3.** Diels–Alder (DA) reaction of 2-methylfuran and *N*-phenylmaleimide.

The bottleneck of deracemization by the dynamic crystallization method is whether or not the target substrate crystallizes as a stable conglomerate. The occurrence of racemic mixtures crystallizing as a conglomerate is approximately 5–10% [59–62]. However, some substrates form conglomerates or chiral crystals at a very high rate due to the effects of molecular shapes and intermolecular interactions. We succeeded in the asymmetric DA reaction utilizing the fact that the DA adduct of 2-methylfuran and *N*-phenylmaleimide was a conglomerate [41]. However, to investigate the generality of this methodology, we synthesized and analyzed a variety of DA adducts with various substituents on the nitrogen atom (Figure 4).

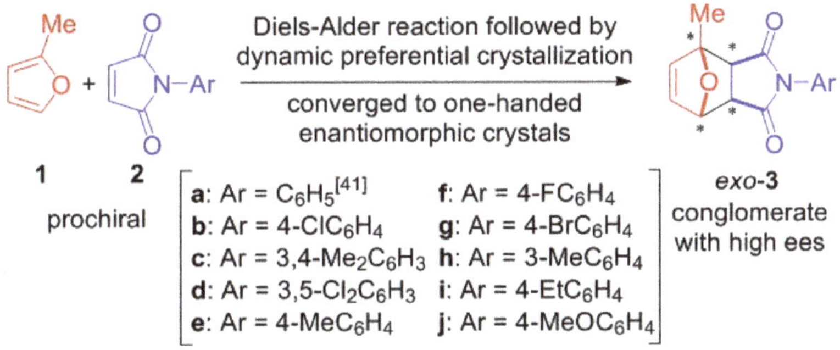

**Figure 4.** Asymmetric Diels–Alder reaction of 2-methylfuran and various *N*-arylmaleimides converged to one-handed enantiomers.

Analysis of the crystal structures of the adducts revealed that they exhibited a very high probability of affording conglomerates. For these substrates, we achieved an asymmetric DA reaction without using an external chiral source and clarified the relationship between the crystal structure and the enantiomeric purity of the products.

## 2. Results and Discussion

In order to investigate the effect of the substituent on the nitrogen atom of the DA adduct on the crystal structure and the enantiomeric purity of the crystal in dynamic crystallization, the adducts *exo*-**3b–j** were synthesized by changing the substituent as shown in Table 1.

Table 1. Space groups of *exo*-3.

| *exo*-3 | R | Space Group |
|---|---|---|
| 3a | $C_6H_5$ | $P2_12_12_1$ [a] |
| 3b | 4-$ClC_6H_4$ | $P2_12_12_1$, (73:27) [b] |
| 3c | 3,4-$Me_2C_6H_3$ | $P2_12_12_1$ |
| 3d | 3,5-$Cl_2C_6H_3$ | $P2_1$, (76:24) [b] |
| 3e | 4-$MeC_6H_4$ | $P2_12_12_1$, polymorphism [c] |
| 3f | 4-$FC_6H_4$ | $P2_1/c$ |
| 3g | 4-$BrC_6H_4$ | $P2_1/n$ |
| 3h | 3-$MeC_6H_4$ | ND [d] |
| 3i | 4-$EtC_6H_4$ | ND [d] |
| 3j | 4-$MeOC_6H_4$ | ND [d] |

[a] Reference No. 41. [b] Disordered crystal consisting of the ratio of both enantiomers indicated in parentheses. [c] Determined by PXRD. [d] Not determined; however, SHG property was inactive by 1064 nm line from an Nd-YAG pulsed laser.

In the DA reaction between 2-methylfuran 1 and maleimides 2b–j, the formation of *endo*-isomers was confirmed at the beginning of the reaction as in the case of the reaction with 2a [41]. When the reaction time was extended, the products predominantly converged to *exo*-isomers, because the crystalline *exo*-adducts were excluded from the reaction system in solution. Therefore, the crystals of the *exo*-adducts were analyzed by single crystal X-ray structure analysis. When the racemate of each *exo*-isomer was recrystallized, single crystals suitable for crystal structure analysis were obtained for *exo*-3b–g. The crystal space groups of these seven types of crystals are shown in Table 1. All of them had $2_1$ helices in the crystal lattices. The space group of 3a–3c and 3e was the orthorhombic $P2_12_12_1$, and 3d and 3f–g were in the monoclinic space group. Surprisingly, five of the ten synthesized substrates, 3a–3e, formed conglomerates of a chiral crystal space group.

In all cases, interactions such as CH–π, C=O–HC and O–HC with relatively small energies were present, which controlled the molecular arrangement (Figures S1–S8). These molecules had nearly spherical shapes and were closely packed in the crystal. Even when alcohol or benzene-based solvents were used, solvent molecules were not incorporated into the crystals.

For 3h–j, single crystals suitable for X-ray crystallography could not be obtained, thus the detailed molecular arrangements were unknown. However, the second harmonic generation (SHG) of these crystals was inactive by irradiation with 1064 nm light from an Nd-YAG pulsed laser, indicating that they might be racemic crystals [63,64]. However, if the emission of the SHG is weak, the possibility of a conglomerate may have been overlooked.

For asymmetric synthesis by the proposed method, the product crystals must be conglomerates. For the four new substrates 3b–3e determined to be in a chiral crystal space group from the above crystal structure analysis, we investigated the absolute asymmetric DA reaction via asymmetric amplification by crystallization.

Another requisite for chiral amplification by dynamic crystallization is rapid racemization under crystallization conditions. Adducts 3 have four chiral centers determined uniquely in the one-step concerted reaction and these cannot directly racemize. However, apparent racemization occurs due to the equilibrium reaction with a reverse DA reaction to regenerate achiral furan and maleimide [65] (Figure 5).

Utilizing this reversible DA reaction, a deracemization reaction by dynamic crystallization was developed. Many dynamic crystallization methods require rapid racemization under crystallization conditions. However, attrition-enhanced deracemization, which is a method based on crystal grinding, has been reported in many successful cases even in systems with a low racemization rate compared to dynamic preferential crystallization methods using a temperature gradient or solvent evaporation [55–58]. However, fast deracemization under the racemization conditions suppresses side reactions, and crystals with high enantiomeric purity can be obtained efficiently.

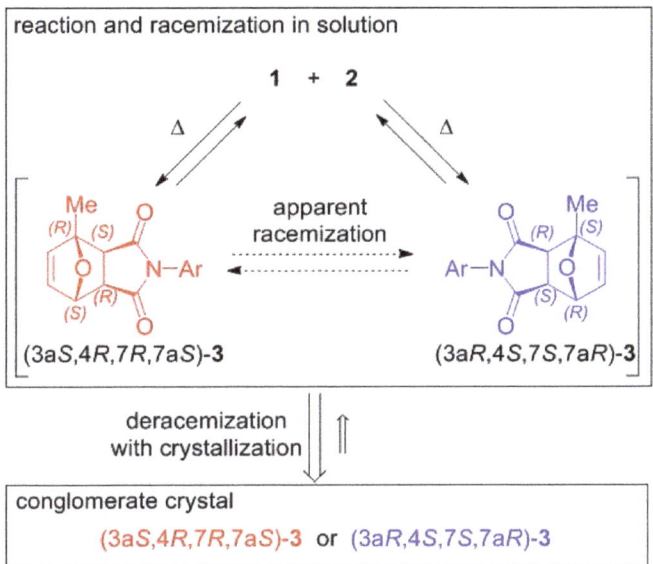

**Figure 5.** Asymmetric synthesis of Diels–Alder reaction products involving racemization via reversible reaction and preferential crystallization.

In the case of the asymmetric synthesis of *exo*-**3a**, we already found out that trifluoroacetic acid (TFA) was the best catalyst for both of the forward and reverse reactions. NMR spectroscopy was used to follow the reversible reaction at 60 °C in deuterated chloroform. Even when the substituents were changed, the catalytic activity of TFA was effective, and the reaction rate was improved in both the DA and *retro*-DA reversible reactions without side reactions. Table 2 shows the half-life of each substrate. For **1a**, as reported in a previous paper, the addition reaction of maleimide reached 50% in about 1 h (Table 2), and an *endo*-adduct was formed at the initial stage of the reaction, which eventually converged to an *exo*-adduct over time [41]. In other cases, almost the same courses in the reactions were observed (Figures S9–S11). The change with time, in these cases, is a reaction in a homogeneous system, but the actual reaction is performed at a higher concentration, where the highly crystalline *exo*-adduct is crystallized and removed from the reaction system.

**Table 2.** The half-life of DA and *retro*-DA reactions with or without trifluoroacetic acid (TFA) [a].

| Substrate | $\tau_{1/2}$ of 2 (h) | | $\tau_{1/2}$ of *exo*-3 (h) | |
|---|---|---|---|---|
| | DA w/o TFA [b] | DA with TFA [c] | *retro*-DA w/o TFA [d] | *retro*-DA with TFA [e] |
| a [f] | 0.92 | 0.25 | 4.5 | 3.0 |
| b | 0.30 | <0.1 | 4.0 | 2.0 |
| c | 0.42 | <0.1 | 6.5 | 3.2 |
| d | 0.30 | <0.1 | 2.0 | 1.2 |

[a] All reactions were monitored by $^1$H NMR spectroscopy. [b] Conditions: maleimide (0.05 M) and 2-methylfuran (0.5 M) in CDCl$_3$ at 60 °C. [c] Conditions: maleimide (0.05 M), 2-methylfuran (0.5 M), and TFA (0.05 M) in CDCl$_3$ at 60 °C. [d] Conditions: *exo*-3 (0.05 M) in CDCl$_3$ at 60 °C. [e] Conditions: *exo*-3 (0.03 M) and TFA (0.03 M) in CDCl$_3$ at 60 °C. [f] Reference No. 41.

The reaction in the NMR tube was also examined for the reverse reaction. The degradation of *exo*-**3** at low concentration (0.05 M) in deuterated chloroform at 60 °C was followed. Maleimide **2** and methylfuran **1** were quantitatively regenerated in all cases. In order to perform highly efficient deracemization, it was necessary to accelerate both the forward and reverse reactions. Specifically, it was necessary to search for a catalyst that greatly accelerated the reverse reaction.

When various maleimides **2a–d** were reacted with 2-methylfuran in the presence of TFA, the DA reaction was accelerated about 4 times for most substrates. Regarding the reverse reaction, it was found that when *exo*-**3a–d** were reacted with TFA in the same concentration of 0.03 M, the reaction was accelerated by 1.5 to 2 times compared to the reaction without TFA (Table 2, Figures S9–S11). Table 2 shows the results at 60 °C, but the actual reaction can be run at 80 °C, whereupon the rate is expected to be several times faster, and a sufficient reversible reaction rate is ensured.

Once the formation of conglomerates and the progress of racemization were confirmed, the asymmetric synthesis involving the dynamic crystallization process was examined. In a sealed tube, *N*-arylmaleimide **2** (100 mg), 2-methylfuran **1** (15.0 equiv), TFA (0–1.0 equiv) as the catalyst, heptane (1.0 mL) as a solvent, and glass beads (2 mm$\Phi$, 250 mg) were added to crush the crystals and the mixture was stirred at 80 °C for several days. The DA reaction proceeded immediately after the start of the reaction, and within a few minutes, crystals of the adducts precipitated and the reaction solution was suspended. After that, deracemization occurred by continuously stirring the obtained suspension. The change in ee value of *exo*-**3** by deracemization was monitored by HPLC using a CHIRALPAK IA (Daicel Ind.) column.

In our previous paper, when 0.5 equiv of TFA was used, the enantiomeric purity started to increase after six days from the start of the reaction and reached 90% ee after 14 days (Figure 6) [41]. Thereafter, the suspension was filtered to isolate the crystals, and *exo*-**3a** was obtained with a yield of 80% and 90% ee. The plot of enantiomeric purity versus time showed a non-linear curve. This sigmoid-like increase in enantiomeric purity is typical for Viedma ripening, and the population balance model [66,67] and existing formulas were extended in view of the effects of Ostwald ripening and autocatalytic enantioselective crystal growth. Theoretical analysis by fitting [68,69] has also been performed. On the other hand, when no glass beads were used, deracemization did not occur.

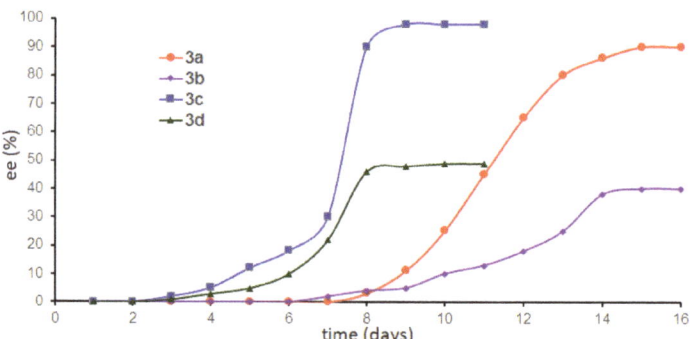

**Figure 6.** Asymmetric DA reaction followed by attrition-enhanced deracemization applied to prochiral **2a–d** (100 mg), 2-methylfuran **1** (5.0 equiv), TFA (0.5 equiv), heptane (1.00 mL), and glass beads (250 mg) at 80 °C in a sealed tube.

Based on the results of the asymmetric reaction of **2a** to *exo*-**3a**, asymmetric DA reactions for **2b–2e** with **1** were also examined leading to conglomerates *exo*-**3b–3e**. As in the case of the reaction of **2a**, 2-methylfuran **1**, various *N*-arylmaleimides **2b–2e**, 0.5 equiv of TFA as a catalyst for promoting racemization, heptane as a solvent, and glass beads for grinding the precipitated solids were stirred in a sealed tube at 80 °C for several days while tracking the change of the ee value of the solid *exo*-**3** (Table 3 and Figure 6).

**Table 3.** Asymmetric DA reaction of **3a-d** [a].

| 3 | Time (day) [b] | Yield (%) [c] | Ee (%) [d] |
|---|---|---|---|
| **3a** [e] | 6–14 | 80 | 90 |
| **3b** | 7–15 | 70 | 40 |
| **3c** | 3–9 | 81 | 98 |
| **3d** | 4–8 | 78 | 49 |

[a] Conditions: prochiral **2** (100 mg), 2-methylfuran **1** (5.0 equiv), TFA (0.5 equiv), heptane (1.00 mL), and glass beads (250 mg) were stirred at 80 °C in a sealed tube. [b] Time required for asymmetric expression and amplification. [c] Yields of crystalline **3** after filtration. [d] Enantiomeric excess of crystals of **3**. [e] Reference No. 41.

When **2b** was used, crystals of DA adduct were precipitated immediately after the reaction, and the continuous stirring while suspending the solids led to an increase in the enantiomeric purity of *exo*-**3b** from the 7th day, reaching 40% ee after 15 days. However, no further asymmetric amplification occurred. This is attributable to the crystal structure of *exo*-**3b**. Disordered packing was indicated by single-crystal structure analysis, and the ratio of the enantiomers was 73:27, a result that exactly reflected the ee value of the converged enantiomorphic crystals.

In the case of **2c**, the crystallinity of the produced *exo*-**3c** was also good, and a stable suspension by glass beads and the stir bar was obtained. The ee value of *exo*-**3c** increased after 3 days and reached 98% after 9 days, achieving the most efficient asymmetric amplification among all the substrates.

When **2d** was used, immediately after the reaction, crystals were precipitated, a suspension was obtained, and asymmetric amplification started 4 days later. Eight days later, the ee reached 49%, after which further amplification did not occur. The reason for this limitation is that, similar to the case of *exo*-**3b**, due to the crystal packing of *exo*-**3d**, the enantiomer contained in the single crystal is disordered in a ratio of 74:24, which exactly reflected the maximum ee value of *exo*-**3d**.

The asymmetric synthesis of *exo*-**3e** using maleimide **2e** was also examined, but *exo*-**3e** was obtained as a racemate without asymmetric amplification. The powdered X-ray crystal structure analysis of the obtained solid was different from the analysis pattern simulated from the conglomerate crystal $P2_12_12_1$ (Figure 7). Conglomerate crystals of *exo*-**3e** gradually changed to a racemic crystal system during suspension with glass beads. The crystal does not contain a crystallization solvent and it showed a polymorphism with a phase transition.

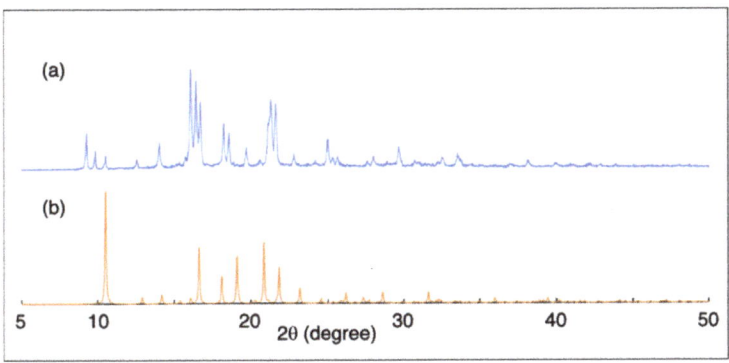

**Figure 7.** PXRD pattern of *exo*-**3e**: (**a**) powder of *exo*-**3e** after grinding, (**b**) calculated pattern from a single crystal of *exo*-**3e** of the $P2_12_12_1$ space group.

It was unable to control the handedness of the crystals obtained after the DA reaction; solids of both types of handedness were obtained in approximately the same number of times for every ten experiments. However, we were able to control the handedness by starting the attrition-enhanced deracemization from a DA adduct with low ee (5% ee). The deracemization started immediately and the same handedness of the enantiomer as the slightly excess stereoisomer could be efficiently obtained.

## 3. Conclusions

Asymmetric DA reactions from prochiral starting materials involving dynamic preferential crystallization were achieved under achiral conditions. Since the DA reaction of 2-methylfuran and various maleimide derivatives proceeded reversibly, *exo*-adducts were obtained as major products as the reaction proceeded. Nine *exo*-adducts were newly synthesized by the DA reaction using maleimides with various substituents on the nitrogen atom. Single crystal X-ray structure analysis revealed that four derivatives gave conglomerates.

When 2-methylfuran and each maleimide in the presence of a catalytic amount of TFA were reacted in a sealed tube, DA adducts precipitated as crystals. The mixture was continuously ground and stirred using glass beads. Deracemization of *exo*-type DA adducts occurred and asymmetric amplification was observed for four substrates. Each final enantiomeric purity was greatly influenced by the crystal structure, and when enantiomers were included in the disorder, they reached an enantiomeric purity reflecting the ratio. In addition, it was found that the chirality of the 3,5-dimethylphenyl derivative was deracemized to 98% ee. We have developed an absolute asymmetric DA reaction that can obtain enantiomerically active DA adducts without using an external asymmetric source.

## 4. Experimental

General Information. NMR spectra were recorded in $CDCl_3$ solutions on a Bruker DPX 300 and DPX 400 spectrometers for $^1$H- and $^{13}$C-NMR. Chemical shifts are reported in parts per million (ppm) relative to TMS as an internal standard. IR spectra were recorded on a JASCO FT/IR-230 spectrometer. HPLC analyses were performed on a JASCO HPLC system (JASCO PU-1580 pump, DG-1580-53, LG-2080-02, MD-2015, UV-2075 and CD-2095 detector). Single crystal X-ray structure analysis was conducted using a SMART APEX II (Bruker AXS) and APEX II ULTRA (Bruker AXS). Powder X-ray crystallographic analysis was performed using D8 ADVANCE (BRUKER AXS). Commercially available *N*-phenylmaleimide and 2-methylfuran were used without further purification. Other maleimides **2b–j** were provided according to the reported procedure [70,71]. (Figures S18–S35)

Synthesis of Exo-**3a–j**.

The corresponding maleimides **2** (1.0 g) and 2-methylfuran **1** (15 equiv) were added to 10 mL of hexane, and the mixture was stirred at 60 °C for 24 h. Thereafter, the solvent and extra amount of methylfuran were removed under reduced pressure, and the crude crystalline products were recrystallized from chloroform/hexane to isolate *exo*-**3**. The structures of known adducts **3a** were determined by comparing their spectral data to literature values. The adducts, **3e–g**, and **3j** are commercially available; however, these materials were easily obtained by the above method.

(3a*S**,4*R**,7*S**,7a*R**)-4-Methyl-2-phenyl-3a,4,7,7a-tetrahydro-1*H*-4,7-epoxyisoindole-1,3(2*H*)-dione (*exo*-**3a**) [41]

Colorless prism; 96% yield; mp: 144–146 °C; $^1$H NMR ($CDCl_3$) δ 1.80 (s, 3H), 2.88 (d, *J* = 6.6 Hz, 1H), 3.14 (d, *J* = 6.3 Hz, 1H), 5.32, (d, *J* = 1.5 Hz, 1H), 6.38 (d, *J* = 5.7 Hz, 1H), 6.57 (dd, *J* = 1.5 and 5.4 Hz, 1H), 7.27–7.30 (m, 2H), 7.37–7.51 (m,3H); $^{13}$C NMR ($CDCl_3$) δ 15.7, 49.5, 50.6, 81.1, 88.6, 126.5, 128.7, 129.1, 131.7, 137.1, 140.7, 174.0, 175.3. (Figures S36 and S37) The enantiomeric purity of the solid was determined by HPLC using a CHIRALPAK IA (Daicel Ind.); column. $t_{R(1)}$ = 20 min for (+)-**3a**, $t_{R(2)}$ = 30.5 min for (−)-**3a**. Eluent: hexane/EtOH = 80:20 (v/v); flow rate: 0.7 mL/min.

(3a*S**,4*R**,7*S**,7a*R**)-2-(4-Chlorophenyl)-4-methyl-3a,4,7,7a-tetrahydro-1*H*-4,7-epoxyisoindole-1,3(2*H*)-dione (*exo*-**3b**)

Colorless crystal; m.p. 120 °C; 90% yield; $^1$H NMR ($CDCl_3$) δ 1.78 (s, 3H), 2.85 (d, *J* = 6.5 Hz, 1H), 3.11 (d, *J* = 6.5 Hz, 1H), 5.29 (d, *J* = 1.8 Hz, 1H), 6.37 (d, *J* = 5.6 Hz, 1H), 6.55 (dd, *J* = 5.6, 1.6 Hz, 1H), 7.23–7.26 (m, 2H), 7.42–7.44 (m, 2H); $^{13}$C NMR ($CDCl_3$) δ 15.7, 49.5, 50.6, 81.1, 88.6, 127.7, 129.2, 130.2,

134.4, 137.0, 140.7, 173.7, 175.0; IR (cm$^{-1}$, KBr) 1701; HRMS (ESI-MS) m/z calcd for C$_{15}$H$_{12}$ClNO$_3$ + H 290.0578, found 290.0577. (Figures S38 and S39) The enantiomeric purity of the solid was determined by HPLC using a CHIRALPAK IA (Daicel Ind.); column. $t_{R(1)}$ = 20 min for (+)-**3b**, $t_{R(2)}$ = 22 min for (−)-**3b**. Eluent: hexane/EtOH = 90: 10 (v/v); flow rate: 1.0 mL/min. (Figures S12 and S13)

(3a*S**,4*R**,7*S**,7a*R**)-2-(3,4-Dimethylphenyl)-4-methyl-3a,4,7,7a-tetrahydro-1*H*-4,7-epoxyisoindole-1,3(2*H*)-dione (*exo*-**3c**)

Colorless crystal; m.p. 140 °C; 91% yield; $^1$H NMR (CDCl$_3$) δ 1.78 (s,3H), 2.27 (s, 3H), 2.27 (s, 3H), 2.84 (d, *J* = 6.5 Hz, 1H), 3.10 (d, *J* = 6.5 Hz, 1H), 5.30 (d, *J* = 1.6 Hz, 1H), 6.36 (d, *J* = 5.6 Hz, 1H), 6.54 (dd, *J* = 5.6, 1.6 Hz, 1H), 6.96–7.01 (m, 2H), 7.21–7.26 (m, 1H); $^{13}$C NMR (CDCl$_3$) δ 15.7, 19.5, 19.8, 49.4, 50.6, 81.0, 88.5, 123.9, 127.5, 129.2, 130.2, 137.0, 137.6, 137.7, 140.7, 174.2, 175.6; IR (cm$^{-1}$, KBr) 1708; HRMS (ESI-MS) m/z calcd for C$_{17}$H$_{17}$NO$_3$ + H 284.1281, found 284.1276. (Figures S40 and S41) The enantiomeric purity of the solid was determined by HPLC using a CHIRALPAK IA (Daicel Ind.); column. $t_{R(1)}$ = 29.5 min for (+)-**3c**, $t_{R(2)}$ = 33 min for (−)-**3c**. Eluent: hexane/EtOH = 80: 20 (v/v); flow rate: 0.5 mL/min. (Figures S14 and S15)

(3a*S**,4*R**,7*S**,7a*R**)-2-(3,5-Dichlorophenyl)-4-methyl-3a,4,7,7a-tetrahydro-1*H*-4,7-epoxyisoindole-1,3(2*H*)-dione (*exo*-**3d**)

Colorless crystal; m.p. 136 °C; 99% yield; $^1$H NMR (CDCl$_3$) δ 1.79 (s, 3H), 2.87 (d, *J* = 6.5 Hz, 1H), 3.13 (d, *J* = 6.5 Hz, 1H), 5.30 (d, *J* = 1.8 Hz, 1H), 6.38 (d, *J* = 5.6 Hz, 1H), 6.57 (dd, *J* = 5.6, 1.6 Hz, 1H), 7.26 (m, 2H), 7.39–7.40 (m, 1H); $^{13}$C NMR (CDCl$_3$) δ 15.7, 49.5, 50.6, 81.2, 88.7, 125.1, 128.8, 133.3, 135.2, 137.1, 140.7, 173.2, 174.5; IR (cm$^{-1}$, KBr) 1712. (Figures S42 and S43) The enantiomeric purity of the solid was determined by HPLC using a CHIRALPAK IA (Daicel Ind.); column. $t_{R(1)}$ = 13.5 min for (+)-**3d**, $t_{R(2)}$ = 16 min for (−)-**3d**. Eluent: hexane/EtOH = 90 : 10 (v/v); flow rate: 1.0 mL/min. (Figures S16 and S17)

(3a*S**,4*R**,7*S**,7a*R**)-4-Methyl-2-(4-tolyl)-3a,4,7,7a-tetrahydro-1*H*-4,7-epoxyisoindole-1,3(2*H*)-dione (*exo*-**3e**)

Colorless crystal; m.p. 132 °C; 90% yield; $^1$H NMR (CDCl$_3$) δ 1.79 (s, 3H), 2.38 (s, 3H), 2.86 (d, *J* = 6.6 Hz, 1H), 3.12 (d, *J* = 6.6 Hz, 1H), 5.31 (d, *J* = 1.8 Hz, 1H), 6.37 (d, *J* = 5.5 Hz, 1H), 6.56 (dd, *J* = 5.6, 1.7 Hz, 1H), 7.13–7.28 (m, 4H); $^{13}$C NMR (CDCl$_3$) δ 15.7, 21.2, 49.4, 50.6, 81.0, 88.5, 126.3, 129.1, 129.7, 137.0, 138.7, 140.7, 174.1, 175.4; IR (cm$^{-1}$, KBr) 1707. (Figures S44 and S45)

(3a*S**,4*R**,7*S**,7a*R**)-2-(4-Fluorophenyl)-4-methyl-3a,4,7,7a-tetrahydro-1*H*-4,7-epoxyisoindole-1,3(2*H*)-dione (*exo*-**3f**)

Colorless crystal; m.p. 133 °C; 88% yield; $^1$H NMR (CDCl$_3$) δ 1.78 (s, 3H), 2.86 (d, *J* = 6.5 Hz, 1H), 3.11 (d, *J* = 6.5 Hz, 1H), 5.29 (d, *J* = 1.8 Hz, 1H), 6.37 (d, *J* = 5.6 Hz, 1H), 6.55 (dd, *J* = 5.7, 1.7 Hz, 1H), 7.13–7.29 (m, 4H); $^{13}$C NMR (CDCl$_3$) δ 15.7, 49.4, 50.6, 81.1, 88.6, 116.0, 116.2, 127.6, 128.3, 128.4, 137.0, 140.7, 160.9, 163.4, 173.9, 175.2. (Figures S46 and S47)

(3a*S**,4*R**,7*S**,7a*R**)-2-(4-Bromophenyl)-4-methyl-3a,4,7,7a-tetrahydro-1*H*-4,7-epoxyisoindole-1,3(2*H*)-dione (*exo*-**3g**)

Colorless crystal; m.p. 127 °C; 94% yield; $^1$H NMR (CDCl$_3$) δ 1.78 (s, 3H), 2.86 (d, *J* = 6.5 Hz, 1H), 3.12 (d, *J* = 6.5 Hz, 1H), 5.29 (d, *J* = 1.8 Hz, 1H), 6.37 (d, *J* = 5.6 Hz, 1H), 6.56 (dd, *J* = 5.7, 1.7 Hz, 1H), 7.18-7.20 (m, 2H), 7.58–7.60 (m, 2H); $^{13}$C NMR (CDCl$_3$) δ 15.7, 49.5, 50.6, 81.1, 88.6, 122.5, 128.0, 130.6, 132.2, 137.0, 140.7, 173.6, 174.9; IR (cm$^{-1}$, KBr) 1705. (Figures S48 and S49)

(3a*S**,4*R**,7*S**,7a*R**)-4-Methyl-2-(3-tolyl)-3a,4,7,7a-tetrahydro-1*H*-4,7-epoxyisoindole-1,3(2*H*)-dione (*exo*-**3h**)

Colorless crystal; m.p. 128 °C; 93% yield; $^1$H NMR (CDCl$_3$) δ 1.79 (s, 3H), 2.38 (s, 3H), 2.86 (d, *J* = 6.5 Hz, 1H), 3.12 (d, *J* = 6.5 Hz, 1H), 5.31 (d, *J* = 1.6 Hz, 1H), 6.37 (d, *J* = 5.6 Hz, 1H), 6.55 (dd, *J* = 5.6, 1.6 Hz, 1H), 7.05–7.37 (m, 4H); $^{13}$C NMR (CDCl$_3$) δ 15.7, 21.3, 49.5, 50.6, 81.1, 88.6, 123.6, 127.1, 128.9, 129.6, 131.6, 137.0, 139.2, 140.7, 174.1, 175.4; IR (cm$^{-1}$, KBr) 1707; HRMS (ESI-MS) m/z calcd for C$_{16}$H$_{15}$NO$_3$ - H 268.0979, found 288.0992. (Figures S50 and S51)

(3a*S*\*,4*R*\*,7*S*\*,7a*R*\*)-2-(4-Ethylphenyl)-4-methyl-3a,4,7,7a-tetrahydro-1*H*-4,7-epoxyisoindole-1,3(2*H*)-dione (*exo*-**3i**)

Colorless crystal; m.p. 114 °C; 88% yield; $^1$H NMR (CDCl$_3$) δ 1.24 (t, *J* = 7.7 Hz, 3H), 1.79 (s, 3H), 2.67 (q, *J* = 7.6 Hz, 2H), 2.85 (d, *J* = 6.5 Hz, 1H), 3.11 (d, *J* = 6.5 Hz, 1H), 5.30 (d, *J* = 1.8 Hz, 1H), 6.36 (d, *J* = 6.5 Hz, 1H), 6.55 (dd, *J* = 5.6, 1.6 Hz, 1H), 7.16-7.30 (m, 4H); $^{13}$C NMR (CDCl$_3$) δ 15.3, 15.7, 28.6, 49.5, 50.6, 81.1, 88.6, 126.4, 128.6, 129.3, 137.0, 140.7, 144.9, 174.2, 175.5; IR (cm$^{-1}$, KBr) 1702; HRMS (ESI-MS) *m/z* calcd for C$_{17}$H$_{17}$NO$_3$ + H 284.1281, found 284.1277. (Figures S52 and S53)

(3a*S*\*,4*R*\*,7*S*\*,7a*R*\*)-2-(4-Methoxyphenyl)-4-methyl-3a,4,7,7a-tetrahydro-1*H*-4,7-epoxyisoindole-1,3(2*H*)-dione (*exo*-**3j**)

Colorless crystal; m.p. 120–121 °C; 94% yield; $^1$H NMR (CDCl$_3$) δ 1.78 (s, 3H), 2.84 (d, *J* = 6.5 Hz, 1H), 3.10 (d, *J* = 6.5 Hz, 1H), 3.82 (s, 3H), 5.29 (d, *J* = 1.8 Hz, 1H), 6.36 (d, *J* = 5.7 Hz, 1H), 6.54 (dd, *J* = 5.6, 1.6 Hz, 1H), 6.94–7.26 (m, 4H); $^{13}$C NMR (CDCl$_3$) δ 15.7, 49.4, 50.5, 55.4, 81.0, 88.5, 114.4, 124.4, 127.7, 137.0, 140.7, 159.5, 174.3, 175.5. (Figures S54 and S55)

Single crystal X-ray structure analysis of (3a*S*,4*R*,7*S*,7a*R*)-2-(4-chlorophenyl)-4-methyl-3a,4,7,7a-tetrahydro-1*H*-4,7-epoxyisoindole-1,3(2*H*)-dione (*exo*-**3b**)

Disordered crystal of the ratio of 73:27, exhibiting (-)-CD sign at 254 nm for major isomer. Colorless prism (0.20 × 0.20 × 0.01 mm$^3$), orthorhombic space group $P2_12_12_1$, *a* = 6.5485(4) Å, *b* = 12.3747(6) Å, *c* = 16.8764(9) Å, *V* = 1367.59(13) Å$^3$, *Z* = 4, λ (CuK*α*) = 1.54178 Å, *ρ* = 1.407 g/cm$^3$, *μ* (CuK*α*) = 2.539 cm, 3954 reflections measured (T = 173 K, 4.430° < θ < 68.264°), nb of independent data collected: 2213 nb of independent data used for refinement: 2127 in the final least-squares refinement cycles on F$^2$, the model converged at $R_1$ = 0.0505, $wR_2$ = 0.1993 [*I* > 2σ(*I*)], $R_1$ = 0.0522, $wR_2$ = 0.1408 (all data), and GOF = 1.065, H-atom parameters constrained, absolute Flack parameter = 0.505(8). (CCDC 1985809). (Figure S1)

Single crystal X-ray structure analysis of (3a*S*,4*R*,7*S*,7a*R*)-2-(4-chlorophenyl)-4-methyl-3a,4,7,7a-tetrahydro-1*H*-4,7-epoxyisoindole-1,3(2*H*)-dione (*exo*-**3b**)

Enantiomerically pure crystal provided by optical resolution using HPLC, exhibiting (-)-CD sign at 254 nm. Colorless prism (0.20 × 0.20 × 0.10 mm$^3$), orthorhombic space group $P2_12_12_1$, *a* = 6.5975(2) Å, *b* = 12.3565(4) Å, *c* = 16.7223(6) Å, *V* = 1363.24(8) Å$^3$, *Z* = 4, λ (CuK*α*) = 1.54178 Å, *ρ* = 1.412 g/cm$^3$, *μ* (CuK*α*) = 2.547 cm, 12970 reflections measured (T = 173 K, 4.449° < θ < 68.213°), nb of independent data collected: 2427, nb of independent data used for refinement: 2397 in the final least-squares refinement cycles on F$^2$, the model converged at $R_1$ = 0.0369, $wR_2$ = 0.0985 [*I* > 2σ(*I*)], $R_1$ = 0.0371, $wR_2$ = 0.0989 (all data), and GOF = 1.059, H-atom parameters constrained, absolute Flack parameter = 0.051(3). (CCDC 1985810). (Figure S2)

Single crystal X-ray structure analysis of (3a*S*,4*R*,7*S*,7a*R*)-2-(3,4-dimethylphenyl)-4-methyl-3a,4,7,7a-tetrahydro-1*H*-4,7-epoxyisoindole-1,3(2*H*)-dione (*exo*-**3c**)

Exhibiting (-)-CD sign at 254 nm. Colorless prism (0.20 × 0.10 × 0.10 mm$^3$), orthorhombic space group $P2_12_12_1$, *a* = 7.9707(3) Å, *b* = 12.9386(5) Å, *c* = 13.7305(5) Å, *V* = 1416.02(9) Å$^3$, *Z* = 4, λ (CuK*α*) = 1.54178 Å, *ρ* = 1.329 g/cm$^3$, *μ* (CuK*α*) = 0.741 cm, 21885 reflections measured (T = 173 K, 4.696° < θ < 68.241°), nb of independent data collected: 2593, nb of independent data used for refinement: 2576 in the final least-squares refinement cycles on F$^2$, the model converged at $R_1$ = 0.0334, $wR_2$ = 0.0855 [*I* > 2σ(*I*)], $R_1$ = 0.0335, $wR_2$ = 0.0856 (all data), and GOF = 1.065, H-atom parameters constrained, absolute Flack parameter = 0.099(16). (CCDC 1985811). (Figure S3)

Single crystal X-ray structure analysis of (3a*S*,4*R*,7*S*,7a*R*)-2-(3,5-dichlorophenyl)-4-methyl-3a,4,7,7a-tetrahydro-1*H*-4,7-epoxyisoindole-1,3(2*H*)-dione (*exo*-**3d**)

Disordered crystal of the ratio of 76:24, exhibiting (-)-CD sign at 254 nm for major isomer. Colorless prism (0.30 × 0.20 × 0.20 mm$^3$), monoclinic space group $P2_1$, *a* = 5.4032(5) Å, *b* = 13.9697(10) Å, *c* = 9.6673(8) Å, *β* = 105.179(4)°, *V* = 704.24(10) Å$^3$, *Z* = 2, λ (CuK*α*) = 1.54178 Å, *ρ* = 1.529 g/cm$^3$, *μ* (CuK*α*) = 4.237 cm, 2350 reflections measured (T = 173 K, 5.7024° < θ < 68.068°), nb of independent data collected: 1457, nb of independent data used for refinement: 1452 in the final least-squares refinement cycles on $F^2$, the model converged at $R_1 = 0.0532$, $wR_2 = 0.1391$ [$I > 2\sigma(I)$], $R_1 = 0.0532$, $wR_2 = 0.1392$ (all data), and GOF = 1.125, H-atom parameters constrained, absolute Flack parameter = 0.12(3). (CCDC 1985812). (Figure S4)

Single crystal X-ray structure analysis of (3a$R$,4$S$,7$R$,7a$S$)-2-(3,5-dichlorophenyl)-4-methyl-3a,4,7,7a-tetrahydro-1$H$-4,7-epoxyisoindole-1,3(2$H$)-dione (*exo*-(+)-**3d**)

Enantiomerically pure crystal provided by optical resolution using HPLC. Colorless prism (0.30 × 0.20 × 0.20 mm$^3$), monoclinic space group $P2_1$, $a = 5.3713(9)$ Å, $b = 13.958(2)$ Å, $c = 9.5751(14)$ Å, $\beta = 103.609(5)°$, $V = 697.72(19)$ Å$^3$, $Z = 2$, $\lambda$ (CuK$\alpha$) = 1.54178 Å, $\rho = 1.529$ g/cm$^3$, $\mu$ (CuK$\alpha$) = 4.277 cm, 10993 reflections measured (T = 173 K, 4.752° < $\theta$ < 72.198°), nb of independent data collected: 2558, nb of independent data used for refinement: 2519 in the final least-squares refinement cycles on $F^2$, the model converged at $R_1 = 0.0429$, $wR_2 = 0.1034$ [$I > 2\sigma(I)$], $R_1 = 0.0431$, $wR_2 = 0.1036$ (all data), and GOF = 1.101, H-atom parameters constrained, absolute Flack parameter = 0.110(5). (CCDC 1985813). (Figure S5)

Single crystal X-ray structure analysis of analysis of (3a$R$,4$S$,7$R$,7a$S$)-4-methyl-2-(4-tolyl)-3a,4,7,7a-tetrahydro-1$H$-4,7-epoxyisoindole-1,3(2$H$)-dione (*exo*-**3e**)

Colorless prism (0.20 × 0.10 × 0.10 mm$^3$), orthorhombic space group $P2_12_12_1$, $a = 9.3007(6)$ Å, $b = 10.6635(6)$ Å, $c = 13.7010(9)$ Å, $V = 1358.84(15)$ Å$^3$, $Z = 4$, $\lambda$ (CuK$\alpha$) = 1.54178 Å, $\rho = 1.316$ g/cm$^3$, $\mu$ (CuK$\alpha$) = 0.746 cm, 4911 reflections measured (T = 173 K, 6.314° < $\theta$ < 68.278°), nb of independent data collected: 2263, nb of independent data used for refinement: 2225 in the final least-squares refinement cycles on $F^2$, the model converged at $R_1 = 0.0458$, $wR_2 = 0.1278$ [$I > 2\sigma(I)$], $R_1 = 0.0462$, $wR_2 = 0.1284$ (all data), and GOF = 1.048, H-atom parameters constrained, absolute Flack parameter = 0.39(7). (CCDC 1985814). (Figure S6)

Single crystal X-ray structure analysis of analysis of (3a$S^*$,4$R^*$,7$S^*$,7a$R^*$)-2-(4-fluorophenyl)-4-methyl-3a,4,7,7a-tetrahydro-1$H$-4,7-epoxyisoindole-1,3(2$H$)-dione (*exo*-**3f**)

Colorless prism (0.20 × 0.10 × 0.03 mm$^3$), monoclinic space group $P2_1/c$, $a = 9.9813(13)$ Å, $b = 13.0236(15)$ Å, $c = 9.7580(8)$ Å, $\beta = 101.379(9)°$, $V = 1243.5(2)$ Å$^3$, $Z = 4$, $\lambda$ (CuK$\alpha$) = 1.54178 Å, $\rho = 1.460$ g/cm$^3$, $\mu$ (CuK$\alpha$) = 0.945 cm, 2238 reflections measured (T = 173 K, 5.655° < $\theta$ < 68.328°), nb of independent data collected: 2238, nb of independent data used for refinement: 1687 in the final least-squares refinement cycles on $F^2$, the model converged at $R_1 = 0.0687$, $wR_2 = 0.1787$ [$I > 2\sigma(I)$], $R_1 = 0.0915$, $wR_2 = 0.1875$ (all data), and GOF = 1.172, H-atom parameters constrained. (CCDC 1985815). (Figure S7)

Single crystal X-ray structure analysis of (3a$S^*$,4$R^*$,7$S^*$,7a$R^*$)-2-(4-bromophenyl)-4-methyl-3a,4,7,7a-tetrahydro-1$H$-4,7-epoxyisoindole-1,3(2$H$)-dione (*exo*-**3g**).

Colorless prism (0.20 × 0.20 × 0.20 mm$^3$), monoclinic space group $P2_1/n$, $a = 20.838(6)$ Å, $b = 6.485(2)$ Å, $c = 20.853(6)$ Å, $\beta = 104.364(3)°$, $V = 2729.9(15)$ Å$^3$, $Z = 8$, $\lambda$ (MoK$\alpha$) = 0.71073 Å, $\rho = 1.626$ g/cm$^3$, $\mu$ (MoK$\alpha$) = 3.018 cm, 5224 reflections measured (T = 173 K, 1.593° < $\theta$ < 27.570°), nb of independent data collected: 5224, nb of independent data used for refinement: 3679 in the final least-squares refinement cycles on $F^2$, the model converged at $R_1 = 0.0511$, $wR_2 = 0.1118$ [$I > 2\sigma(I)$], $R_1 = 0.0928$, $wR_2 = 0.1310$ (all data), and GOF = 1.025, H-atom parameters constrained. (CCDC 1985818). (Figure S8)

Reaction conditions for asymmetric DA reaction via dynamic crystallization

In a sealed tube ($L = 200$ mm, $\Phi = 25$ mm), $N$-arylmaleimide **2** (100 mg), 2-methylfuran **1** (15 eq.), TFA (0–1.0 eq), and heptane (1.0 mL) were stirred with or without glass beads (250 mg) using a stir bar at 80 °C. The crystalline adduct **3** appeared after a few minutes, and the solution was kept in suspension by stirring at 600 rpm for several days at 80 °C. The change of ee value of crystalline **3** was monitored by HPLC using CHIRALPAK IA (Daicel Ind.) column; eluent: *n*-hexane/EtOH. Finally, crystalline *exo*-**3** was isolated by filtration. The same procedure was performed for all substrates.

**Supplementary Materials:** The following are available online at http://www.mdpi.com/2073-8994/12/6/910/s1, Figure S1: Single crystal X-Ray crystallographic analysis of *exo*-**3b** (disordered), Figure S2: Single crystal X-Ray crystallographic analysis of *exo*-**3b** (enantiopure), Figure S3: Single crystal X-Ray crystallographic analysis of *exo*-**3c** (enantiopure), Figure S4: Single crystal X-Ray crystallographic analysis of *exo*-**3d** (disordered), Figure S5: Single crystal X-Ray crystallographic analysis of *exo*-**3d** (enantiopure), Figure S6: Single crystal X-Ray crystallographic analysis of *exo*-**3e** (enantiopure), Figure S7: Single crystal X-Ray crystallographic analysis of *exo*-**3f**, Figure S8: Single crystal X-Ray crystallographic analysis of *exo*-**3g**, Figure S9: Time course for DA reaction of **1** (0.5 M) and **2b–d** (0.05 M) at 60 °C in CDCl$_3$ monitored by $^1$H NMR, Figure S10: Time course for reverse-DA reaction of *exo*-**3b–d** (0.05 M) at 60 °C in CDCl$_3$ monitored by $^1$H NMR, Figure S11: Time course for reverse-DA reaction of *exo*-**3b–d** (0.03 M) in the presence of TFA (0.03 eq) at 60 °C in CDCl$_3$ monitored by $^1$H NMR, Figure S12: HPLC analysis of racemic *exo*-**3b**, Figure S13: HPLC analysis of 40% ee of *exo*-(-)-**3b**, Figure S14. HPLC analysis of racemic *exo*-**3c**, Figure S15. HPLC analysis of 98% ee of *exo*-(-)-**3c**, Figure S16. HPLC analysis of racemic of *exo*-**3d**, Figure S17. HPLC analysis of 49% ee of *exo*-(+)-**3d**, Figure S18: $^1$H NMR spectrum of maleimide **2b**, Figure S19: $^{13}$C NMR spectrum of maleimide **2b**, Figure S20: $^1$H NMR spectrum of maleimide **2c**, Figure S21: $^{13}$C NMR spectrum of maleimide **2c**, Figure S22: $^1$H NMR spectrum of maleimide **2d**, Figure S23: $^{13}$C NMR spectrum of maleimide **2d**, Figure S24: $^1$H NMR spectrum of maleimide **2e**, Figure S25: $^{13}$C NMR spectrum of maleimide **2e**, Figure S26: $^1$H NMR spectrum of maleimide **2f**, Figure S27: $^{13}$C NMR spectrum of maleimide **2f**, Figure S28: $^1$H NMR spectrum of maleimide **2g**, Figure S29: $^{13}$C NMR spectrum of maleimide **2g**, Figure S30: $^1$H NMR spectrum of maleimide **2h**, Figure S31: $^{13}$C NMR spectrum of maleimide **2h**, Figure S32: $^1$H NMR spectrum of maleimide **2i**, Figure S33: $^{13}$C NMR spectrum of maleimide **2i**, Figure S34: $^1$H NMR spectrum of maleimide **2j**, Figure S35: $^{13}$C NMR spectrum of maleimide **2j**, Figure S36: $^1$H NMR spectrum of *exo*-**3a**, Figure S37: $^{13}$C NMR spectrum of *exo*-**3a**, Figure S38: $^1$H NMR spectrum of *exo*-**3b**, Figure S39: $^{13}$C NMR spectrum of *exo*-**3b**, Figure S40: $^1$H NMR spectrum of *exo*-**3c**, Figure S41: $^{13}$C NMR spectrum of *exo*-**3c**, Figure S42: $^1$H NMR spectrum of *endo*-**3d**, Figure S43: $^{13}$C NMR spectrum of *endo*-**3d**, Figure S44: $^1$H NMR spectrum of *exo*-**3e**, Figure S45: $^{13}$C NMR spectrum of *exo*-**3e**, Figure S46: $^1$H NMR spectrum of *endo*-**3f**, Figure S47: $^{13}$C NMR spectrum of *endo*-**3f**, Figure S48: $^1$H NMR spectrum of *exo*-**3g**, Figure S49: $^{13}$C NMR spectrum of *exo*-**3g**, Figure S50: $^1$H NMR spectrum of *endo*-**3h**, Figure S51: $^{13}$C NMR spectrum of *endo*-**3h**, Figure S52: $^1$H NMR spectrum of *exo*-**3i**, Figure S53: $^{13}$C NMR spectrum of *exo*-**3i**, Figure S54: $^1$H NMR spectrum of *endo*-**3j**, Figure S55: $^{13}$C NMR spectrum of *endo*-**3j**.

**Author Contributions:** Conceptualization, M.S.; methodology, N.U., S.T., W.S., M.S.; validation, Y.Y. and T.M.; formal analysis, N.U., S.T., W.S.; investigation, N.U., S.T., W.S.; data curation, Y.Y., T.M., M.S.; writing—original draft preparation, N.U., S.T., M.S.; writing—review and editing, M.S.; supervision, M.S.; project administration, M.S.; funding acquisition, M.S. All authors have read and agreed to the published version of the manuscript.

**Funding:** This work was supported by Grants-in-Aid for Scientific Research (No. 19H02708) from the Ministry of Education, Culture, Sports, Science, and Technology (MEXT) of the Japanese Government. Uemura acknowledges financial support from the Frontier Science Program of the Graduate School of Science and Engineering, Chiba University.

**Conflicts of Interest:** The authors declare no conflict of interest.

## References

1. Diels, O.; Alder, K. Synthesen in der Hydroaromatischen Reihe. *Justus Liebigs Ann. Chem.* **1928**, *460*, 98–122. [CrossRef]
2. Diels, O.; Alder, K. Synthesen in der Hydroaromatischen Reihe, IV. Mitteilung: Über die Anlagerung von Maleinsäure-anhydrid an Arylierte Diene, Triene und Fulvene. *Ber. Dtsch. Chem. Ges.* **1929**, *62*, 2081–2087. [CrossRef]
3. Nicolaou, K.C.; Snyder, S.A.; Montagnon, T.; Vassilikogiannakis, G. The Diels-Alder Reaction in Total Synthesis. *Angew. Chem. Int. Ed.* **2002**, *41*, 1668–1698. [CrossRef]
4. Takao, K.-I.; Munakata, R.; Tadano, K. Recent advances in natural product synthesis by using intramolecular Diels-Alder reactions. *Chem. Rev.* **2005**, *105*, 4779–4807. [CrossRef]
5. Min, L.; Liu, X.; Li, C.-C. Total Synthesis of Natural Products with Bridged Bicyclo[m.n.1] Ring Systems via Type II [5 + 2] Cycloaddition. *Acc. Chem. Res.* **2020**, *53*, 703–718. [CrossRef]
6. Breunig, M.; Yuan, P.; Gaich, T. An Unexpected Transannular [4 + 2] Cycloaddition during the Total Synthesis of (+)-Norcembrene. *Angew. Chem. Int. Ed.* **2020**, *59*, 5521–5525. [CrossRef]
7. Chavan, S.P.; Kadam, A.L.; Gonnade, R.G. Enantioselective Formal Total Synthesis of (-)-Quinagolide. *Org. Lett.* **2019**, *21*, 9089–9093. [CrossRef]
8. Burns, A.S.; Rychnovsky, S.D. Total Synthesis and Structure Revision of (-)-Illisimonin A, a Neuroprotective Sesquiterpenoid from the Fruits of Illicium simonsii. *J. Am. Chem. Soc.* **2019**, *141*, 13295–13300. [CrossRef]

9. Zhou, S.; Xia, K.; Leng, X.; Li, A. Asymmetric Total Synthesis of Arcutinidine, Arcutinine, and Arcutine. *J. Am. Chem. Soc.* **2019**, *141*, 13718–13723. [CrossRef]
10. Maurya, V.; Appayee, C. Enantioselective Total Synthesis of Potent 9β-11-Hydroxyhexahydrocannabinol. *J. Org. Chem.* **2020**, *85*, 1291–1297. [CrossRef]
11. Corey, E.J.; Shibata, T.; Lee, T.W. Asymmetric Diels-Alder Reactions Catalyzed by a Triflic Acid Activated Chiral Oxazaborolidine. *J. Am. Chem. Soc.* **2002**, *124*, 3808–3809. [CrossRef] [PubMed]
12. Choy, W.; Reed, L.A.; Masamune, S. Asymmetric Diels-Alder Reaction: Design of Chiral Dienophiles. *J. Org. Chem.* **1983**, *48*, 1137–1139. [CrossRef]
13. Oppolzer, W. Asymmetric Diels-Alder and Ene Reactions in Organic Synthesis. New Synthetic Methods (48). *Angew. Chem. Int. Ed.* **1984**, *23*, 876–889. [CrossRef]
14. Kagan, H.B.; Riant, O. Catalytic Asymmetric Diels-Alder Reactions. *Chem. Rev.* **1992**, *92*, 1007–1019. [CrossRef]
15. Mehta, G.; Uma, R. Stereoelectronic Control in Diels–Alder Reaction of Dissymmetric 1,3-Dienes. *Acc. Chem. Res.* **2000**, *33*, 278–286. [CrossRef] [PubMed]
16. Wilson, R.M.; Jen, W.S.; MacMillan, D.W.C. Enantioselective Organocatalytic Intramolecular Diels–Alder Reactions. The Asymmetric Synthesis of Solanapyrone D. *J. Am. Chem. Soc.* **2005**, *127*, 11616–11617. [CrossRef]
17. Zhou, Y.; Lin, L.; Liu, X.; Hu, X.; Lu, Y.; Zhang, X.; Feng, X. Catalytic Asymmetric Diels-Alder Reaction/[3,3] Sigmatropic Rearrangement Cascade of 1-Thiocyanatobutadienes. *Angew. Chem. Int. Ed.* **2018**, *57*, 9113–9116. [CrossRef]
18. Li, M.; Carreras, V.; Jalba, A.; Ollevier, T. Asymmetric Diels-Alder Reaction of α,β-Unsaturated Oxazolidin-2-one Derivatives Catalyzed by a Chiral Fe(III)-Bipyridine Diol Complex. *Org. Lett.* **2018**, *20*, 995–998. [CrossRef]
19. Zheng, J.; Lin, L.; Fu, K.; Zheng, H.; Liu, X.; Feng, X. Synthesis of Optically Pure Spiro[cyclohexane-oxindoline] Derivatives via Catalytic Asymmetric Diels-Alder Reaction of Brassard-Type Diene with Methyleneindolines. *J. Org. Chem.* **2015**, *80*, 8836–8842. [CrossRef]
20. Chauhan, M.S.; Kumar, P.; Singh, S. Synthesis of MacMillan catalyst modified with ionic liquid as a recoverable catalyst for asymmetric Diels-Alder reaction. *RSC Adv.* **2015**, *5*, 52636–52641. [CrossRef]
21. Walborsky, H.; Barash, L.; Davis, T. Communications- Partial Asymmetric Syntheses: The Diels-Alder Reaction. *J. Am. Chem. Soc.* **1988**, *110*, 1238–1256.
22. Evans, D.A.; Chapman, K.T.; Bisaha, J. Asymmetric Diels-Alder Cycloaddition Reactions with Chiral α,β-Unsaturated N-Acyloxazolidinones. *J. Am. Chem. Soc.* **1988**, *110*, 1238–1256. [CrossRef]
23. Robiette, R.; Cheboub-Benchaba, K.; Peeters, D.; Marchand-Brynaert, J. Design of a New and Highly Effective Chiral Auxiliary for Diels-Alder Reaction of 1-Aminodiene. *J. Org. Chem.* **2003**, *68*, 9809–9812. [CrossRef] [PubMed]
24. Lakner, F.J.; Negrete, G.R. A new and convenient chiral auxiliary for asymmetric Diels-Alder cycloadditions in environmentally benign solvents. *Synlett* **2002**, *4*, 643–645. [CrossRef]
25. Hachiya, S.; Kasashima, Y.; Yagishita, F.; Mino, T.; Masu, H.; Sakamoto, M. Asymmetric Transformation by Dynamic Crystallization of Achiral Succinimides. *Chem. Commun.* **2013**, *49*, 4776–4778. [CrossRef]
26. Addadi, L.; Lahav, M. *Origin of Optical Activity in Nature*; Walker, D.C., Ed.; Elsevier: New York, NY, USA, 1979.
27. Mason, S.F. Origins of Biomolecular Handedness. *Nature* **1984**, *311*, 19–23. [CrossRef]
28. Bonner, W.A. The Origin and Amplification of Biomolecular Chirality. *Orig. Life Evol. Biosph.* **1991**, *21*, 59. [CrossRef]
29. Avalos, M.; Babiano, R.; Cintas, P.; Jiménez, J.L.; Palacios, J.C.; Barron, L.D. Absolute Asymmetric Synthesis under Physical Fields: Facts and Fictions. *Chem. Rev.* **1998**, *98*, 2391–2404. [CrossRef]
30. Feringa, B.L.; van Delden, R. Absolute Asymmetric Synthesis: The Origin, Control, and Amplification of Chirality. *Angew. Chem. Int. Ed.* **1999**, *38*, 3418–3438. [CrossRef]
31. Tsogoeva, S.B.; Wei, S.; Freund, M.; Mauksch, M. Deracemization with reversible Mannich type reaction. *Angew. Chem. Int. Ed.* **2009**, *48*, 590–594. [CrossRef]
32. Flock, A.M.; Reucher, C.M.M.; Bolm, C. Enantioenrichment by Iterative Retro-Aldol/Aldol ReactionCatalyzed by an Achiral or Racemic Base. *Chem. Eur. J.* **2010**, *16*, 3918–3921. [CrossRef] [PubMed]
33. Yagishita, F.; Ishikawa, H.; Onuki, T.; Hachiya, S.; Mino, T.; Sakamoto, M. Total spontaneous resolution by deracemization of isoindolinones. *Angew. Chem. Int. Ed.* **2012**, *51*, 13023–13025. [CrossRef] [PubMed]

34. Steendam, R.R.E.; Verkade, J.M.M.; Van Benthem, T.J.B.; Meekes, H.; Van Enckevort, W.J.P.; Raap, J.; Rutjes, F.P.J.T.; Vlieg, E. Emergence of Single-molecule Chirality from Achiral Reactants. *Nat. Commun.* **2014**, *5*, 5543.
35. Kaji, Y.; Uemura, N.; Kasashima, Y.; Ishikawa, H.; Yoshida, Y.; Mino, T.; Sakamoto, M. Asymmetric Synthesis of an Amino Acid Derivative from Achiral Aroyl Acrylamide by Reversible Michael Addition and Preferential Crystallization. *Chem. Eur. J.* **2016**, *22*, 16429–16432. [CrossRef] [PubMed]
36. Uemura, N.; Sano, K.; Matsumoto, A.; Yoshida, Y.; Mino, T.; Sakamoto, M. Absolute Asymmetric Synthesis of an Aspartic Acid Derivative from Prochiral Maleic Acid and Pyridine under Achiral Conditions. *Chem. Asian J.* **2019**, *14*, 4150–4153. [CrossRef]
37. Kawasaki, T.; Takamatsu, N.; Aiba, S.; Tokunaga, Y. Spontaneous Formation and Amplification of an Enantioenriched α-Amino Nitrile: A Chiral Precursor for Strecker Amino Acid Synthesis. *Chem. Commun.* **2015**, *51*, 14377–14380. [CrossRef]
38. Takamatsu, N.; Aiba, S.; Yamada, T.; Tokunaga, Y.; Kawasaki, T. Highly Stereoselective Strecker Synthesis Induced by a Slight Modification of Benzhydrylamine from Achiral to Chiral. *Chem. Eur. J.* **2018**, *24*, 1304–1310. [CrossRef]
39. Sakamoto, M.; Shiratsuki, K.; Uemura, N.; Ishikawa, H.; Yoshida, Y.; Kasashima, Y.; Mino, T. Asymmetric Synthesis by Using Natural Sunlight under Absolute Achiral Conditions. *Chem. Eur. J.* **2017**, *23*, 1717–1721. [CrossRef]
40. Ishikawa, H.; Uemura, N.; Yagishita, Y.; Baba, N.; Yoshida, Y.; Mino, T.; Kasashima, Y.; Sakamoto, M. Asymmetric Synthesis Involving Reversible Photodimerization of a Prochiral Flavonoid Followed by Crystallization. *Eur. J. Org. Chem.* **2017**, *46*, 6878–6881. [CrossRef]
41. Uemura, N.; Toyoda, S.; Ishikawa, H.; Yoshida, Y.; Mino, T.; Kasashima, Y.; Sakamoto, M. Asymmetric Diels–Alder Reaction Involving Dynamic Enantioselective Crystallization. *J. Org. Chem.* **2018**, *83*, 9300–9304. [CrossRef]
42. Viedma, C. Chiral Symmetry Breaking during Crystallization: Complete Chiral Purity Induced by Nonlinear Autocatalysis and Recycling. *Phys. Rev. Lett.* **2005**, *94*, 065504. [CrossRef] [PubMed]
43. Coquerel, G. Crystallization of Molecular Systems from Solution: Phase Diagrams, Supersaturation and Other Basic Concepts. *Chem. Soc. Rev.* **2014**, *43*, 2286–2300. [CrossRef] [PubMed]
44. Viedma, C.; Ortiz, J.E.; de Torres, T.; Izumi, T.; Blackmond, D.G.; Viedma, C.; Ortiz, J.E.; de Torres, T.; Izumi, T.; Blackmond, D.G. Evolution of Solid Phase Homochirality for a Proteinogenic Amino Acid. *J. Am. Chem. Soc.* **2008**, *130*, 15274–15275. [CrossRef] [PubMed]
45. Noorduin, W.L.; Bode, A.A.C.; van der Meiden, M.; Meekes, H.; van Etteger, A.F.; van Enckevort, W.J.P.; Christianen, P.C.M.; Kaptein, B.; Kellogg, R.M.; Rasing, T.; et al. Complete Chiral Symmetry Breaking of an Amino Acid Derivative Directed by Circularly Polarized Light. *Nature Chem.* **2009**, *1*, 729–732. [CrossRef]
46. Gherase, D.; Conroy, D.; Matar, O.K.; Blackmond, D.G. Experimental and Theoretical Study of the Emergence of Single Chirality in Attrition-Enhanced Deracemization. *Cryst. Growth Des.* **2014**, *14*, 928–937. [CrossRef]
47. Sogutoglu, L.-C.; Steendam, R.R.E.; Meekes, H.; Vlieg, E.; Rutjes, F.P.J.T. Viedma Ripening: A Reliable Crystallisation Method to Reach Single Chirality. *Chem. Soc. Rev.* **2015**, *44*, 6723–6732. [CrossRef]
48. Steendam, R.R.E.; Kulka, M.W.; Meekes, H.; van Enckevort, W.J.P.; Raap, J.; Vlieg, E.; Rutjes, F.P.J.T. One-Pot Synthesis, Crystallization and Deracemization of Isoindolinones from Achiral Reactants. *Eur. J. Org. Chem.* **2015**, 7249–7252. [CrossRef]
49. Nguyen, T.P.T.; Cheung, P.S.M.; Werber, L.; Gagnon, J.; Sivakumar, R.; Lennox, C.; Sossin, A.; Mastai, Y.; Cuccia, L.A. Directing the Viedma ripening of ethylenediammonium sulfate using "Tailor-made" chiral additives. *Chem. Commun.* **2016**, *52*, 12626–12629. [CrossRef]
50. Sivakumar, R.; Askari, M.S.; Woo, S.; Madwar, C.; Ottenwaelder, X.; Bohle, D.S.; Cuccia, L.A. Homochiral crystal generation via sequential dehydration and Viedma ripening. *Cryst. Eng. Comm.* **2016**, *18*, 4277–4280. [CrossRef]
51. Breveglieri, F.; Maggioni, G.M.; Mazzotti, M. Deracemization of NMPA via Temperature Cycles. *Cryst. Growth Des.* **2018**, *18*, 1873–1881. [CrossRef]
52. Engwerda, A.H.J.; Maassen, R.; Tinnemans, P.; Meekes, H.; Rutjes, F.P.J.T.; Vlieg, E. Attrition-Enhanced Deracemization of the Antimalaria Drug Mefloquine. *Angew. Chem. Int. Ed.* **2019**, *58*, 1670–1673. [CrossRef] [PubMed]
53. Houk, K.N.; Blackmond, D.G. Isotopically Directed Symmetry Breaking and Enantioenrichment in Attrition-Enhanced Deracemization. *J. Am. Chem. Soc.* **2020**, *142*, 3873–3879.
54. Ishikawa, H.; Ban, K.; Uemura, N.; Yoshida, Y.; Mino, T.; Kasashima, Y.; Sakamoto, M. Attrition-Enhanced Deracemization of Axially Chiral Nicotinamides. *Eur. J. Org. Chem.* **2020**, *8*, 1001–1005. [CrossRef]

55. Havinga, E. Spontaneous Formation of Optically Active Substances. *Biochim. Biophys. Acta* **1954**, *13*, 171–174. [CrossRef]
56. Frank, F.C. On Spontaneous Asymmetric Synthesis. *Biochim. Biophys. Acta* **1953**, *11*, 459–463. [CrossRef]
57. Yoshioka, R. Racemization, Optical Resolution and Crystallization-induced Asymmetric Transformation of Amino Acids and Pharmaceutical Intermediates. *Top. Curr. Chem.* **2007**, *269*, 83–132.
58. Sakamoto, M.; Mino, T. Asymmetric Reaction Using Molecular Chirality Controlled by Spontaneous Crystallization. *Eur. J. Org. Chem.* **2017**, 6878–6881. [CrossRef]
59. Jacques, J.; Collet, A.; Wilen, S.H. *Enantiomers, Racemates and Resolution*; Krieger: Malabar, FL, USA, 1994.
60. Coquerel, G. *Chiral Discrimination in the Solid State: Applications to Resolution and Deracemization*; Springer: Tokyo, Japan, 2015; pp. 393–420.
61. Kellogg, R.M. *How to Use Pasteur's Tweezers*; Springer: Tokyo, Japan, 2015; pp. 421–443.
62. Sakamoto, M.; Mino, T. *Total Resolution of Racemates by Dynamic Preferential Crystallization*; Springer: Tokyo, Japan, 2015; pp. 445–462.
63. Petralli-Mallow, T.; Wong, T.M.; Byers, J.D.; Yee, H.I.; Hicks, J.M. Circular Dichroism Spectroscopy at Interfaces: A Surface Second Harmonic Generation Study. *J. Phys. Chem.* **1993**, *97*, 1383–1388. [CrossRef]
64. Fischer, P.; Hache, F. Nonlinear Optical Spectroscopy of Chiral Molecules. *Chirality* **2005**, *17*, 421–437. [CrossRef]
65. Kotha, S.; Banerjee, S. Recent developments in the *retro*-Diels–Alder reaction. *RSC Adv.* **2013**, *3*, 7642–7666. [CrossRef]
66. Saito, Y.; Hyuga, H. Chiral Crystal Growth under Grinding. *J. Phys. Soc. Jpn.* **2008**, *77*, 113001/1. [CrossRef]
67. Martin, I.; Marco, M.A. A Population Balance Model for Chiral Resolution via Viedma Ripening. *Cryst. Growth Des.* **2011**, *11*, 4611.
68. Peter, J.S. Kinetics and Thermodynamics of Efficient Chiral Symmetry Breaking in Nearly Racemic Mixtures of Conglomerate Crystals. *Cryst. Growth Des.* **2011**, *11*, 1957–1965.
69. Xiouras, C.; Van Cleemput, E.; Kumpen, A.; Ter Horst, J.H.; Van Gerven, T.; Stefanidis, G.D. Towards Deracemization in the Absence of Grinding through Crystal Transformation, Ripening, and Racemization. *Cryst. Growth Des.* **2017**, *17*, 882–890. [CrossRef]
70. Matuszak, N.; Muccioli, G.G.; Labar, G.; Lambert, D.M. Synthesis and in Vitro Evaluation of N-Substituted Maleimide Derivatives as Selective Monoglyceride Lipase Inhibitors. *Bioorg. Med. Chem. Lett.* **2010**, *20*, 1510–1515. [CrossRef]
71. Eloh, K.; Demurtas, M.; Mura, M.G.; Deplano, A.; Onnis, V.; Sasanelli, N.; Maxia, A.; Caboni, P. Potent Nematicidal Activity of Maleimide Derivatives on Meloidogyne incognita. *J. Agric. Food Chem.* **2016**, *64*, 4876–4881. [CrossRef]

© 2020 by the authors. Licensee MDPI, Basel, Switzerland. This article is an open access article distributed under the terms and conditions of the Creative Commons Attribution (CC BY) license (http://creativecommons.org/licenses/by/4.0/).

*Review*

# Symmetry Breaking and Photomechanical Behavior of Photochromic Organic Crystals

Daichi Kitagawa [1], Christopher J. Bardeen [2],*  and Seiya Kobatake [1],*

1. Department of Applied Chemistry, Graduate School of Engineering, Osaka City University, 3-3-138 Sugimoto, Sumiyoshi-ku, Osaka 558-8585, Japan; kitagawa@osaka-cu.ac.jp
2. Department of Chemistry, University of California, 501 Big Springs Road, Riverside, CA 92521, USA
* Correspondence: christopher.bardeen@ucr.edu (C.J.B.); kobatake@a-chem.eng.osaka-cu.ac.jp (S.K.); Tel.: +1-951-827-2723 (C.J.B.); +81-6-6605-2797 (S.K.)

Received: 27 August 2020; Accepted: 7 September 2020; Published: 9 September 2020

**Abstract:** Photomechanical materials exhibit mechanical motion in response to light as an external stimulus. They have attracted much attention because they can convert light energy directly to mechanical energy, and their motions can be controlled without any physical contact. This review paper introduces the photomechanical motions of photoresponsive molecular crystals, especially bending and twisting behaviors, from the viewpoint of symmetry breaking. The bending (right–left symmetry breaking) and twisting (chiral symmetry breaking) of photomechanical crystals are based on both intrinsic and extrinsic factors like molecular orientation in the crystal and illumination conditions. The ability to design and control this symmetry breaking will be vital for generating new science and new technological applications for organic crystalline materials.

**Keywords:** photomechanical; crystal; right–left symmetry breaking; chiral symmetry breaking

## 1. Introduction

Symmetry is a very important concept in various fields, from natural sciences like mathematics, physics, chemistry, biology, geology, and astronomy to engineering fields such as architecture and urban design. For instance, the 1979 and 2008 Nobel Prizes in Physics were awarded to research on symmetry, specifically a unified symmetry description of electromagnetic and weak interactions and the discovery of the mechanism of spontaneous breaking of symmetry [1–5]. The well-known Woodward–Hoffmann rules in chemistry that rationalize pericyclic reactions rely on the fact that the symmetry of the molecular orbitals of the electrons involved in the reaction must be preserved during the reaction [6]. The 1981 Nobel Prize in Chemistry was awarded to the Woodward–Hoffmann rules and the frontier molecular orbital theory reported by Kenichi Fukui [7]. In biology, the homochirality of amino acids that almost always exist in the left-handed form (L-amino acids) is well known. The origin of this homochirality is not known but may be related to symmetry breaking. Moreover, in architecture, symmetry also plays an important role. In Islamic architecture, the elegance of the mosque is due to its symmetry and golden ratio. Thus, our life is closely related to symmetry.

The concept of symmetry is also important in biological and materials science research. For example, Kuroda et al. reported that the zygotic left–right asymmetry pathway in snails is dictated by its chiral blastomere arrangement [8]. Briefly, they physically twisted the blastomere and showed that the right-handed and left-handed conch was determined by the difference in the shape of the blastomere (Figure 1). Ishii et al. also reported on the control of chiral supramolecular nanoarchitectures by macroscopic mechanical rotations [9]. They revealed that the macroscopic mechanical rotation of a rotary evaporator could induce enantioselective H-aggregation of achiral phthalocyanines: counterclockwise rotation resulted in right-handed aggregation, but clockwise rotation gave left-handed aggregation (Figure 2). These results indicate that the chirality of physical outputs can be controlled by applying external mechanical stimuli that break the symmetry. In this paper, we introduce photomechanical molecular crystals that respond to light as an external stimulus and show that this perturbation can generate chiral mechanical motion as an output.

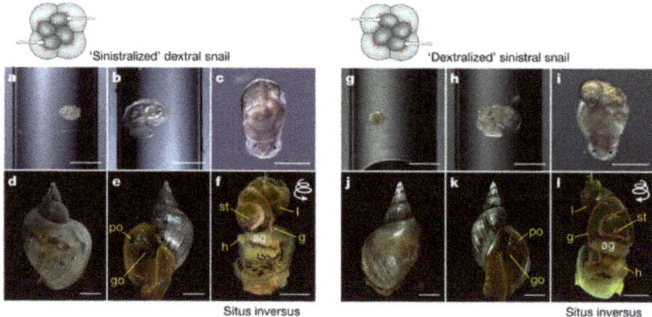

**Figure 1.** Generation of right-handed and left-handed snails, depending on the physical twist of the blastomere. After the third-cleavage manipulations, both sinistralized dextral embryos (**a**) and dextralized sinistral embryos (**g**) were raised to adult snails. Development was observed at trochophore (**a,g**), veliger (**b,h**) and juvenile snail (**c,i**) stages. Adult snails were pictured dorsally (**d,j**) and ventrally (**e,k**). The shell was removed to observe the position of internal organs (**f,l**, dorsal view). ag, albumen gland; g with dotted red line, gut; go, female genital opening; h, heart; l with white coil, liver; st, stomach; po, pulmonary sac opening. Scale bars: **a–c**, **g–i**, 0.5 mm; **d–f**, **j–l**, 5 mm. Reproduced from [8] with permission of Springer Nature, copyright 2009.

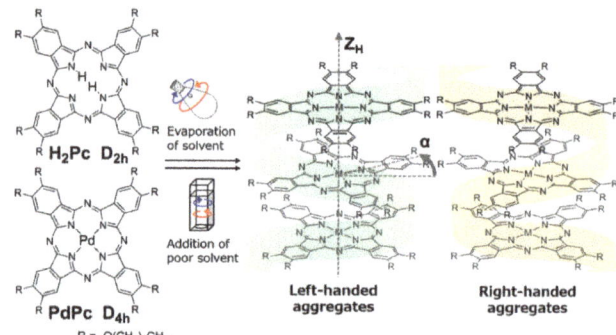

**Figure 2.** Control of chiral supramolecular nanoarchitectures by macroscopic mechanical rotations. Reproduced from [9] with permission of Wiley-VCH Verlag GmbH & Co. KGaA, Weinheim, copyright 2013.

## 2. History and Background of Photomechanical Molecular Crystals

Photomechanical materials exhibiting mechanical motions upon photoirradiation generally consist of photoresponsive molecules. When photoresponsive molecules undergo geometrical changes due to photochemical reactions, the individual molecular motions act in concert, resulting in the mechanical motion of the material itself. Liquid crystalline polymers and molecular crystals made from photoresponsive molecules are well known and have been intensively investigated so far.

The history of photomechanical molecular crystals starts from research in 1982, reported by Abakumov and Nevodchikov et al. [10]. The crystals, composed of a semiquinone complex of platinum group metals, exhibited a photomechanical bending of the crystal by as much as 45° upon irradiation with visible or near-infrared light. The crystal reverted to its original shape within 0.1 s when the light exposure was stopped. The bending behavior was due to the radical-mediated formation of dimerized Rh-Rh bonds in the crystal. However, at that time, this photomechanical behavior did not attract much attention. After a while, in the early 2000s, it was noted again. Bardeen et al. reported on the photochemically driven expansion of crystalline nanorods composed of an anthracene derivative, 9-tert-butyl anthroate (9-TBAE) (Figure 3) [11]. 9-TBAE undergoes a [4 + 4] photodimerization in the crystalline state, which results in a 15% increase in rod length without fragmentation. Kobatake and Irie et al. reported on the rapid and photoreversible shape changes of photochromic diarylethene crystals, including transforming from a square shape to a lozenge shape, expansion and contraction, and bending (Figure 4) [12]. After these remarkable findings, many researchers joined this research field, and it has been revealed that crystals of various photoresponsive molecules can exhibit photomechanical behavior such as expansion and contraction, bending, twisting, coiling, rolling, and so on (Figure 5) [13–43]. As can be seen, bending is an especially common motion.

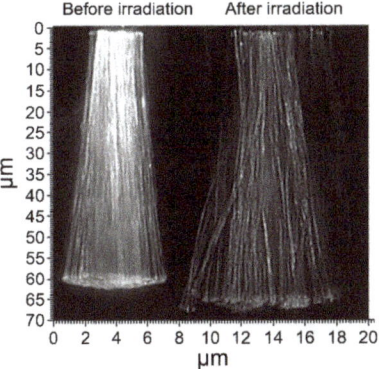

**Figure 3.** Fluorescence image of a bundle of nanorods, consisting of 9-tert-butyl anthroate (9-TBAE), before and after irradiation with a 365 nm light. The rods lengthen by an average of 15% as measured along a single rod. Reproduced from [11] with permission of the American Chemical Society, copyright 2006.

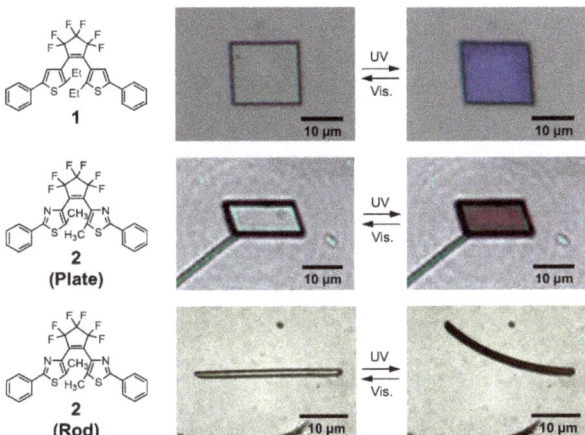

**Figure 4.** Photoreversible crystal shape changes of diarylethene derivatives **1** and **2**. Adapted from [12] with permission of Springer Nature, copyright 2007.

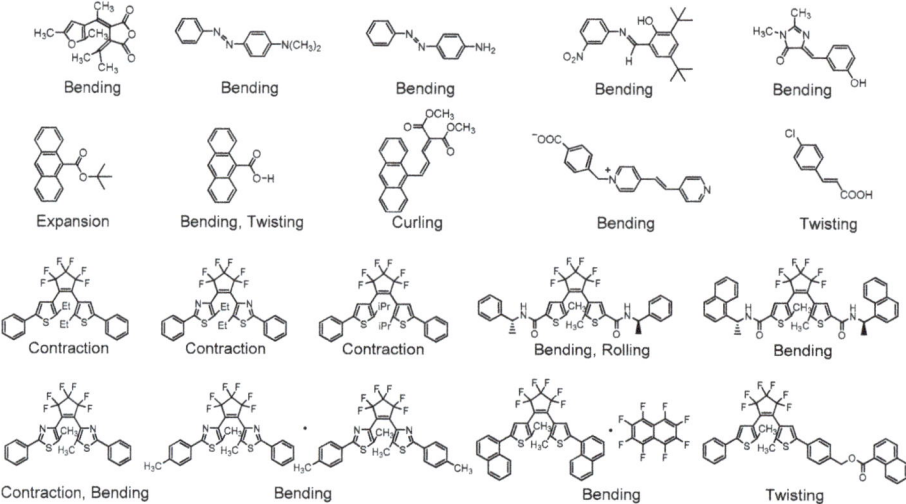

**Figure 5.** Photochromic compounds exhibiting photomechanical behaviors in crystals.

## 3. Photomechanical Bending Motion of Molecular Crystals: Right–Left Symmetry Breaking

Photomechanical motion of photoresponsive organic crystals is ascribed to the strain generated by the photoreactions in the crystal. The mechanism can be explained with the example of photomechanical crystal bending as follows. Upon ultraviolet (UV) light irradiation from one side of the crystal, photoresponsive molecules undergo various reactions such as intramolecular ring-closing and opening reactions, cis–trans isomerization, intermolecular [2 + 2] photodimerization, [4 + 4] photodimerization, and so on, which results in the formation of a bimorph or bimetal structure between reactants and photoproducts. This asymmetry arises because the photoreactions proceed more on the side where light is irradiated. The strain induced by this bimorph or bimetal structure leads to the bending deformation of the crystal itself (Figure 6) [44]. From the viewpoint of symmetry, photomechanical behavior is induced by the breaking of symmetry of the molecular structure and arrangement in the crystal due to the progress of the photoreaction. Moreover, the bending direction can be controlled by moving the illumination source. Most bending crystals based on the bimorph or bimetal mechanism exhibit this control of symmetry breaking by an extrinsic factor. Thus, it can be said that the symmetry breaking plays an important role even in the simplest photomechanical behavior.

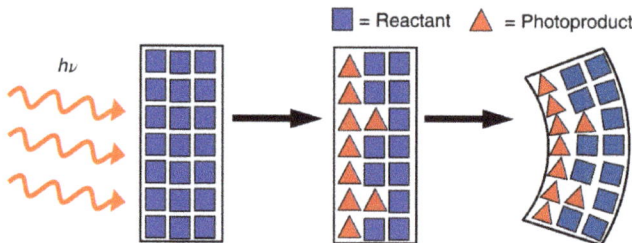

**Figure 6.** Schematic illustration of a bimorph or bimetal structure composed of reactants and photoproducts. Adapted from [44] with permission of John Wiley & Sons, Ltd., copyright 2017.

As mentioned in the previous section, bending is the most common photomechanical motion. In addition to illumination direction, the bending behavior is influenced by the molecular crystal structure itself. A good illustration of this can be seen by taking diarylethene molecular crystals as examples. Diarylethene molecules undergo a 6π ring-closing and ring-opening reaction between the colorless open-ring isomer and the colored closed-ring isomer. During this photoisomerization reaction, the long axis of the molecule shrinks, the short axis of the molecule extends, and the thickness of the molecule decreases, as shown in Figure 7 [45]. This small but apparent molecular structural change results in photomechanical motion. When a rod-shaped crystal of a diarylethene derivative, 1-(5-methyl-2-(4-(p-vinylbenzoyloxymethyl)phenyl)-4-thiazolyl)-2-(5-methyl-2-phenyl-4-thiazolyl) perfluorocyclopentene (4), is irradiated with UV light, the crystal bends toward the light source and straightens out after irradiation with visible light (Figure 8a) [23]. On the other hand, the crystal of a diarylethene derivative, 1,2-bis(2-methyl-5-(4-(1-naphthoyloxymethyl) phenyl)-3-thienyl)perfluorocyclopentene (5), shows different behavior. Upon irradiation with UV light, the crystal bends away from the light source and returns to its initial shape by visible light irradiation (Figure 8b) [21]. The different bending directions are ascribed to different molecular packings in the crystals. Therefore, molecular packing also plays an important role as an intrinsic factor in determining the bending direction.

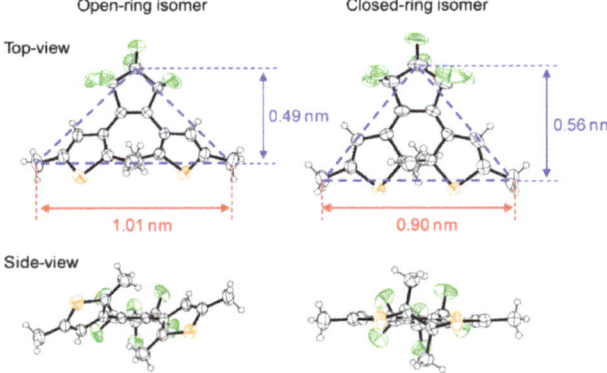

**Figure 7.** Molecular structural change of a diarylethene derivative, 1,2-bis(2,5-dimethyl-3-thienyl)-perfluorocyclopentene (3), accompanied with the photochromic reaction. Reproduced from [45] with permission of the American Chemical Society, copyright 2014.

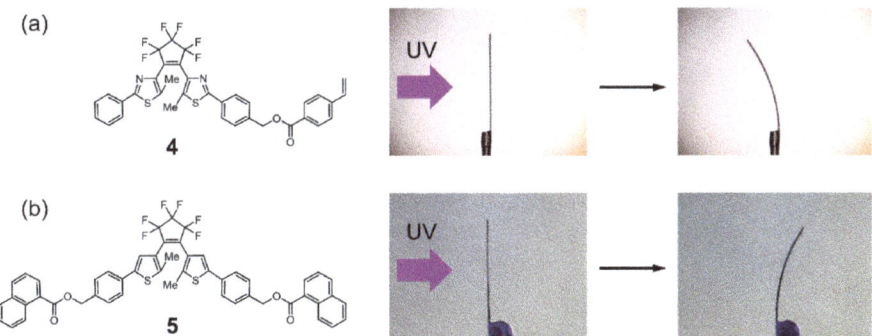

**Figure 8.** Photomechanical crystal bending behaviors of diarylethene derivatives 4 and 5 (a) bending toward the UV light source (adapted from [23] with permission of the European Society for Photobiology, the European Photochemistry Association, and The Royal Society of Chemistry, copyright 2014) and (b) bending away from the UV light source at 365 nm (adapted from [21] with permission of the American Chemical Society, copyright 2013).

One question is whether such bending behavior can be observed for crystals of any size. It is a long-known fact for chemists that when the size of a crystal is large, the crystal usually disintegrates as the photoreaction proceeds. To gain insight into the size dependence of the photomechanical bending behavior, the dependence of the bending velocity on the crystal thickness was investigated [21,23]. The initial bending velocity ($V_{init}$) increases as the crystal thickness decreases, and the relationship between the $V_{init}$ and the crystal thickness was well explained by Timoshenko's bimetal model. Based on this analysis, it was found that a crystal with a thickness of a few micrometers bent well. In larger crystals, the strain generated by the photoisomerization reaction cannot be relaxed by deformation of the crystal, and the crystal tends to fracture. Furthermore, it was found that $V_{init}$ increased in proportion to the light irradiation intensity, suggesting that the photomechanical bending was directly proportional to the amount of molecular reactions that have occurred [46]. This finding inspired us to examine the effect of illumination conditions on photomechanical behavior. When the crystal of a diarylethene derivative, 1,2-bis(5-methyl-2-phenyl-4-thiazolyl)perfluorocyclopentene (2), was irradiated with a 365 nm light, it bent toward the incident light. However, upon irradiation with a 380 nm light, the crystal bent away from the light source at first, then bent back toward the incident light after prolonged irradiation (Figure 9) [47]. This result is related to the difference in depth of the photochromic reaction from the crystal surface for different illumination wavelengths. Similar results could be observed when polarized UV light was used as the incident light [48]. Thus, the photomechanical bending direction (i.e., right–left symmetry breaking) can be controlled by both intrinsic and extrinsic factors, such as the molecular packing in the crystal and the illumination conditions (source direction and wavelength).

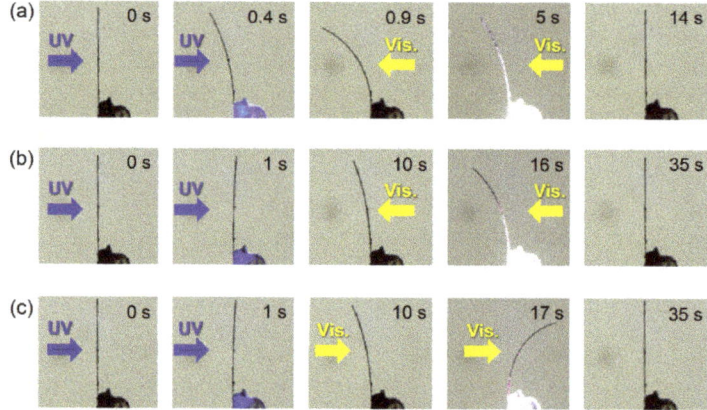

**Figure 9.** Different photomechanical bending behavior of a diarylethene **2** crystal depending on the illumination wavelength. (**a**) The crystal was irradiated with a 365 nm light from the left side and then irradiated with visible light from the right side. (**b**) The crystal was irradiated with a 380 nm light from the left side and then irradiated with visible light from the right side. (**c**) The crystal was irradiated with a 380 nm light from the left side and then irradiated with visible light from the left side. Reproduced from [47] with permission of the PCCP Owner Societies.

## 4. Crystal Twisting: Chiral Symmetry Breaking by Intrinsic Factors

In the research field of crystal growth, twisted crystal growth is an interesting topic. Twisted crystal growth has been observed in crystals of a variety of substances including elements, minerals, simple salts, organic molecules, and polymers. Figure 10 shows the photographs of the twisted crystals of natural quartz and $K_2Cr_2O_7$ as examples of twisted crystal growth. Shtukenberg and Kahr et al. have written a review paper on the mechanism of growth-actuated twisting of single crystals [49]. Since various factors can play a role in growth-actuated twisting, such as chemical composition, size, shape, and growth conditions, there is no universal mechanism for the twisting. However, in all cases, it can be concluded that the crystal twists as a result of the chiral symmetry breaking due to various factors such as surface charge, electrostatics, piezoelectricity, screw dislocation, twinning, surface stress, and so on. These are all examples of symmetry breaking due to intrinsic environmental factors in the absence of a chemical reaction.

**Figure 10.** Photographs of a representative twisted crystal growth of (**a**) natural quartz and (**b**) $K_2Cr_2O_7$. Reproduced from [49] with permission of Wiley-VCH Verlag GmbH & Co. KGaA, Weinheim, copyright 2014.

Recently, Sureshan et al. reported on chirality-controlled twisting of dipeptide crystals by a thermal topochemical reaction [50]. The presence of chiral reactants could induce crystal twisting during the reaction under heating, and the direction of the twisting (i.e., right-handed or left-handed) depended on the chirality of the precursor dipeptide (Figure 11). This is an example of stimuli-responsive twisting. The heating acts as a stimulus and leads to chiral symmetry breaking, resulting in twisting. A twisting motion due to photochemical reactions has also been reported. Bardeen et al. reported that microribbon crystals composed of 9-anthracenecarboxylic acid (9AC) exhibited twisting upon UV irradiation [26]. After the light was turned off, they relaxed back to their original shapes over the course of minutes (Figure 12). This is T-type (thermally reversible) photomechanical twisting. Kitagawa and Kobatake et al. reported that microribbon crystals made from a diarylethene derivative also exhibited reversible photomechanical crystal twisting [22]. In this case, the crystal could be reversibly switched back and forth between twisted and straight with alternating irradiation with UV and visible light (Figure 13). This is P-type (photochemically reversible) photomechanical twisting. These photomechanical crystal twisting behaviors were observed under spatially uniform light irradiation. The twisting motion is induced by strain in the diagonal direction relative to the crystal's long axis, which depends on the molecular orientation with respect to the long axis of the plate. Al-Kaysi and Bardeen et al. elucidated that controlling crystal morphology resulted in different photomechanical behaviors [51]. When crystals of 9-methylanthracene (9MA) were prepared by seeded growth using a surfactant, hexagonal microplates were obtained that exhibited photomechanical curling behavior (Figure 14a). On the other hand, rectangular microribbons prepared by a floating drop method (i.e., dropping the organic solvent containing 9MA into water) showed helical twisting behavior upon UV irradiation (Figure 14b). The hexagonal microplates and rectangular microribbons of 9MA are the same polymorph, but have different internal molecular orientations (i.e., the crystal growth directions of the hexagonal microplates and rectangular microribbons are different).

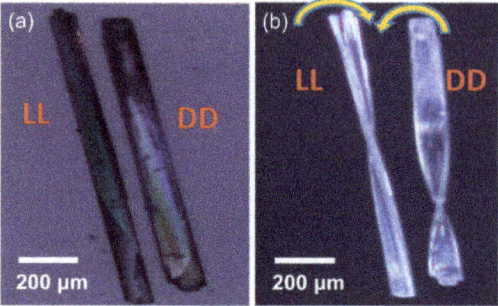

**Figure 11.** Twisting of crystals consisting of chiral dipeptides (N$_3$-$_L$-Ala-$_L$-Val-NHCH$_2$C≡CH) upon heating as an external stimulus (**a**) before heating and (**b**) after heating at 85 °C for 1 day. Adapteed from [50] with permission of National Academy of Sciences, copyright 2018.

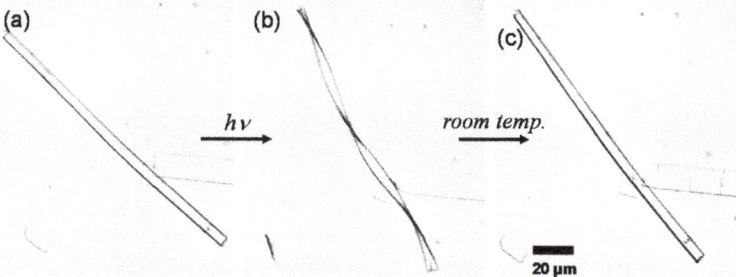

**Figure 12.** Photomechanical twisting behavior of a 9AC crystal upon UV irradiation (**a**) before photoirradiation and (**b**) immediately after irradiation. The twisted crystal returns to the original shape over the course of minutes by removing the incident light. (**c**) Nine minutes after removing the light. Adapted from [26] with permission of the American Chemical Society, copyright 2011.

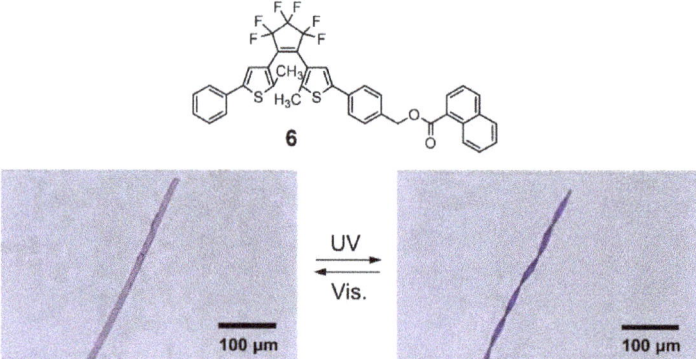

**Figure 13.** Photoreversible photomechanical twisting of a diarylethene **6** crystal upon alternating irradiation with UV and visible light. Adapted from [22] with permission of Wiley-VCH Verlag GmbH & Co. KGaA, Weinheim, copyright 2013.

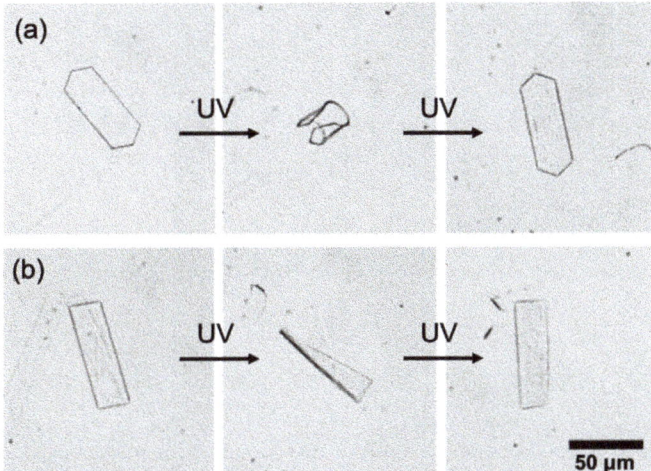

**Figure 14.** Photomechanical curling and twisting of 9-methylanthracene (9MA) crystals, depending on the crystal morphology. (**a**) A 9MA hexagonal microplate and (**b**) a 9MA rectangle microplate. Adapted from [51] with permission of Wiley-VCH Verlag GmbH & Co. KGaA, Weinheim, copyright 2018.

By controlling the molecular orientation within a crystal, it is possible to generate different photomechanical behaviors (Figure 14). On larger scales, the shape of the crystal itself can lead to symmetry breaking and directional motion. Katsonis et al. reported that ribbons made from a liquid crystalline polymer with an azobenzene derivative could exhibit curling, right-handed or left-handed helical twisting, depending on the direction in which they were cut (Figure 15) [52]. This is because the direction of the shrinkage and expansion in the plane of the material can be tuned by the direction of the cutting. This is an excellent study on controlling the photomechanical twisting by controlling the macroscopic structure. Al-Kaysi and Bardeen et al. reported that branched crystals composed of 4-fluoroanthracene-9-carboxylic acid (4F-9AC) could be prepared by a pH-driven reprecipitation method [53]. A branched crystal will rotate in one direction like a ratchet under sequential illumination (Figure 16). This is due to symmetry breaking by the crystal branching. The rotation direction depends on the chirality of the branched crystal shape. In all the cases in this section, the twisting was intrinsic in the sense that some internal structural factor led to twisting. Changing the external perturbation, i.e., the light field, was not used to influence the twisting motion.

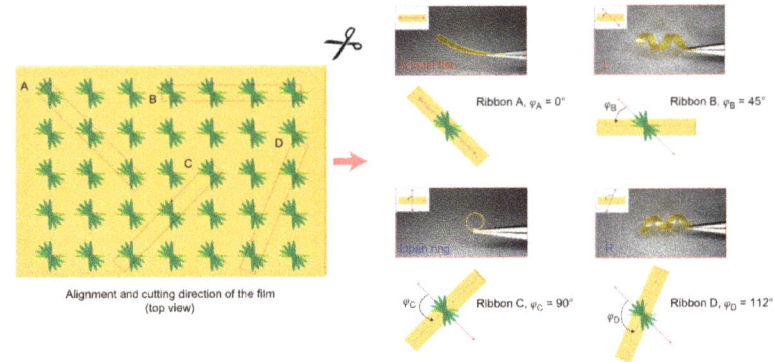

**Figure 15.** Photomechanical behaviors of liquid crystalline azobenzene polymer, depending the cutting direction. Reproduced from [52] with permission of Springer Nature, copyright 2014.

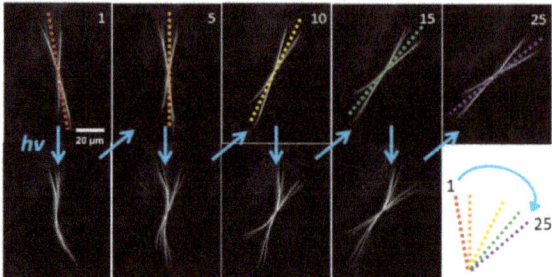

**Figure 16.** Photomechanical ratchet-like motion of X-shaped 4F-9AC crystals upon UV irradiation. Adapted from [53] with permission of Wiley-VCH Verlag GmbH & Co. KGaA, Weinheim, copyright 2016.

## 5. Control of Chiral Crystal Twisting by Control of Light Illumination as an Extrinsic Factor

The control of chiral symmetry breaking by external factors is an important research topic, and many scientists have made efforts in this field [54]. For example, in the introduction, we described how a dynamic external perturbation (rotovap spinning direction) could change the chiral structure of supramolecular aggregates. In photomechanical materials, control of chirality by extrinsic factors (that is, control of photomechanical twisting) has also been investigated.

Light is the main extrinsic perturbation that can be used in the case of molecular crystals. Chirality control can be realized by changing the illumination direction. We demonstrated that ribbon crystals that consist of a diarylethene derivative could exhibit different twisting motions, ranging from a helicoid to a cylindrical helix, depending on the angle of the incident light (Figure 17) [55]. This was ascribed to the preferential excitation of differently oriented molecules within the crystal by different light directions. In other words, by exciting differently oriented molecules, the photoinduced strain tensor in the crystal (and thus the mode of photomechanical deformation) can be controlled by the direction of the UV light irradiation. Note that this control is only possible for crystals where all the molecules have a precise orientation with respect to the lab frame. The detailed mechanism is still under investigation, but this is a unique example of controlling chiral photomechanical twisting by an extrinsic factor.

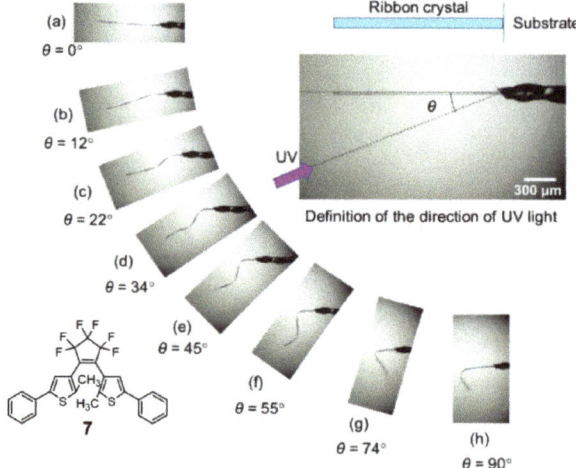

**Figure 17.** Control of photomechanical twisting of a ribbon crystal consisting of diarylethene 7 by illumination direction. Reproduced from [55] with permission of the American Chemical Society, copyright 2018.

## 6. Future Directions

There is now a large body of work showing that photoreactive crystals can generate mechanical motion. However, one major challenge is incorporating these crystals into actuator structures that can do useful work. As examples of applications, molecular crystal devices that give rise to cantilever motion [18], gearwheel rotation [19], and current switching [56] have been demonstrated, as shown in Figure 18. However, nano- and micro-sized crystals can only generate small amounts of work in isolation. It is still challenging to sum up the work of many individual crystals into one large output. As one of the strategies to overcome this issue, embedding crystals in polymer hosts has been demonstrated [57–60]. However, it is still difficult to order the crystals regularly in polymer hosts, and a large output has not been obtained. Recently, we reported that hybrid organic–inorganic materials consisting of a diarylethene derivative and anodic aluminum oxide (AAO) porous template can exhibit photomechanical actuation (Figure 19) [61,62]. These ordered composites combine the photoresponsive properties of the organic with the high elastic modulus of the ceramic, making them a photon-powered analog to piezoelectric actuators.

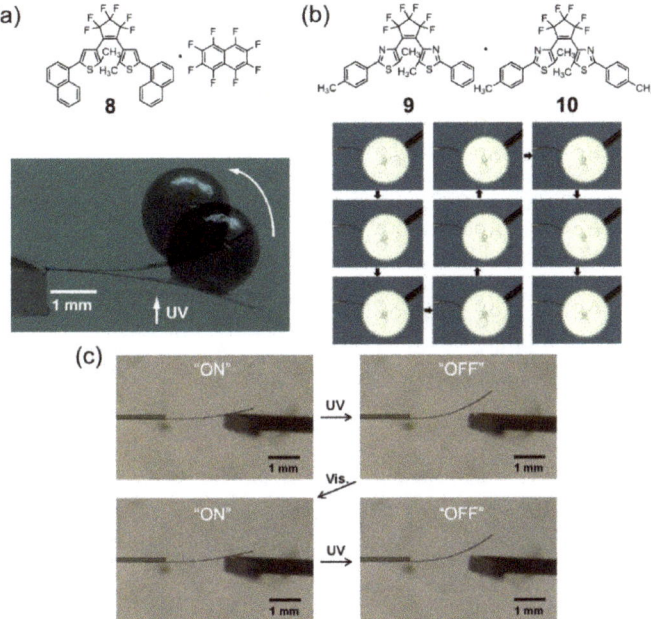

**Figure 18.** Examples of application of photomechanical bending behaviors of diarylethene crystals. (**a**) Molecular crystal cantilever, consisting of **8** and a perfluoronaphthalene co-crystal (reproduced from [18] with permission of the American Chemical Society, copyright 2010); (**b**) gearwheel rotation, working by complicated motions of a diarylethene derivatives **9** and **10** mixed crystal (reproduced from [19] with permission of Wiley-VCH Verlag GmbH & Co. KGaA, Weinheim, copyright 2012); and (**c**) current switching caused by photoreversible bending of a gold-coated diarylethene **2** crystal (reproduced from [56] with permission of The Royal Society of Chemistry).

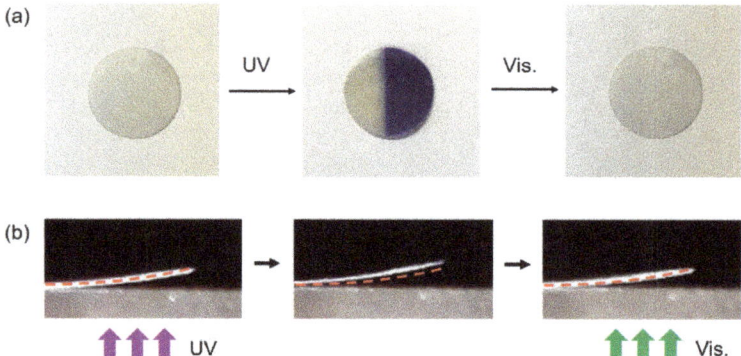

**Figure 19.** (**a**) Photographs of the hybrid organic–inorganic materials, consisting of a diarylethene derivative 7 and anodic aluminum oxide (AAO) porous template, and (**b**) its photomechanical actuation. Reproduced from [61] with permission of the American Chemical Society, copyright 2019.

A second challenge for organic photomechanical crystals is to improve their energy conversion and power efficiency. For example, in most cases, only a few percent of the photochromic molecules react and contribute to the photomechanical response. To increase conversion, negative photochromic compounds are the best candidates. The absorption spectra of negative photochromes undergo a shift to higher energies after photoisomerization, allowing the excitation light to penetrate through the crystal. Bardeen et al. recently reported on the photomechanical behavior of a phenylbutadiene derivative, (*E*)-4-fluorocinnamaldehyde malononitrile ((*E*)-4FCM), that undergoes a negative photochromic reaction based on a [2+2] photocycloaddition in the crystal form [63]. The nanowire bundles composed of (*E*)-4FCM exhibited a rapid expansion and spread by as much as 300% (Figure 20). AAO membranes containing this molecule were capable of lifting approximately four times as much weight as membranes containing a positive photochrome based on diarylethene. Thus, varying and optimizing the molecular photochrome is still one of the essential topics.

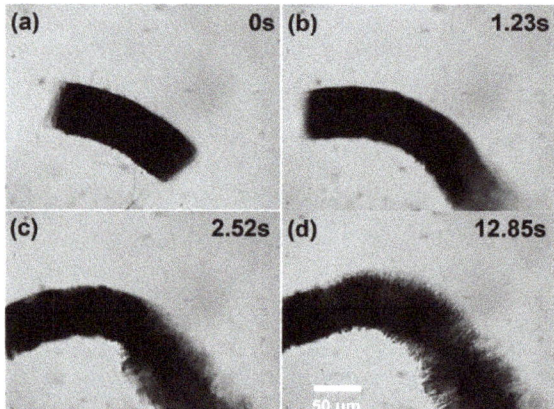

**Figure 20.** Spreading of nanowire bundles consisting of (*E*)-4-fluorocinnamaldehyde malononitrile ((*E*)-4FCM) upon UV irradiation. Reproduced from [63] with permission of The Royal Society of Chemistry.

## 7. Conclusions

In this paper, we introduced various photomechanical motions of photochromic molecular crystals from the viewpoint of symmetry breaking, such as right–left symmetry breaking and chiral symmetry breaking, by both intrinsic and extrinsic factors. Photomechanical molecular crystals have great potential as materials that convert light energy directly to mechanical energy, with large elastic moduli, high energy densities, and fast response times. This is an advantage compared with liquid crystalline polymers. In addition to the challenges described in the previous section, it is necessary to develop a quantitative understanding of how molecular-level structure changes and photochemical reaction kinetics give rise to a macroscopic photomechanical response. Such a predictive understanding will allow us to design photomechanical crystals with properties optimized for different applications. While there are still problems to be addressed, it is clear that this research field is rapidly developing and attracting many researchers. The ability of these high-symmetry structures to undergo symmetry breaking motion in response to light will continue to generate new science and new technological applications for organic crystalline materials.

**Author Contributions:** Conceptualization, D.K. and C.J.B.; writing—original draft preparation, D.K.; writing—review and editing, C.J.B and S.K.; project administration, S.K.; All authors have read and agreed to the published version of the manuscript.

**Funding:** The research was supported by JSPS KAKENHI Grant Number JP26107013 in Scientific Research on Innovative Areas Photosynergetics (S.K.), the United States National Science Foundation, grant DMR-1810514 (C.J.B.), and JSPS KAKENHI Grant Number JP16K17896 in Scientific Research for Young Scientists B (D.K.).

**Conflicts of Interest:** The authors declare no conflict of interest.

## References

1. Glashow, S.L. Partial-symmetries of Weak Interactions. *Nucl. Phys.* **1961**, *22*, 579–588. [CrossRef]
2. Salam, A.; Ward, J.C. Electromagnetic and Weak Interactions. *Phys. Lett.* **1964**, *13*, 168–171. [CrossRef]
3. Weinberg, S. A Model of Leptons. *Phys. Rev. Lett.* **1967**, *19*, 1264–1266. [CrossRef]
4. Nambu, Y. Axial Vector Current Conservation in Weak Interactions. *Phys. Rev. Lett.* **1960**, *4*, 380. [CrossRef]
5. Kobayashi, M.; Maskawa, T. CP-Violation in the Renormalizable Theory of Weak Interaction. *Prog. Theor. Phys.* **1973**, *49*, 652–657. [CrossRef]
6. Hoffmann, R.; Woodward, R.B. Orbital Symmetry Control of Chemical Reactions. *Science* **1970**, *167*, 825–831. [CrossRef]
7. Fukui, K.; Yonezawa, T.; Shingu, H. A Molecular Orbital Theory of Reactivity in Aromatic Hydrocarbons. *J. Chem. Phys.* **1952**, *20*, 722. [CrossRef]
8. Kuroda, R.; Endo, B.; Abe, M.; Shimizu, M. Chiral Blastomere Arrangement Dictates Zygotic Left-right Asymmetry Pathway in Snails. *Nature* **2009**, *462*, 790–794. [CrossRef]
9. Kuroha, M.; Nambu, S.; Hattori, S.; Kitagawa, Y.; Niimura, K.; Mizuno, Y.; Hamba, F.; Ishii, K. Chiral Supramolecular Nanoarchitectures from Macroscopic Mechanical Rotations: Effects on Enantioselective Aggregation Behavior of Phthalocyanines. *Angew. Chem. Int. Ed.* **2019**, *58*, 18454–18459. [CrossRef]
10. Abakumov, G.A.; Nevodchikov, V.I. Thermomechanical and Photomechanical Effects Observed on Crystals of a Free-radical Complex. *Dokl. Akad. Nauk Sssr* **1982**, *266*, 1407–1410.
11. Al-Kaysi, R.O.; Mueller, A.M.; Bardeen, C.J. Photochemically Driven Shape Changes of Crystalline Organic Nanorods. *J. Am. Chem. Soc.* **2006**, *128*, 15938–15939. [CrossRef] [PubMed]
12. Kobatake, S.; Takami, S.; Muto, H.; Ishikawa, T.; Irie, M. Rapid and Reversible Shape Changes of Molecular Crystals on Photoirradiation. *Nature* **2007**, *446*, 778–781. [CrossRef] [PubMed]
13. Koshima, H.; Ojima, N.; Uchimoto, H. Mechanical Motion of Azobenzene Crystals upon Photoirradiation. *J. Am. Chem. Soc.* **2009**, *131*, 6890–6891. [CrossRef] [PubMed]
14. Koshima, H.; Takechi, K.; Uchimoto, H.; Shiro, M.; Hashizume, D. Photomechanical Bending of Salicylideneaniline Crystals. *Chem. Commun.* **2011**, *47*, 11423–11425. [CrossRef]
15. Koshima, H.; Nakaya, H.; Uchimoto, H.; Ojima, N. Photomechanical Motion of Furylfulgide Crystals. *Chem. Lett.* **2012**, *41*, 107–109. [CrossRef]

16. Koshima, H.; Ojima, N. Photomechanical Bending of 4-Aminoazobenzene Crystals. *Dye. Pigm.* **2012**, *92*, 798–801. [CrossRef]
17. Uchida, K.; Sukata, S.I.; Matsuzawa, Y.; Akazawa, M.; de Jong, J.J.D.; Katsonis, N.; Kojima, Y.; Nakamura, S.; Areephong, J.; Meetsma, A.; et al. Photoresponsive Rolling and Bending of Thin Crystals of Chiral Diarylethenes. *Chem. Commun.* **2008**, *3*, 326–328. [CrossRef]
18. Morimoto, M.; Irie, M. A Diarylethene Cocrystal that Converts Light into Mechanical Work. *J. Am. Chem. Soc.* **2010**, *132*, 14172–14178. [CrossRef]
19. Terao, F.; Morimoto, M.; Irie, M. Light-Driven Molecular-Crystal Actuators: Rapid and Reversible Bending of Rodlike Mixed Crystals of Diarylethene Derivatives. *Angew. Chem. Int. Ed.* **2012**, *51*, 901–904. [CrossRef]
20. Kuroki, L.; Takami, S.; Yoza, K.; Morimoto, M.; Irie, M. Photoinduced Shape Changes of Diarylethene Single Crystals: Correlation between Shape Changes and Molecular Packing. *Photochem. Photobiol. Sci.* **2010**, *9*, 221–225. [CrossRef]
21. Kitagawa, D.; Kobatake, S. Crystal Thickness Dependence of Photoinduced Crystal Bending of 1,2-Bis(2-methyl-5-(4-(1-naphthoyloxymethyl)phenyl)-3-thienyl)perfluorocyclopentene. *J. Phys. Chem. C* **2013**, *117*, 20887–20892. [CrossRef]
22. Kitagawa, D.; Nishi, H.; Kobatake, S. Photoinduced Twisting of a Photochromic Diarylethene Crystal. *Angew. Chem. Int. Ed.* **2013**, *52*, 9320–9322. [CrossRef] [PubMed]
23. Kitagawa, D.; Kobatake, S. Crystal Thickness Dependence of the Photoinduced Crystal Bending of 1-(5-Methyl-2-(4-(p-vinylbenzoyloxymethyl)phenyl)-4-thiazolyl)-2-(5-methyl-2-phenyl-4-thiazolyl) perfluoro-cyclopentene. *Photochem. Photobiol. Sci.* **2014**, *13*, 764–769. [CrossRef] [PubMed]
24. Kitagawa, D.; Iwaihara, C.; Nishi, H.; Kobatake, S. Quantitative Evaluation of Photoinduced Bending Speed of Diarylethene Crystals. *Crystals* **2015**, *5*, 551–561. [CrossRef]
25. Al-Kaysi, R.O.; Bardeen, C.J. Reversible Photoinduced Shape Changes of Crystalline Organic Nanorods. *Adv. Mater.* **2007**, *19*, 1276–1280. [CrossRef]
26. Zhu, L.; Al-Kaysi, R.O.; Bardeen, C.J. Reversible Photoinduced Twisting of Molecular Crystal Microribbons. *J. Am. Chem. Soc.* **2011**, *133*, 12569–12575. [CrossRef]
27. Zhu, L.; Al-Kaysi, R.O.; Dillon, R.J.; Tham, F.S.; Bardeen, C.J. Crystal Structures and Photophysical Properties of 9-Anthracene Carboxylic Acid Derivatives for Photomechanical Applications. *Cryst. Growth Des.* **2011**, *11*, 4975–4983. [CrossRef]
28. Kim, T.; Zhu, L.; Mueller, L.J.; Bardeen, C.J. Dependence of the Solid-state Photomechanical Response of 4-Chlorocinnamic Acid on Crystal Shape and Size. *CrystEngComm* **2012**, *14*, 7792–7799. [CrossRef]
29. Kim, T.; Al-Muhanna, M.K.; Al-Suwaidan, S.D.; Al-Kaysi, R.O.; Bardeen, C.J. Photoinduced Curling of Organic Molecular Crystal Nanowires. *Angew. Chem. Int. Ed.* **2013**, *52*, 6889–6893. [CrossRef]
30. Kitagawa, D.; Kawasaki, K.; Tanaka, R.; Kobatake, S. Mechanical Behavior of Molecular Crystals Induced by Combination of Photochromic Reaction and Reversible Single-Crystal-to-Single-Crystal Phase Transition. *Chem. Mater.* **2017**, *29*, 7524–7532. [CrossRef]
31. Tong, F.; Kitagawa, D.; Dong, X.; Kobatake, S.; Bardeen, C.J. Photomechanical Motion of Diarylethene Molecular Crystal Nanowires. *Nanoscale* **2018**, *10*, 3393–3398. [CrossRef] [PubMed]
32. Hatano, E.; Morimoto, M.; Imai, T.; Hyodo, K.; Fujimoto, A.; Nishimura, R.; Sekine, A.; Yasuda, N.; Yokojima, S.; Nakamura, S.; et al. Photosalient Phenomena that Mimic Impatiens Are Observed in Hollow Crystals of Diarylethene with a Perfluorocyclohexene Ring. *Angew. Chem. Int. Ed.* **2017**, *56*, 12576–12580. [CrossRef] [PubMed]
33. Nakagawa, Y.; Morimoto, M.; Yasuda, N.; Hyodo, K.; Yokojima, S.; Nakamura, S.; Uchida, K. Photosalient Effect of Diarylethene Crystals of Thiazoyl and Thienyl Derivatives. *Chem. Eur. J.* **2019**, *25*, 7874–7880. [CrossRef] [PubMed]
34. Taniguchi, T.; Fujisawa, J.; Shiro, M.; Koshima, H.; Asahi, T. Mechanical Motion of Chiral Azobenzene Crystals with Twisting upon Photoirradiation. *Chem. Eur. J.* **2016**, *22*, 7950–7958. [CrossRef] [PubMed]
35. Bushuyev, O.S.; Singleton, T.A.; Barrett, C.J. Fast, Reversible, and General Photomechanical Motion in Single Crystals of Various Azo Compounds Using Visible Light. *Adv. Mater.* **2013**, *25*, 1796–1800. [CrossRef]
36. Bushuyev, O.S.; Tomberg, A.; Friscic, T.; Barrett, C.J. Shaping Crystals with Light: Crystal-to-Crystal Isomerization and Photomechanical Effect in Fluorinated Azobenzenes. *J. Am. Chem. Soc.* **2013**, *135*, 12556–12559. [CrossRef]

37. Samanta, R.; Kitagawa, D.; Mondal, A.; Bhattacharya, M.; Annadhasan, M.; Mondal, S.; Chandrasekar, R.; Kobatake, S.; Reddy, C.M. Mechanical Actuation and Patterning of Rewritable Crystalline Monomer–Polymer Heterostructures via Topochemical Polymerization in a Dual-Responsive Photochromic Organic Material. *ACS Appl. Mater. Interfaces* **2020**, *12*, 16856–16863. [CrossRef]
38. Samanta, R.; Ghosh, S.; Devarapalli, R.; Reddy, C.M. Visible Light Mediated Photopolymerization in Single Crystals: Photomechanical Bending and Thermomechanical Unbending. *Chem. Mater.* **2018**, *30*, 577–581. [CrossRef]
39. Naumov, P.; Kowalik, J.; Solntsev, K.M.; Baldridge, A.; Moon, J.-S.; Kranz, C.; Tolbert, L.M. Topochemistry and Photomechanical Effects in Crystals of Green Fluorescent Protein-like Chromophores: Effects of Hydrogen Bonding and Crystal Packing. *J. Am. Chem. Soc.* **2010**, *132*, 5845–5857. [CrossRef]
40. Nath, N.K.; Pejov, L.; Nichols, S.M.; Hu, C.; Saleh, N.; Kahr, B.; Naumov, P. Model for Photoinduced Bending of Slender Molecular Crystals. *J. Am. Chem. Soc.* **2014**, *136*, 2757–2766. [CrossRef]
41. Nath, N.K.; Runcevski, T.; Lai, C.Y.; Chiesa, M.; Dinnebier, R.E.; Naumov, P. Surface and Bulk Effects in Photochemical Reactions and Photomechanical Effects in Dynamic Molecular Crystals. *J. Am. Chem. Soc.* **2015**, *137*, 13866–13875. [CrossRef] [PubMed]
42. Gupta, P.; Karothu, D.P.; Ahmed, E.; Naumov, P.; Nath, N.K. All-in-One: Thermally Twistable, Photobendable, Elastically Deformable and Self-Healable Soft Crystal. *Angew. Chem. Int. Ed.* **2018**, *57*, 8498–8502. [CrossRef] [PubMed]
43. Halabi, J.M.; Ahmed, E.; Catalano, L.; Karothu, D.P.; Rezgui, R.; Naumov, P. Spatial Photocontrol of the Optical Output from an Organic Crystal Waveguide. *J. Am. Chem. Soc.* **2019**, *141*, 14966–14970. [CrossRef] [PubMed]
44. Zhu, L.; Tong, F.; Al-Kaysi, R.O.; Bardeen, C.J. Photomechanical Effects in Photochromic Crystals. In *Photomechanical Materials, Composites, and Systems*; White, T.J., Ed.; John Wiley & Sons, Inc.: Hoboken, NJ, USA, 2017; Chapter 7; pp. 233–274.
45. Irie, M.; Fukaminato, T.; Matsuda, K.; Kobatake, S. Photochromism of diarylethene molecules and crystals: Memories, switches, and actuators. *Chem. Rev.* **2014**, *114*, 12174–12277. [CrossRef] [PubMed]
46. Hirano, A.; Hashimoto, T.; Kitagawa, D.; Kono, K.; Kobatake, S. Dependence of Photoinduced Bending Behavior of Diarylethene Crystals on Ultraviolet Irradiation Power. *Cryst. Growth Des.* **2017**, *17*, 4819–4825. [CrossRef]
47. Kitagawa, D.; Tanaka, R.; Kobatake, S. Dependence of Photoinduced Bending Behavior of Diarylethene Crystals on Irradiation Wavelength of Ultraviolet Light. *Phys. Chem. Chem. Phys.* **2015**, *17*, 27300–27305. [CrossRef] [PubMed]
48. Hirano, A.; Kitagawa, D.; Kobatake, S. Photomechanical Bending Behavior of Photochromic Diarylethene Crystals Induced under Polarized Light. *CrystEngComm* **2019**, *21*, 2495–2501. [CrossRef]
49. Shtukenberg, A.G.; Punin, Y.O.; Gujral, A.; Kahr, B. Growth Actuated Bending and Twisting of Single Crystals. *Angew. Chem. Int. Ed.* **2014**, *53*, 672–699. [CrossRef]
50. Rai, R.; Krishnan, B.P.; Sureshan, K.M. Chirality-Controlled Spontaneous Twisting of Crystals Due to Thermal Topochemical Reaction. *Proc. Natl. Acad. Sci. USA* **2018**, *115*, 2896–2901. [CrossRef]
51. Tong, F.; Xu, W.; Al-Haidar, M.; Kitagawa, D.; Al-Kaysi, R.O.; Bardeen, C.J. Photomechanically Induced Magnetic Field Response by Controlling Molecular Orientation in 9-Methylanthracene Microcrystals. *Angew. Chem. Int. Ed.* **2018**, *57*, 7080–7084. [CrossRef]
52. Iamsaard, S.; Aßhoff, S.J.; Matt, B.; Kudernac, T.; Cornelissen, J.J.; Fletcher, S.P.; Katsonis, N. Conversion of Light into Macroscopic Helical Motion. *Nat. Chem.* **2014**, *6*, 229–235. [CrossRef] [PubMed]
53. Zhu, L.; Al-Kaysi, R.O.; Bardeen, C.J. Photoinduced Ratchet-Like Rotational Motion of Branched Molecular Crystals. *Angew. Chem. Int. Ed.* **2016**, *55*, 7073–7076. [CrossRef] [PubMed]
54. Hickenboth, C.R.; Moore, J.S.; White, S.R.; Sottos, N.R.; Baudry, J.; Wilson, S.R. Biasing Reaction Pathways with Mechanical Force. *Nature* **2007**, *446*, 423–427. [CrossRef]
55. Kitagawa, D.; Tsujioka, H.; Tong, F.; Dong, X.; Bardeen, C.J.; Kobatake, S. Control of Photomechanical Crystal Twisting by Illumination Direction. *J. Am. Chem. Soc.* **2018**, *140*, 4208–4212. [CrossRef] [PubMed]
56. Kitagawa, D.; Kobatake, S. Photoreversible Current ON/OFF Switching by the Photoinduced Bending of Gold-Coated Diarylethene Crystals. *Chem. Commun.* **2015**, *51*, 4421–4424. [CrossRef] [PubMed]
57. Lan, T.; Chen, W. Hybrid Nanoscale Organic Molecular Crystals Assembly as a Photon-Controlled Actuator. *Angew. Chem. Int. Ed.* **2013**, *52*, 6496–6500. [CrossRef] [PubMed]

58. Yu, Q.; Yang, X.; Chen, Y.; Yu, K.; Gao, J.; Liu, Z.; Cheng, P.; Zhang, Z.; Aguila, B.; Ma, S. Fabrication of Light-Triggered Soft Artificial Muscles via a Mixed-Matrix Membrane Strategy. *Angew. Chem. Int. Ed.* **2018**, *57*, 10192–10196. [CrossRef]
59. Sahoo, S.C.; Nath, N.K.; Zhang, L.; Semreen, M.H.; Al-Tel, T.H.; Naumov, P. Actuation Based on Thermo/photosalient Effect: A Biogenic Smart Hybrid Driven by Light and Heat. *RSC Adv.* **2014**, *4*, 7640–7647. [CrossRef]
60. Koshima, H.; Matsudomi, M.; Uemura, Y.; Kimura, F.; Kimura, T. Light-driven Bending of Polymer Films in Which Salicylidenephenylethylamine Crystals are Aligned Magnetically. *Chem. Lett.* **2013**, *42*, 1517–1519. [CrossRef]
61. Dong, X.; Tong, F.; Hanson, K.M.; Al-Kaysi, R.O.; Kitagawa, D.; Kobatake, S.; Bardeen, C.J. Hybrid Organic–Inorganic Photon-Powered Actuators Based on Aligned Diarylethene Nanocrystals. *Chem. Mater.* **2019**, *31*, 1016–1022. [CrossRef]
62. Dong, X.; Guo, T.; Kitagawa, D.; Kobatake, S.; Palffy-Muhoray, P.; Bardeen, C.J. Effects of Template and Molecular Nanostructure on the Performance of Organic–Inorganic Photomechanical Actuator Membranes. *Adv. Funct. Mater.* **2020**, *30*, 1902396. [CrossRef]
63. Tong, F.; Xu, W.; Guo, T.; Lui, B.F.; Hayward, R.C.; Palffy-Muhoray, P.; Al-Kaysi, R.O.; Bardeen, C.J. Photomechanical Molecular Crystals and Nanowire Assemblies Based on the [2+2] Photodimerization of a Phenylbutadiene Derivative. *J. Mater. Chem. C* **2020**, *8*, 5036–5044. [CrossRef]

© 2020 by the authors. Licensee MDPI, Basel, Switzerland. This article is an open access article distributed under the terms and conditions of the Creative Commons Attribution (CC BY) license (http://creativecommons.org/licenses/by/4.0/).

*Review*

# Vapochromism of Organic Crystals Based on Macrocyclic Compounds and Inclusion Complexes

Toshikazu Ono * and Yoshio Hisaeda *

Department of Chemistry and Biochemistry, Graduate School of Engineering, Center for Molecular Systems (CMS), Kyushu University, 744 Motooka, Nishi-ku, Fukuoka 819-0395, Japan
* Correspondence: tono@mail.cstm.kyushu-u.ac.jp (T.O.); yhisatcm@mail.cstm.kyushu-u.ac.jp (Y.H.)

Received: 28 October 2020; Accepted: 18 November 2020; Published: 19 November 2020

**Abstract:** Vapochromic materials, which change color and luminescence when exposed to specific vapors and gases, have attracted considerable attention in recent years owing to their potential applications in a wide range of fields such as chemical sensors and environmental monitors. Although the mechanism of vapochromism is still unclear, several studies have elucidated it from the viewpoint of crystal engineering. In this mini-review, we investigate recent advances in the vapochromism of organic crystals. Among them, macrocyclic molecules and inclusion complexes, which have apparent host–guest interactions with analyte molecules (specific vapors and gases), are described. When the host compound is properly designed, its cavity size and symmetry change in response to guest molecules, influencing the optical properties by changing the molecular inclusion and recognition abilities. This information highlights the importance of structure–property relationships resulting from the molecular recognition at the solid–vapor interface.

**Keywords:** vapochromism; fluorescence; macrocycles; inclusion crystals; host–guest chemistry

## 1. Introduction

The development of chemosensors has been the subject of intensive research for potential applications in various fields covering human health, industries, and security fields. Gas chromatography, high-performance liquid chromatography, and electrochemical sensing are commonly used for the detection of small molecules and volatile organic compounds. However, most of these methods fail to meet the requirements of simple operation and are expensive. In this context, vapochromic materials that undergo color and/or fluorescence changes in response to specific gases and vapors have been a promising phenomenon. Colorimetric sensor arrays with pattern recognition capabilities have been widely used to detect and discriminate multiple chemically similar samples.

In the last decades, vapochromic materials based on organic dyes, metal complexes, metal organic frameworks, and covalent organic frameworks have attracted a lot of attention and various researches have been carried out. Various metal complexes are known to show vapochromism in the solid-states due to the significant changes in the metal-to-metal interaction and coordination bonding modes caused by the adsorption and desorption of vapor molecules. Several examples of metal-containing vapochromic materials such as Pt(II) or Au(I) have been reported and compiled in review papers [1–6]. On the other hand, metal-free vapochromic materials have been recently considered. The purpose of this mini-review is to investigate recent advances in the fields of vapochromism/vapofluorochromism of organic materials, especially focusing on host–guest compounds with distinct mechanisms and where structural identification has been achieved.

## 2. Vapochromic Materials Based on Macrocyclic Compounds

Macrocyclic compounds have long been known as host compounds, and cyclodextrins [7], calixarenes [8], thiacalixarenes [9], pillararenes [10], and cucurbiturils [11] have been reported. Some of them incorporate guest molecules into the cavity of their cyclic framework via intermolecular interactions to form a host–guest complex in solution and the solid-states. For example, the room-temperature phosphorescence from guest molecules has been observed when cyclodextrin derivatives and halogenated naphthalene are used as host compounds and guest molecules, respectively [12]. This is because the formation of the host–guest complex can suppress thermal deactivation. Thus, the formation of complexes, such as host–guest complexes, has been important for expressing hidden optical functions.

Macrocyclic compounds have internal cavities based on their molecular shape. If the size of the guest molecule does not match the size of the cavities, it will not be incorporated into the cavities. If the guest molecule matches the size of the cavities, it will be selectively incorporated into the cavities. These molecular recognition capabilities have been used to retain and remove substances and for sustained release materials. Based on this phenomenon, various sensor materials have been developed as molecular recognition membranes by combining with electrochemical sensors [13]. Research on vapochromism, in which the host compound is a solid and the guest compound is a vapor (gas), attracts great attention because of the absorption and luminescence response before and after the adsorption of the guest molecule. These studies have widely used porous materials as hosts that can adsorb gases and small organic compounds as guests. Among them, studies in which the crystal structures have been identified before and after guest adsorption are essential to discuss the mechanism of vapochromism. Recently, vapochromic materials based on pillar[n]arene derivatives, which are host compounds with a cyclic structure, have been of great interest.

Pillar[n]arenes, reported by Ogoshi in 2008 [14], are macrocyclic molecules composed of 1,4-di-alkoxybenzene, which are called pillar[5]arene or pillar[6]arene, etc., depending on the repeating unit "n". Research progress with these compounds has been rapidly expanded and has been summarized in some review papers regarding supramolecular complexes and host–guest chemistry [10,15,16]. In this mini-review, we present various vapochromic materials based on host–guest chemistry inspired by the unique structures of pillar[n]arenes and their related compounds.

The difference between the pillar[5]arene and the pillar[6]arene is the size of the cavity. The cavity size of pillar[5]arenes and pillar[6]arenes is ca. 4.7 and ca. 6.7 Å, respectively. The pillar[5]arene tends to incorporate linear chain hydrocarbons via CH–π interactions, whereas the pillar[6]arene tends to accommodate larger molecules such as aromatic molecules and cyclic alkanes. This cavity size-dependent selectivity of the guest molecule makes it possible to fractionate positional substitutes in organic compounds. For example, it has been reported that these materials can be used to separate styrene from ethylbenzene and to separate three xylene isomers [17,18]. Separation of linear and branched alkanes is demonstrated using the host–guest chemistry of pillar[n]arenes as well [19]. However, the color of the solid powder of a typical pillar[n]arene is white, making it difficult to visually detect the uptake event of the guest molecules, for example, by color-change. To overcome these problems, Ogoshi et al. [20] used a novel macrocyclic molecule in which an electron-accepting molecule, benzoquinone, was introduced into the molecular structure of the pillar[5]arene (1) (Figure 1). This material (1) was brown-color in the absence of any guest molecules. The color of the material changed from brown to light-red when linear alkane vapors were introduced, indicating that the alkane vapors, which had been challenging to detect due to the color change, were successfully detected with the naked eye. Uptake of methanol vapor induced a different color change, from brown to black. It was also found that branched and cyclic alkane vapors did not show a color change when exposed to them. Furthermore, when a mixture of linear, cyclic, and branched alkane vapors was used, it was found that the gas selectively adsorbed linear alkane vapors and showed a color change. In other words, it is possible to detect the presence of linear alkane gas molecules in the mixed alkane gas by the color change. Recently, Ogoshi et al. [21] investigated vapochromic behaviors of pillar[6]arene with one benzoquinone unit to detect various small aromatic guests vapors.

**Figure 1.** Vapochromic behavior of **1** against various alkane vapors and methanol vapors. Reproduced with permission from Reference [20]. Copyright 2017 American Chemical Society.

Vapochromic materials have been reported to be scalable to various guest molecules based on molecular design. Huang et al. [22] changed the analytes to volatile aliphatic aldehydes and investigated their vapochromic behaviors using **1** (Figure 2). Aliphatic aldehydes are selected as target compounds because they are generally highly reactive in nature and are potentially hazardous volatile organic compounds that are considered unfavorable effects on the environmental and human health. Exposure of guest-free crystals to various types of aliphatic aldehyde vapors resulted in quantitative adsorption of these aldehyde vapors and different color changes were observed. Comparison of the single-crystal structures revealed that different types of aldehyde vapors led to change in the relative positions of the adjacent 1,4-diethoxybenzene and benzoquinone units. As a result, the charge-transfer (CT) interactions between the electron-rich 1,4-diethoxybenzene and electron-deficient benzoquinone units were altered, resulting in vapochromic behaviors with different color changes. These behaviors were regenerated by the heat treatment and were able to repeatedly observe guest adsorption and desorption without any particular loss of performance.

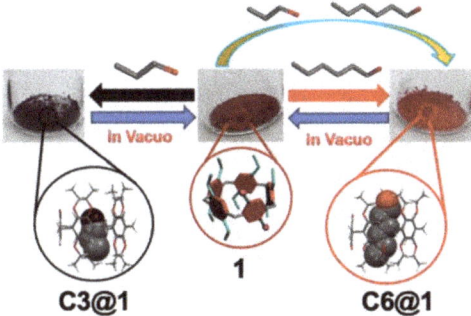

**Figure 2.** Vapochromic behavior of **1** against aliphatic aldehydes vapors. Reproduced with permission from Reference [22]. Copyright 2018 American Chemical Society.

Subsequently, Ogoshi et al. [23] investigated the creation of a vapochromic material using a unique mechanism that changes its liquid-to-solid state when exposed to specific vapors based on pillar[6]arene derivatives (**2**) (Figure 3). By attaching 12 *n*-hexyl ($C_6H_{13}$) chains to pillar[6]arene, this system was transformed into a room-temperature structural liquid, that is, a nanoscale system with some order but no periodic structure. This clear liquid has the characteristic of becoming turbid solid when exposed to guest vapors for several seconds. In particular, the state changes are expected to apply to new vapor detection systems due to the selectivity of vapors, such as cyclohexane. Powder X-ray diffraction (PXRD) measurements supported the fact that the linear *n*-hexyl group was initially contained in the

cavity of pillar[6]arene, but the uptake of the cyclohexane guest vapor results in the de-threading of the
$n$-hexyl group. This is because the cavity size of pillar[6]arene is more suitable for cyclohexane than
linear alkyl chains. The guest inclusion behaviors were also confirmed by proton NMR measurements.

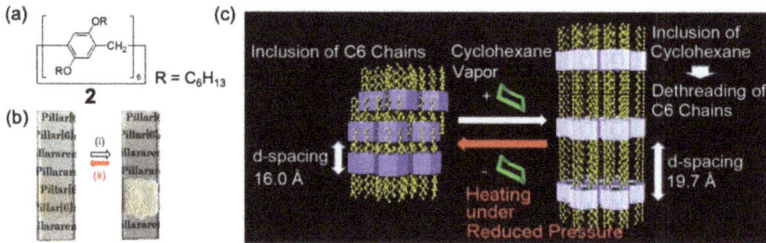

**Figure 3.** (a) Chemical structure of **2**. (b) Photographs of **2** by (i) exposing structural liquid **2** to cyclohexane vapor at 25 °C for 30 min and (ii) heating the solid **2** at 80 °C under reduced pressure for 30 min. (c) Schematic illustration of the guest vapor-induced state change of **2**. Reproduced with permission from Reference [23]. Copyright 2019 American Chemical Society.

To improve the vapochromic properties of pillar[n]arenes, direct modification of fluorophore in their molecular skeleton has also been reported. Huang et al. [24] proposed a novel macrocyclic molecule, in which an anthracene moiety was introduced in a part of pillar[5]arene (**3**) (Figure 4). Upon exposure of this material to vapors of various ketone with different alkyl chains (C3–C8), the compound showed unique vapochromic/vapofluorochromic behavior, as shown in Figure 4c. For example, after exposure to C4, C5, and C6, the anthracene moieties exist as monomers in the solid-state. In contrast, after exposure to C7 and C8, the anthracene moieties effectively form excimers in the solid-states. Single-crystal structures and PXRD experiments suggested that two types of anthracene assemblies, H-aggregation for fluorescence of excimer and J-aggregation for fluorescence of monomer, were found to exhibit different vapochromic properties.

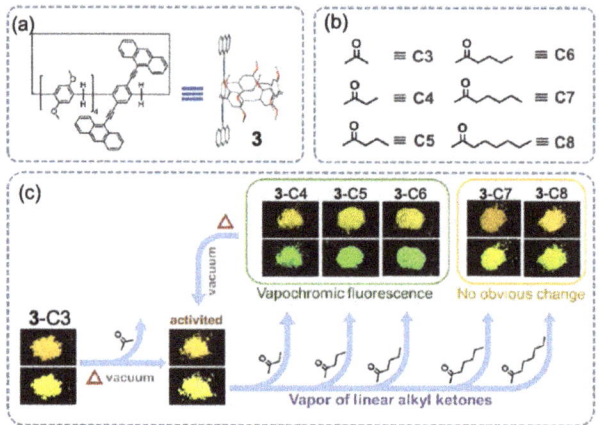

**Figure 4.** (a) Chemical structure and crystal structure of **3**. (b) Chemical structures of ketones (C3–C8). (c) Schematic illustration of activation and vapochromic/vapofluorochromic behaviors of **3**. Reproduced with permission from Reference [24]. Copyright 2019 American Chemical Society.

Another example of pillararenes modification with a fluorescent material is the so-called BowtieArene, which consists of a tetraphenylethylene (TPE) fluorophore and two pillar[5]arenes [25].

This figure-of-eight macrocycle material exhibits unique multi-stimuli-responsive fluorescence accompanied by vapor absorption with host–guest interactions.

It has also been reported that vapochromic properties can be found using pillar[n]arene without fluorescent tags. For example, Ogoshi et al. [26] reported that pillar[6]arene showed vapofluorochromism against benzonitrile, methyl benzoate, and divinylbenzene by crystal-state host–guest complexation at the solid–vapor interface. Li et al. [27] proposed the co-crystallization approach to demonstrate vapochromic materials (Figure 5). The red-colored co-crystals (4•5) were prepared by slow evaporation of the tetrahydrofuran-saturated solution of 4 and 5. The red coloration was caused by intermolecular CT interactions between the electron-rich 4 and electron-deficient 5 with absorption at 452 nm. However, after the co-crystals (4•5) were smashed to power and activated by heating at 70 °C under vacuum for 12 h, the materials (4•5α) color changed to white. Exposure of 4•5α to various vapors of haloalkane resulted in a color changes from white to a red or orange color. Such a combination of macrocycle/host–guest chemistry and co-crystal engineering offers great potential for further research.

**Figure 5.** (a) Chemical structures of 4 and 5. (b) Crystal structure of 4•5. (c) Pictures of 4•5 and 4•5α. (d) Vapochromic behavior of 4•5α against various vapors. Reproduced with permission from Reference [27]. Copyright 2016 Wiley-VCH.

Following these trends, it is possible to design various vapochromic materials in which changes in absorption and luminescence properties are induced by the uptake of guests, adding a new twist to the macrocyclic that function as host molecules. In particular, those with an apparent crystal structure allow for a detailed evaluation of host–guest interactions. Research is being carried out on how host molecules with what structures exhibit vapochromic properties.

## 3. Vapochromic Materials Based on Inclusion Crystals

The guest inclusion phenomenon is frequently observed in many organic compounds' crystallization process, regardless of the cyclic compounds mentioned above. This is because organic compounds have complex structures in three dimensions. For example, organic compounds with bulky and rigid substituents tend to include guest molecules because they do not pack easily without guests [28–31]. This suggests that a phenomenon may be found in which the packing structure changes depending on the type of guest, resulting in a change in optical properties. Indeed, inclusion crystals consisting of organic dyes as host molecules and various guest molecules have been reported. Most of them have been studied to modulate solid-state optical properties by changing the packing structure depending on the type of guest molecules. If this phenomenon allows the preparation of host–guest complexes through the solid-gas interface, it can cause vapochromism. The evaluation of the crystal structure of the host–guest complexes provides important information to discuss the mechanism of vapochromism.

In 2010, Naota et al. [32] proposed the first vapochromic organic crystals based on naphthalenediimide (NDI) derivatives (Figure 6). They designed S-shaped NDI derivatives bearing two pyrrole-imine tethers via an alkyl chain (6). The NDI moiety is an electron-deficient π-conjugated molecule that forms CT complex with electron-rich pyrrole-imine moieties and exhibits absorption and coloration in the visible light region. It has been found that the coloration of this material changes when various solvent molecules are incorporated as guests into the material's cavities due to a shift in the relative position of the NDI moiety and the pyrrole-imine tethers, resulting in a change in the CT interaction and coloration.

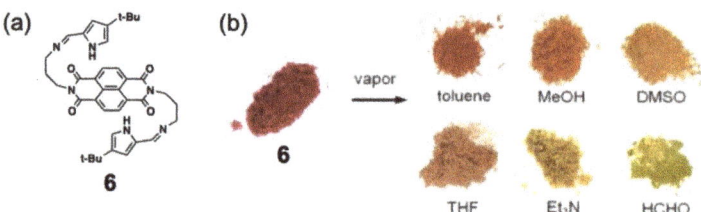

**Figure 6.** (**a**) Chemical structures of **6**. (**b**) Vapochromic behavior of **6** against various vapors.

In 2016, we investigated the vapochromic behavior of NDI derivatives bearing two tris(pentafluorophenyl)borane units (**7**) (Figure 7) [33,34]. The luminescence intensity of the supramolecular host was weak, but it increased by 76, 46, and 37 times under saturated vapor pressure in response to vapors of toluene, benzene, and *m*-xylene, respectively. On the other hand, no increase in luminescence intensity was observed for the vapors of methanol, ethanol, acetone, dichloromethane, chloroform, hexane, and cyclohexane, although adsorption of molecules was observed. The guest-dependent color and fluorescence changes which were observed, attributed to intermolecular CT interactions between the electron-deficient NDI, and the electron-rich aromatic guest molecules were observed. These results show that vapochromic/vapofluorochromic behaviors against aromatic compounds can be achieved by host–guest interaction at the solid–vapor interface.

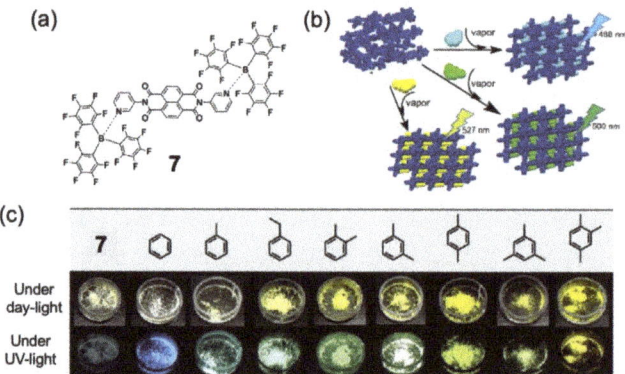

**Figure 7.** (**a**) Chemical structure of **7**. (**b**) Guest inclusion behavior of **7** against vapors of guest molecules. (**c**) Photographs of **7** and **7** with exposure of guest vapors under daylight (upper row) and under ultraviolet-light (365 nm, bottom row). Reproduced with permission from Reference [34]. Copyright 2016 Wiley-VCH.

Based on these findings, we systematically varied the side chains of NDI derivatives (**8–10**) and evaluated the vapochromic behaviors (Figure 8) [35,36]. Among them, NDI with a 2-benzophenone unit (**8**) showed a vapochromic/vapofluorochromic behavior toward toluene, *p*-xylene, 4-fluorotoluene, and anisole. The detection limits for toluene and *p*-xylene vapors are 10 parts-per-billion (ppm) and 20 ppm, respectively. From the single-crystal structures obtained from recrystallization, it was found that the guest molecules were incorporated in the space of the host framework to form a CT complex. This suggests that the absorption and fluorescence behavior is dependent on the type of guest molecule incorporated in the crystal. On the other hand, NDI with a diphenylmethane unit and NDI with a 3,5-di-tert-butylbenzene unit as substituents did not exhibit vapochromic behavior. This means that modification of side-chain structure of NDIs can be devised to achieve novel vapochromic/vapofluorochromic materials without significantly changing the electronic properties of the compounds. Modification of the solid-state structure plays an important role in the development of molecular recognition of the solid–vapor interface between sensor materials and the analytes. In addition, two sensor ensembles (**7** and **8**) have been combined to create a mini-sensor array used to identify small aromatic compounds containing benzene, toluene, and xylene isomers with the naked eye.

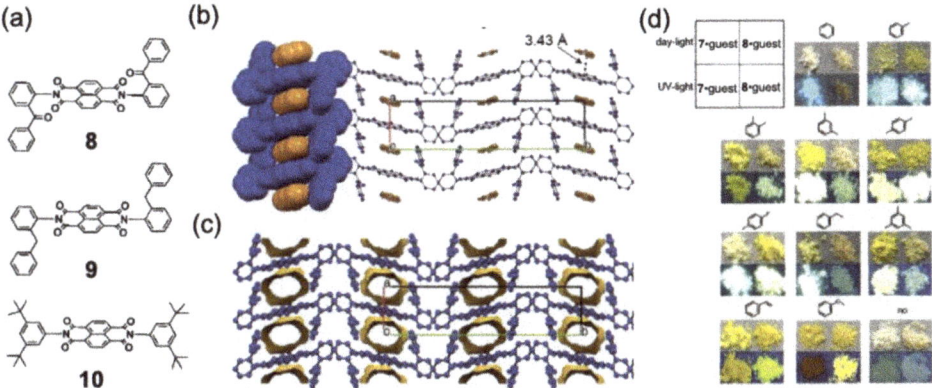

**Figure 8.** (**a**) Chemical structures of **8**–**10**. (**b**) Crystal structure of **8** including anisole as a guest molecule. (**c**) Visualization of calculated voids and contact surface maps of **8**. (**d**) Vapochromic bahavior of **7** and **8**, and **7** and **8** after exposing guest vapors for 1 day.

The introduction of bulky substituents of dyes is essential to create porous crystals that can act as vapochromic materials. For example, Yang et al. investigated porous organic crystals by modifying the H-shaped pentiptycene scaffold of dyes and have been applied as vapochromic materials [37,38]. The pentiptycene unit is well known as a rigid-bulky substituent [39] so that aggregation-induced fluorescence quenching is minimized and pore volume accessible to guest molecules is created.

Porous crystals can also be prepared by dendron modification of dyes and have been applied as vapochromic materials. In 2020, Takeda et al. [40] studied porous organic crystals consisting of π-conjugated molecules with multibranched carbazole units as dendritic propeller sites (**11**) (Figure 9). It was found that these crystals can capture and release water molecules from the atmosphere. Also, the apparent color of the crystalline powder changes is readily detectable with the naked eye. For example, at room-temperature (25 °C), the crystals are yellow when the humidity is below 40%, but they turn entirely red when they reach 50%. This color change is reversible in response to humidity changes, and the humidity and temperature in which the color change occurs are close to our living environment. Thus, crystals can be used as high-performance sensors or adsorbates under water-containing conditions.

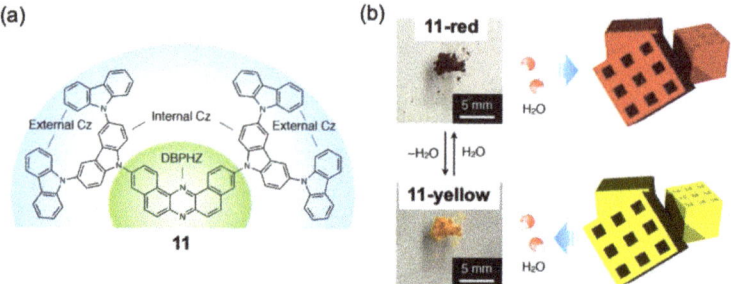

**Figure 9.** (a) Chemical structure of **11**. (b) Photographs of vaphochromic behavior of **11** upon uptake/release of water molecules. Reproduced with permission from Reference [40]. Copyright 2020 Nature publishing group.

Vapochromism of organic crystals using intramolecular proton transfer has been reported. Uekusa et al. [41] suggested that 5-aminoisophthalic acid (**12**) exhibited a reversible color change upon exposure to solvent vapors via a pseudopolymorphic transformation between pink hemihydrate and yellow anhydrous crystals (Figure 10). When exposed to solvent vapors such as methanol, ethanol, and acetonitrile, the white crystals gradually turn into yellow anhydrous crystals, and they revert to hemihydrate crystals in the presence of water. Crystal structural analysis suggested that in the crystal structure of the hemihydrate form, carboxylate anion ($COO^-$) and ammonium cation ($NH^{3+}$) were formed through proton transfer to form a zwitterion. On the other hand, in the non-hydrated state, the amino group remains in the form of $NH_2$ conformation, and the compound **12** is in a nonionic state with no proton transfer. The protonation and deprotonation of the carboxylate group has been assigned by observing the twisting of the group against the plane of the aromatic ring. Stimulated by solvent vapors, the compound **12** undergoes the dehydration/hydration conversion, resulting in a vapochromic behavior with a significant impact on the proton transfer behavior. The similar strategy was also reported by the same group using pimemidic acid [42].

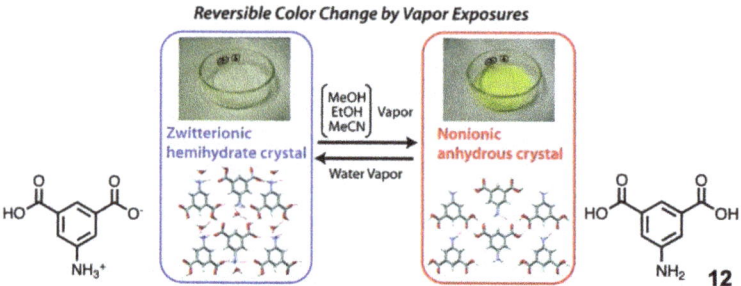

**Figure 10.** Vapochromic behavior of **12**. Reproduced with permission from Reference [41]. Copyright 2019 American Chemical Society.

In 2019, we investigated [43] the vapochromism of organic crystals using intermolecular proton transfer of acid-base complexes. To do this, three crystals, salts, cocrystal, and salt-cocrystal continuum, were designed by $pK_a$ values of acids and bases (Figure 11). The structure–property relationships of crystals were well studied to realize the regulation of proton transfer dynamics between acid-base complexes by photoluminescence color change. An acid-base complex consisting of a pyridine-modified pyrrolopyrrole dye with salicylic acid (**13**•**14**), categorized as a salt-cocrystal continuum, undergoes vapochromism against vapors of dichloromethane. The luminescent color change is attributed to modulation of the degree of protonation by alternating the crystal packing environment upon inclusion and desorption of dichloromethane. When the complex was fumed with triethylamine (TEA) vapor for 1 h, the emission color changed to blue. The emission color change originated from deprotonation of **13** by TEA vapor.

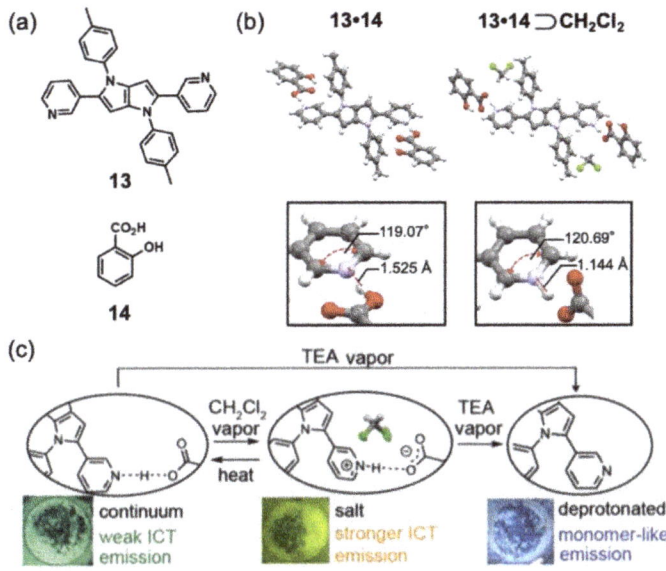

**Figure 11.** (a) Chemical structures of **13** and **14**. (b) Crystal structure of acid-base complex (**13**•**14**) and its CH$_2$Cl$_2$ included structure. (c) Schematic illustration of emission color tuning by various vapor molecules.

## 4. Conclusions

Vapochromic materials based on host–guest chemistry at the solid–gas interface have been extensively studied. In the first part, we focused on the pillararene derivatives among the macrocyclic molecules as hosts. We introduced their guest-recognition abilities depending on the size of the intrinsic ring framework. Modification of the macrocyclic molecules allows the rational design of vapochromic materials and studies on creating rational vapochromic materials and their mechanisms. In the latter part, we introduced various vapochromic materials, focusing on the fact that the inclusion crystal is a host–guest complex. The importance of introducing appropriately bulky substituents into the dye molecules and a new mechanism of vapochromic materials, in which proton transfer changes in response to guest vapors, were introduced. It is expected that the creation of vapochromic materials will continue through various molecular materials and complexes. We believe that the future of these studies has the potential to be applied to artificial olfactory systems that can outperform the olfactory capabilities of living organisms.

**Author Contributions:** Both authors contributed to this publication. All authors have read and agreed to the published version of the manuscript.

**Funding:** This research was funded by JSPS KAKENHI, grant numbers JP17H04875, JP20H04675, and JP20K21212.

**Conflicts of Interest:** The authors declare no conflict of interest.

## References

1. Kreno, L.E.; Leong, K.; Farha, O.K.; Allendorf, M.; Van Duyne, R.P.; Hupp, J.T. Metal-organic framework materials as chemical sensors. *Chem. Rev.* **2012**, *112*, 1105–1125. [CrossRef] [PubMed]
2. Zhang, X.; Li, B.; Chen, Z.-H.; Chen, Z.-N. Luminescence vapochromism in solid materials based on metal complexes for detection of volatile organic compounds (VOCs). *J. Mater. Chem.* **2012**, *22*, 11427–11441. [CrossRef]
3. Wenger, O.S. Vapochromism in organometallic and coordination complexes: Chemical sensors for volatile organic compounds. *Chem. Rev.* **2013**, *113*, 3686–3733. [CrossRef] [PubMed]
4. Zhou, X.; Lee, S.; Xu, Z.; Yoon, J. Recent Progress on the Development of Chemosensors for Gases. *Chem. Rev.* **2015**, *115*, 7944–8000. [CrossRef] [PubMed]
5. Kato, M.; Ito, H.; Hasegawa, M.; Ishii, K. Soft Crystals: Flexible Response Systems with High Structural Order. *Chem. Eur. J.* **2019**, *25*, 5105–5112. [CrossRef] [PubMed]
6. Li, E.; Jie, K.; Liu, M.; Sheng, X.; Zhu, W.; Huang, F. Vapochromic crystals: Understanding vapochromism from the perspective of crystal engineering. *Chem. Soc. Rev.* **2020**, *49*, 1517–1544. [CrossRef]
7. Crini, G. Review: A History of Cyclodextrins. *Chem. Rev.* **2014**, *114*, 10940–10975. [CrossRef]
8. Ikeda, A.; Shinkai, S. Novel Cavity Design Using Calix[n]arene Skeletons: Toward Molecular Recognition and Metal Binding. *Chem. Rev.* **1997**, *97*, 1713–1734. [CrossRef]
9. Morohashi, N.; Narumi, F.; Iki, N.; Hattori, T.; Miyano, S. Thiacalixarenes. *Chem. Rev.* **2006**, *106*, 5291–5316. [CrossRef]
10. Ogoshi, T.; Yamagishi, T.-A.; Nakamoto, Y. Pillar-shaped macrocyclic hosts pillar [n] arenes: New key players for supramolecular chemistry. *Chem. Rev.* **2016**, *116*, 7937–8002. [CrossRef]
11. Barrow, S.J.; Kasera, S.; Rowland, M.J.; del Barrio, J.; Scherman, O.A. Cucurbituril-Based Molecular Recognition. *Chem. Rev.* **2015**, *115*, 12320–12406. [CrossRef] [PubMed]
12. Li, D.; Lu, F.; Wang, J.; Hu, W.; Cao, X.M.; Ma, X.; Tian, H. Amorphous Metal-Free Room-Temperature Phosphorescent Small Molecules with Multicolor Photoluminescence via a Host-Guest and Dual-Emission Strategy. *J. Am. Chem. Soc.* **2018**, *140*, 1916–1923. [CrossRef] [PubMed]
13. Pirondini, L.; Dalcanale, E. Molecular recognition at the gas–solid interface: A powerful tool for chemical sensing. *Chem. Soc. Rev.* **2007**, *36*, 695–706. [CrossRef] [PubMed]
14. Ogoshi, T.; Kanai, S.; Fujinami, S.; Yamagishi, T.-A.; Nakamoto, Y. para-Bridged Symmetrical Pillar[5]arenes: Their Lewis Acid Catalyzed Synthesis and Host–Guest Property. *J. Am. Chem. Soc.* **2008**, *130*, 5022–5023. [CrossRef]
15. Ogoshi, T.; Tsuchida, H.; Kakuta, T.; Yamagishi, T.A.; Taema, A.; Ono, T.; Sugimoto, M.; Mizuno, M. Ultralong Room-Temperature Phosphorescence from Amorphous Polymer Poly(Styrene Sulfonic Acid) in Air in the Dry Solid State. *Adv. Funct. Mater.* **2018**, *28*. [CrossRef]
16. Jie, K.; Zhou, Y.; Li, E.; Huang, F. Nonporous Adaptive Crystals of Pillararenes. *Acc. Chem. Res.* **2018**, *51*, 2064–2072. [CrossRef]
17. Jie, K.; Liu, M.; Zhou, Y.; Little, M.A.; Bonakala, S.; Chong, S.Y.; Stephenson, A.; Chen, L.; Huang, F.; Cooper, A.I. Styrene Purification by Guest-Induced Restructuring of Pillar[6]arene. *J. Am. Chem. Soc.* **2017**, *139*, 2908–2911. [CrossRef]
18. Jie, K.; Liu, M.; Zhou, Y.; Little, M.A.; Pulido, A.; Chong, S.Y.; Stephenson, A.; Hughes, A.R.; Sakakibara, F.; Ogoshi, T.; et al. Near-Ideal Xylene Selectivity in Adaptive Molecular Pillar[n]arene Crystals. *J. Am. Chem. Soc.* **2018**, *140*, 6921–6930. [CrossRef]
19. Kakuta, T.; Yamagishi, T.-A.; Ogoshi, T. Stimuli-Responsive Supramolecular Assemblies Constructed from Pillar[n]arenes. *Acc. Chem. Res.* **2018**, *51*, 1656–1666. [CrossRef]
20. Ogoshi, T.; Shimada, Y.; Sakata, Y.; Akine, S.; Yamagishi, T.-A. Alkane-Shape-Selective Vapochromic Behavior Based on Crystal-State Host–Guest Complexation of Pillar[5]arene Containing One Benzoquinone Unit. *J. Am. Chem. Soc.* **2017**, *139*, 5664–5667. [CrossRef]
21. Wada, K.; Kakuta, T.; Yamagishi, T.-A.; Ogoshi, T. Obvious vapochromic color changes of a pillar[6]arene containing one benzoquinone unit with a mechanochromic change before vapor exposure. *Chem. Commun.* **2020**, *56*, 4344–4347. [CrossRef] [PubMed]

22. Li, E.; Jie, K.; Zhou, Y.; Zhao, R.; Zhang, B.; Wang, Q.; Liu, J.; Huang, F. Aliphatic Aldehyde Detection and Adsorption by Nonporous Adaptive Pillar[4]arene[1]quinone Crystals with Vapochromic Behavior. *Acs Appl. Mater. Interfaces* **2018**, *10*, 23147–23153. [CrossRef] [PubMed]
23. Ogoshi, T.; Maruyama, K.; Sakatsume, Y.; Kakuta, T.; Yamagishi, T.-A.; Ichikawa, T.; Mizuno, M. Guest Vapor-Induced State Change of Structural Liquid Pillar[6]arene. *J. Am. Chem. Soc.* **2019**, *141*, 785–789. [CrossRef] [PubMed]
24. Li, Q.; Zhu, H.; Huang, F. Alkyl Chain Length-Selective Vapor-Induced Fluorochromism of Pillar[5]arene-Based Nonporous Adaptive Crystals. *J. Am. Chem. Soc.* **2019**, *141*, 13290–13294. [CrossRef]
25. Lei, S.-N.; Xiao, H.; Zeng, Y.; Tung, C.-H.; Wu, L.-Z.; Cong, H. BowtieArene: A Dual Macrocycle Exhibiting Stimuli-Responsive Fluorescence. *Angew. Chem. Int. Ed.* **2020**, *59*, 10059–10065. [CrossRef]
26. Ogoshi, T.; Hamada, Y.; Sueto, R.; Kojima, R.; Sakakibara, F.; Nagata, Y.; Sakata, Y.; Akine, S.; Ono, T.; Kakuta, T.; et al. Vapoluminescence Behavior Triggered by Crystal-State Complexation between Host Crystals and Guest Vapors Exhibiting No Visible Fluorescence. *Cryst. Growth Des.* **2020**. [CrossRef]
27. Li, B.; Cui, L.; Li, C. Macrocycle Co-Crystals Showing Vapochromism to Haloalkanes. *Angew. Chem. Int. Ed.* **2020**. [CrossRef]
28. Bishop, R. Designing new lattice inclusion hosts. *Chem. Soc. Rev.* **1996**, *25*, 311–319. [CrossRef]
29. Fei, Z.; Kocher, N.; Mohrschladt, C.J.; Ihmels, H.; Stalke, D. Single Crystals of the Disubstituted Anthracene 9, 10-(Ph2P□S) 2C14H8 Selectively and Reversibly Detect Toluene by Solid-State Fluorescence Emission. *Angew. Chem. Int. Ed.* **2003**, *42*, 783–787. [CrossRef]
30. Ooyama, Y.; Nagano, S.; Okamura, M.; Yoshida, K. Solid-State Fluorescence Changes of 2-(4-Cyanophenyl)-5-[4-(diethylamino)phenyl]-3H-imidazo[4,5-a]naphthalene upon Inclusion of Organic Solvent Molecules. *Eur. J. Org. Chem.* **2008**, *2008*, 5899–5906. [CrossRef]
31. Hinoue, T.; Miyata, M.; Hisaki, I.; Tohnai, N. Guest-Responsive Fluorescence of Inclusion Crystals with π-Stacked Supramolecular Beads. *Angew. Chem. Int. Ed.* **2012**, *51*, 155–158. [CrossRef] [PubMed]
32. Takahashi, E.; Takaya, H.; Naota, T. Dynamic vapochromic behaviors of organic crystals based on the open-close motions of S-shaped donor-acceptor folding units. *Chem. Eur. J.* **2010**, *16*, 4793–4802. [CrossRef] [PubMed]
33. Ono, T.; Sugimoto, M.; Hisaeda, Y. Multicomponent Molecular Puzzles for Photofunction Design: Emission Color Variation in Lewis Acid–Base Pair Crystals Coupled with Guest-to-Host Charge Transfer Excitation. *J. Am. Chem. Soc.* **2015**, *137*, 9519–9522. [CrossRef] [PubMed]
34. Hatanaka, S.; Ono, T.; Hisaeda, Y. Turn-On Fluorogenic and Chromogenic Detection of Small Aromatic Hydrocarbon Vapors by a Porous Supramolecular Host. *Chem. Eur. J.* **2016**, *22*, 10346–10350. [CrossRef]
35. Ono, T.; Tsukiyama, Y.; Taema, A.; Hisaeda, Y. Inclusion Crystal Growth and Optical Properties of Organic Charge-transfer Complexes Built from Small Aromatic Guest Molecules and Naphthalenediimide Derivatives. *Chem. Lett.* **2017**, *46*, 801–804. [CrossRef]
36. Ono, T.; Tsukiyama, Y.; Hatanaka, S.; Sakatsume, Y.; Ogoshi, T.; Hisaeda, Y. Inclusion crystals as vapochromic chemosensors: Fabrication of a mini-sensor array for discrimination of small aromatic molecules based on side-chain engineering of naphthalenediimide derivatives. *J. Mater. Chem. C* **2019**, *7*, 9726–9734. [CrossRef]
37. Matsunaga, Y.; Yang, J.S. Multicolor Fluorescence Writing Based on Host-Guest Interactions and Force-Induced Fluorescence-Color Memory. *Angew. Chem. Int. Ed.* **2015**, *54*, 7985–7989. [CrossRef]
38. Hsu, L.-Y.; Maity, S.; Matsunaga, Y.; Hsu, Y.-F.; Liu, Y.-H.; Peng, S.-M.; Shinmyozu, T.; Yang, J.-S. Photomechanochromic vs. mechanochromic fluorescence of a unichromophoric bimodal molecular solid: Multicolour fluorescence patterning. *Chem. Sci.* **2018**, *9*, 8990–9001. [CrossRef]
39. Yang, J.-S.; Swager, T.M. Porous Shape Persistent Fluorescent Polymer Films: An Approach to TNT Sensory Materials. *J. Am. Chem. Soc.* **1998**, *120*, 5321–5322. [CrossRef]
40. Yamagishi, H.; Nakajima, S.; Yoo, J.; Okazaki, M.; Takeda, Y.; Minakata, S.; Albrecht, K.; Yamamoto, K.; Badía-Domínguez, I.; Oliva, M.M.; et al. Sigmoidally hydrochromic molecular porous crystal with rotatable dendrons. *Commun. Chem.* **2020**, *3*, 118. [CrossRef]
41. Fujii, K.; Sakon, A.; Sekine, A.; Uekusa, H. Reversible Color Switching of an Organic Crystal Induced by Organic Solvent Vapors. *Cryst. Growth Des.* **2011**, *11*, 4305–4308. [CrossRef]

42. Sakon, A.; Sekine, A.; Uekusa, H. Powder Structure Analysis of Vapochromic Quinolone Antibacterial Agent Crystals. *Cryst. Growth Des.* **2016**, *16*, 4635–4645. [CrossRef]
43. Yano, Y.; Ono, T.; Hatanaka, S.; Gryko, D.T.; Hisaeda, Y. Salt–cocrystal continuum for photofunction modulation: Stimuli-responsive fluorescence color-tuning of pyridine-modified intramolecular charge-transfer dyes and acid complexes. *J. Mater. Chem. C* **2019**, *7*, 8847–8854. [CrossRef]

**Publisher's Note:** MDPI stays neutral with regard to jurisdictional claims in published maps and institutional affiliations.

 © 2020 by the authors. Licensee MDPI, Basel, Switzerland. This article is an open access article distributed under the terms and conditions of the Creative Commons Attribution (CC BY) license (http://creativecommons.org/licenses/by/4.0/).

*Review*

# Elastic Organic Crystals of π-Conjugated Molecules: New Concept for Materials Chemistry

Shotaro Hayashi [1,2]

[1] School of Environmental Science and Engineering, Kochi University of Technology, 185 Miyanokuchi, Tosayamada, Kami City, Kochi 782-8502, Japan; hayashi.shotaro@kochi-tech.ac.jp

[2] Research Center for Molecular Design, Kochi University of Technology, 185 Miyanokuchi, Tosayamada, Kami City, Kochi 782-8502, Japan

Received: 28 September 2020; Accepted: 4 December 2020; Published: 7 December 2020

**Abstract:** It is generally believed that organic single crystals composed of a densely packed arrangement of anisotropic, organic small molecules are less useful as functional materials due to their mechanically inflexible and brittle nature, compared to polymers bearing flexible chains and thereby exhibiting viscoelasticity. Nevertheless, organic crystals have attracted much attention because of their tunable optoelectronic properties and a variety of elegant crystal habits and unique ordered or disordered molecular packings arising from the anisotropic molecular structures. However, the recent emergence of flexible organic crystal materials showing plasticity and elasticity has considerably changed the concept of organic single crystals. In this review, the author summarizes the state-of-the-art development of flexible organic crystal materials, especially functional elastic organic crystals which are expected to provide a foothold for the next generation of organic crystal materials.

**Keywords:** elastic organic crystals; mechanical deformation; π-conjugated molecules; photoluminescence; deformation-induced photoluminescence changes

## 1. Introduction

In general, organic molecular crystals have been believed to be inflexible and brittle (Figure 1a,b), because they have a densely packed, three-dimensional (3D), supramolecular structure and hence do not possess a mechanism for relaxation of the stress loaded. Meanwhile, it is known that several organic molecular crystals display plastic-like bending deformability, i.e., plasticity [1–7]. These crystals are plastically bended by photoirradiation or deformed by external stress (Figure 1a,c). Until 2016, various flexible organic crystal materials were discovered and reported [8–15].

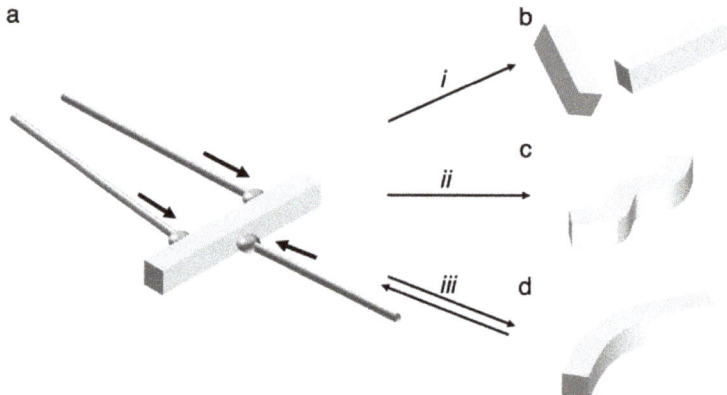

**Figure 1.** Schematic illustration of three points stress for organic crystals. (**a**) Straight organic crystals. *i*: brittle breakage. *ii*: plastically bending deformation. *iii*: elastically bending deformation. (**b**) Brocken crystals. (**c**) Plastically deformed crystals. (**d**) Elastically deformed crystals.

Recently, the examples of ferroelastic and superelastic crystals that can change their shape due to a crystal-to-crystal phase transition caused by the load of stress, and the examples of pseudo-elastic shape recovery of plastically deformed solids have been reported [8–10]. Since that time, we have been interested in elastic organic crystals, i.e., elastically deformable organic crystals that can realize bending deformation and spontaneous shape recovery in response to the load of stress (Figure 1a,d). By the way, the origin of elasticity observed for polymers can be classified into two thermodynamical factors, namely enthalpy and entropy effects [11,12]. In the case of organic molecular crystals, the elasticity should be related to the change of the molecular arrangement in the crystal during deformation. If the entropic elasticity operated in the crystals at the nanoscopic level, by applying an external force, the regularly arranged molecular packing were forced to change to an irregular state and could not return to the original packing according to the second law of thermodynamics. This assumption was not correct. The elastic organic crystals showed enthalpic deformation by applying an external force and returned to the original molecular packing owing to the restoring force.

In this context, we have realized the combination of the elastic flexibility and unique 'optical and electronic functions' of organic molecular crystals [16–24], which is most likely to open up a new field of organic crystal materials chemistry (Figure 2). π-conjugated molecules were reported to demonstrate various optoelectronic functions, such as light absorption and emission, semiconductor performance and so forth, because of the π-electron systems [25–42]. Accordingly, it was expected that endowing a flexible and tough elastic crystal with this π-functionality would enable to develop a new research field which could lead to the development of flexible and wearable optoelectronic single-crystal devices (Figure 2). Furthermore, it was also anticipated to find out additional novel phenomena caused by the mechanical deformation of a dense supramolecular structure, despite anisotropic molecular orientation. In this review, the author summarizes the design (Section 2), potential structures (Section 3), detailed deformation (Section 4), and unique properties (Sections 5 and 6) of elastic organic crystals of π-conjugated molecules.

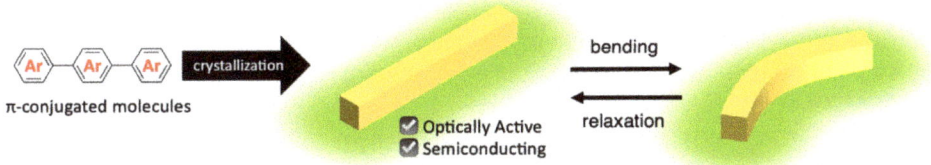

**Figure 2.** Schematic illustration of the concept on "elastic organic crystals of π-conjugated molecules". Flexible organic crystals with π-functionality.

## 2. Design of Elastic Organic Crystals Composed of π-Conjugated Molecules

Here the definite guidelines for the molecular design to afford elastic organic crystals with specific functions are discussed. To endow organic crystals with flexibility, they must have an intermolecular packing so as to relax the stress loaded (Figure 3a) [14,15]. Wire bundles, such as the fibril lamella structure, can exhibit high flexibility, as seen in the macrostructure of muscle fibers (Figure 3b). Therefore, if close attention is paid to the molecular structure that forms these wire bundles, and if the constituent molecules form a slip-stack structure with "rigid and highly planar π-conjugated molecules", the "stable structure" due to molecular sliding accompanies stress loading. The change in the crystal structure accompanying the change in the "metastable structure" would cause deformation and relaxation of the crystal (Figure 3c). To construct a structure that allows expansion and contraction of organic crystals, a method that causes molecular slipping by face-to-face slip-stacking must be devised (Figure 3c,d). It is envisioned that molecular sliding between π-molecular plains affects the microscopic stretchability (Figure 3e).

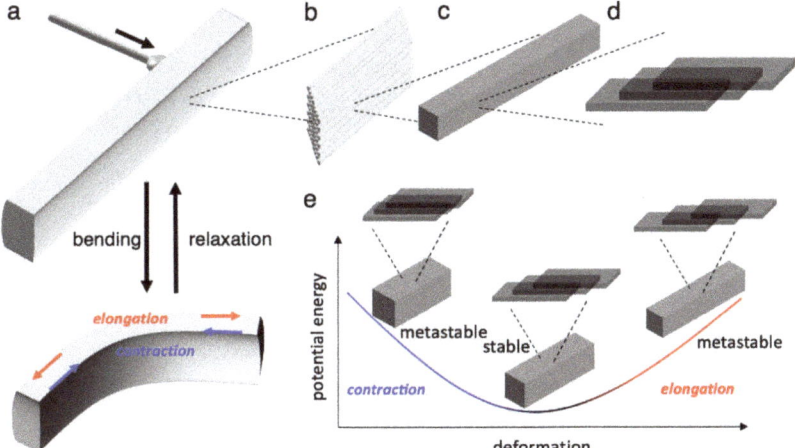

**Figure 3.** Enthalpic elasticity in elastically bendable organic crystals, elastic organic crystals. (**a**) Schematic illustration of bending stress for organic crystals. (**b**) Illustration of the Fibril Lamella structure. (**c**) Fibril in Lamella structure. (**d**) Face-to-face slip-stacked structure. (**e**) Plausible potential energy curve of the elastic organic crystals under bending stress. Length changes at outside and inside in the crystals. Molecular sliding occurs under both elongation and contraction stress.

Predicting and designing the crystal structure of any compound is extremely difficult. However, the skillful design of molecular structures and realization of intended packing structures based on functional group interactions and molecular shapes have been receiving attention in recent years. This methodology is called noncovalent synthesis, as opposed to organic synthesis based on covalent bond conversion. The structural unit is called a supramolecular synthon, like a synthon in organic synthesis [14,43,44]. In recent years, based on the structure concept shown in Figure 4, the author has designed, synthesized, and crystallized π-conjugated molecules to achieve intended crystal structures and create functional elastic crystals.

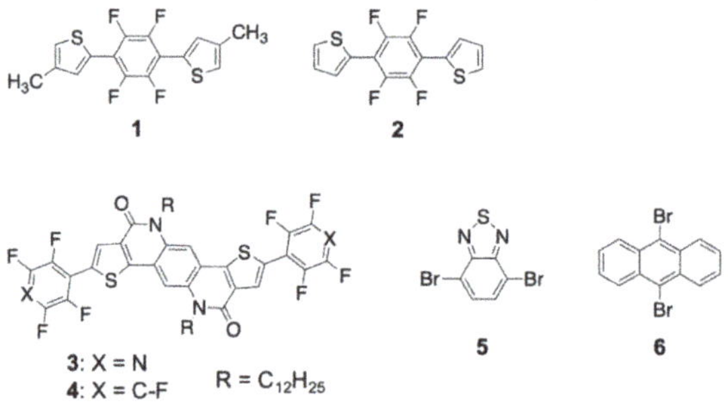

**Figure 4.** Chemical structures for light-emitting elastic organic crystals reported by Hayashi so far. **1** [15], **2** [16], **3** [20], **4** [20], **5** [19], **6** [21].

To form rigid and planar π-conjugated molecules and their slip stack-structures, we have focused on oligothiophenes containing fluoroarene structures **1–4** (Figure 4) [34,37,45,46]. The fluorine atom of the fluoroarene interacts with the sulfur atom of thiophene and forms hydrogen bonds with the adjacent aromatic hydrogen to improve the rigidity (suppression of free rotation) and planarity by multiple intramolecular interactions, and it is expected that face-to-face slip-stacking can be effectively formed by the thiophene-tetrafluorophenylene-thiophene alternating structures **1–4** (Figure 4).

The crystallization of molecules **1–4** crystallized on a millimeter or centimeter scale. From the crystal structure of each molecule, an interatomic distance shorter than the sum of the van der Waals radii of sulfur and fluorine or that of hydrogen and fluorine was observed in the molecule. Therefore, high rigidity and flatness based on the intramolecular S···F interaction and H···F hydrogen bond were suggested. In addition, the molecules pack by the face-to-face slip-stacking structure, and it can be considered that a single crystal with a target crystal structure based on the molecular design is obtained.

Owing to the stress applied to each crystal, it was found that the crystals of **1–4** showed elasticity, whereas the crystal of bis(5-methylthien-2-yl)-2,3,5,6-tetrafluorobenzene was brittle like a normal organic crystal. For example, the crystal of **1** bent when stress was applied, and it quickly returned to its original shape when the stress was released (Figure 5). That is to say, the crystal exhibiting elastic properties was obtained by the proposed design of a molecular crystal structure. In addition, this crystal showed efficient emission characteristics (sky blue emission, $\lambda_{PL}$ = 500 nm, $\varphi_{PL}$ = 24%), and possessed functions unique to the π-electron system [15,16].

**Figure 5.** Photograph of elastic bending of the light-emitting elastic organic crystal of **1**. Reproduced from [15] with permission of Wiley-VCH Verlag GmbH & Co. KGaA, Weinheim, copyright 2016.

The intramolecular H···F hydrogen bond in the crystal structure is stronger than the S···F interaction [34,37,45,46]. Therefore, when the core is difluorophenylene, the intramolecular H···F hydrogen bond with thiophene takes precedence, and the planarity increases. However, increasing the molecular hardness (suppression of free rotation) by four interactions by using a tetrafluorophenylene skeleton as the core is considered to afford the ultimate elasticity. In addition, owing to the face-to-face slip-stacking structure, which is another key to develop elasticity, the π-conjugated molecules form a through-space type of conjugated system, different from roughly isolated solution in solution. The absorption and photoluminescence (PL) bands undergo a large red-shift [15,16,37]. Therefore, it can be considered that such a crystal structure design exerts not only the mechanical properties of the crystal, but also a favorable effect on the optical properties.

## 3. Potential Structures of Elastic Organic Crystals

The author believes that there are some specific methods for developing such elastically deformable properties of organic crystals (Figure 6a) [47–53]. One method is to form a crystal structure by slip-stacking of planar π-conjugated molecules, as described above (Figure 6b) [15–18]. Conversely, as reported by Desiraju and co-workers [14], a method to make the crystal structure easy to move by halogen–halogen interactions can be used (Figure 6c). Because of this interaction, plastically deformable organic crystals have also been discovered. In addition, supramolecular host crystals including guests are likely to develop elasticity (Figure 6d). Ghosh and Reddy [13] reported elastic organic cocrystals formed from caffeine and 4-chloro-3-nitrobenzoic acid in methanol [13]. The guest methanol molecules in the cocrystal contribute to the elastic bending flexibility, which is lost upon desorption of methanol. This result indicates the importance of guest solvent molecules to evolve the elasticity. From this point of view, the supramolecular crystals developed by Ono et al. [54] are likely to periodically contain various solvents, resulting in various properties.

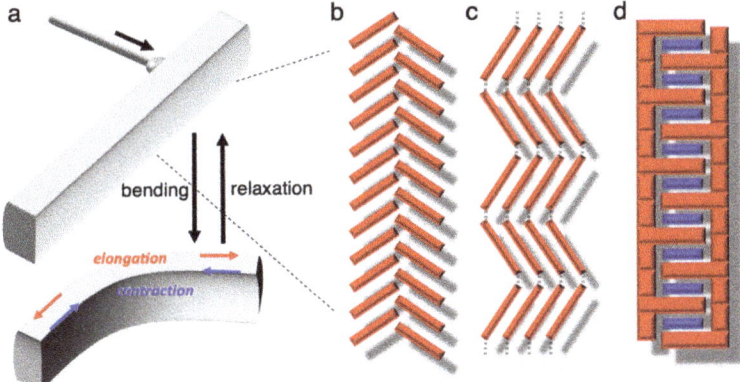

**Figure 6.** (**a**) Schematic illustration of applied stress for the elastic organic crystals. (**b**) llustration of face-to-face slip-stacking of planar conjugated molecules. (**c**) Illustration of zig-zag packing with intermolecular halogen-halogen interactions. (**d**) Illustration of supramolecular host crystal including solvents.

## 4. Detailed Deformation of Elastic Organic Crystals

Elongation and contraction are commonly observed with elastic bending deformation of elastic organic crystals. From observation of the crystal tip, it was found that the outer side elongated and the inner side contracted (Figure 7a). The elongation and contraction in the bent form are the same as the deformation caused by bending of rubber (Figure 7b). This is common deformation behavior for elastomers, and the elongation and contraction ratio of crystals can be calculated from this curvature.

The bending deformation of a plastic material results in a different crystal tip compared with the case of elastic materials (Figure 7c). Elongation and contraction do not occur in plastically deformable organic crystals because of slipping between the crystal layers. Thus, the strain of the elastic organic crystals (ε) is calculated by the following formula:

$$\varepsilon = 2dr, \tag{1}$$

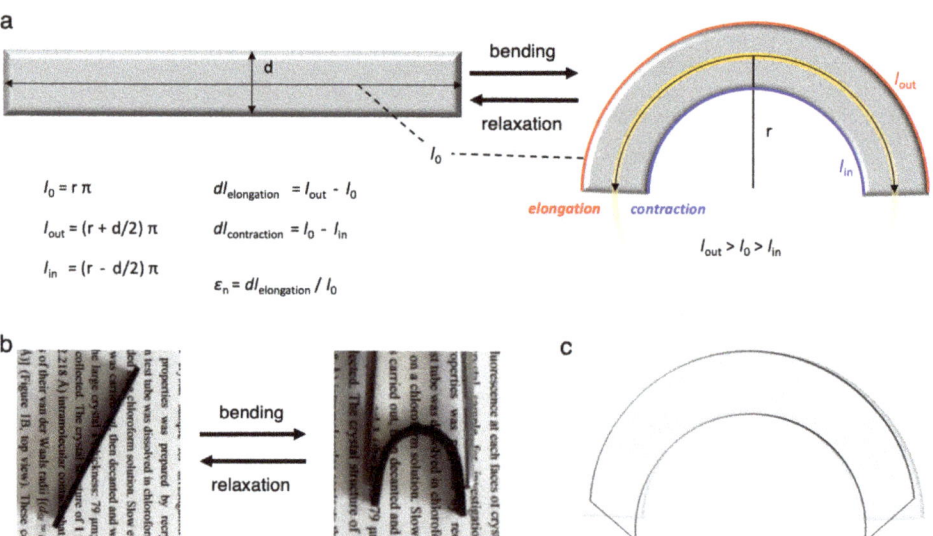

**Figure 7.** (a) Schematic illustration of bending-relaxation of elastic organic crystals. (b) Photograph of straight and bent rubber. (c) Illustration of the difference between elastic and plastic bent materials. Reproduced from [19] with permission of American Chemical Society, copyright 2017.

Because such crystal deformation causes a change in the crystal structure, it is considered to affect the results of structural analysis and physical property measurement. With the development of light-emitting elastic organic crystals, the author first discussed the effect of deformation on the physical properties from the PL spectrum. This will be described later. Conversely, Worthy et al. [55] found that there is a qualitative difference in the crystal structure by investigating from elongation at the outside to contraction at the inside by focusing the beam in X-ray structural analysis of the bent cupper acetylacetonate, Cu(acac)$_2$, crystal. The author proposed a facile method for investigating nanostructure changes of flexible organic crystals by using readily accessible X-ray equipment [56]. X-ray diffraction (XRD) with a curved jig, which applied macroscopic stress–strain (%), revealed reversible and quantitative crystal structure changes under bending stress and relaxation. Importantly, the method provides a way to quantitatively measure reversible structural changes without synchrotron X-ray analysis. In addition, Kenny et al. [57] discussed the relation between the magnetic properties and the bending deformation of this crystal. That is to say, elastic deformation is capable of producing a difference in the crystal structure and physical properties.

Because various mechanical deformations are attractive, it is possible to discover various applications by realizing not only flexibility, but also wearability. The crystals mechanically deform from symmetrical natural form to unnatural non-symmetrical shapes. For the first time, the author discovered not only bending, but also coiling and twisting of elastic organic crystals (Figure 8). Although this crystal showed various deformation behaviors, the deformable behavior differed depending on the crystal structure. This is because the molecular orientation differs for each crystal, but it is very interesting that the mechanical deformation behavior can be controlled by the arrangement. In addition, this deformation behavior is observed macroscopically, but in the future it is desirable that the change is microscopic.

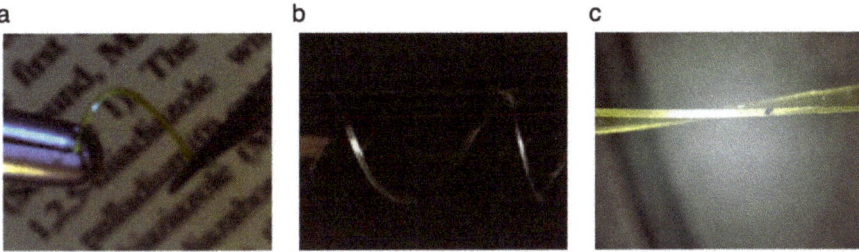

**Figure 8.** (**a**) Photograph of elastically bending. (**b**) Photograph of elastically coiling. (**c**) Photograph of elastically twisting. Reproduced from [15] with permission of Wiley-VCH Verlag GmbH & Co. KGaA, Weinheim, copyright 2016.

Poisson's effect is the ratio of the strain generated in the direction perpendicular to the stress to the strain generated in the stress direction when stress is applied to the body within the elastic limit [21]. In addition, when the crystals are stretched with a jig, changes due to such deformation occur. Poisson's effect of organic crystals is derived from deformation of the crystal lattice. Unlike general elastomers, it is expected that the amount of change differs depending on the direction based on the anisotropic arrangement of the molecules (Figure 9). Therefore, a unique Poisson's effect with a densely arranged molecular arrangement is shown. The calculated Poisson's ratio ($v$) is defined as the ratio of the change in the width per unit width of the material (e.g., plastic or metal) to the change in its length per unit length as a result of strain. For common materials, $v = 0.2$–$0.5$ because of the wide contraction as the material stretches. The Poisson's ratio is calculated by the following formula:

$$v_b = \varepsilon_b/\varepsilon_a, \quad v_c = \varepsilon_c/\varepsilon_a, \tag{2}$$

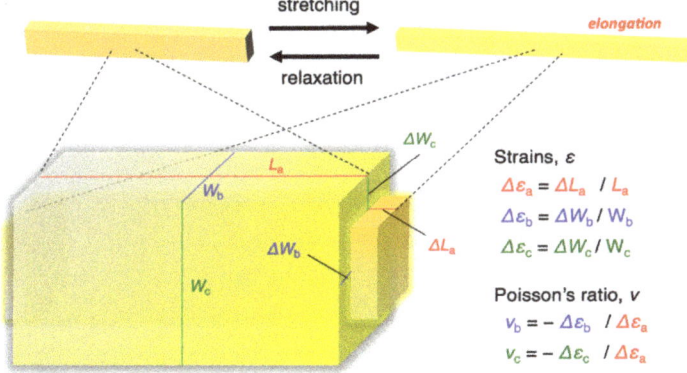

**Figure 9.** Schematic illustration of stretching stress for the elastic organic crystals. Relationships between strain and Poisson's ratio. Reproduced from [21] with permission of Wiley-VCH Verlag GmbH & Co. KGaA, Weinheim, copyright 2020.

The $v$ values of the (011)- and (001)-faces of the crystal of 9,10-dibromoanthracene **6** induced by elongation of the long axis ($\Delta\varepsilon_a$) were determined to be 0.243 and 0.057, respectively (Figure 10a) [21]. Thus, the (011)-face direction is more easily changed than the (001)-face direction. Therefore, the macroscopic anisotropic deformation of an elastic organic crystal can be observed. Conversely, density functional theory (DFT) simulation of the crystal structure suggested that the width lengths of the $b$ and $c$ axes decreased upon elongation of the $a$ axis. DFT simulations supported these results. The lengths of the $b$ and $c$ axes were predicted to increase upon contraction of the $a$ axis. The change of the $c$ axis was larger than that of the $b$ axis, which is consistent with the experimental results. Thus, the change in Poisson's ratio of the $c$ axis is larger than that of the $b$ axis. Simulations of the crystal not only supported this result, but also revealed changes in the intermolecular packing (Figure 10b). Interestingly, this structural change can be investigated experimentally. 1D X-ray diffraction (XRD) measurements of the crystal under applied elongation stress with a (011)-face-up setup geometry gave microscopic insight into the contraction of the short axis of the crystal (Figure 10c). Under the original straight-shape state, two strong diffraction peaks at $2\theta = 5.577°$ and $11.197°$ were observed (Figure 10d,e). These are correspond to $d$ spacings of 15.8465 and 7.9023 Å arising from the (001) and (002) planes of the single crystal. The observed values of the $d$ spacings were consistent with those calculated from the single crystal data of **6**. Under the application of elongation stress, the diffraction peaks shifted to the larger $2\theta$ region (Figure 10d,e), which suggested that the lengths of the layer spacings of the (001) and (002) planes decreased owing to contraction of the $c$ axis upon elongation of the $a$ axis, which was also observed in macroscopic deformation. The diffraction peak from the (001) plane of $5.613°$ ($d$ = 15.7449 Å) under elongation stress. The contraction ratio was 0.64%. When the elongation stress was released, the XRD peaks almost returned to their original positions, which indicated that the shape of the crystal lattice was also recovered (Figure 10d,e). This result suggests that reversible macroscopic crystal shape deformation induced microscopic crystal lattice deformation.

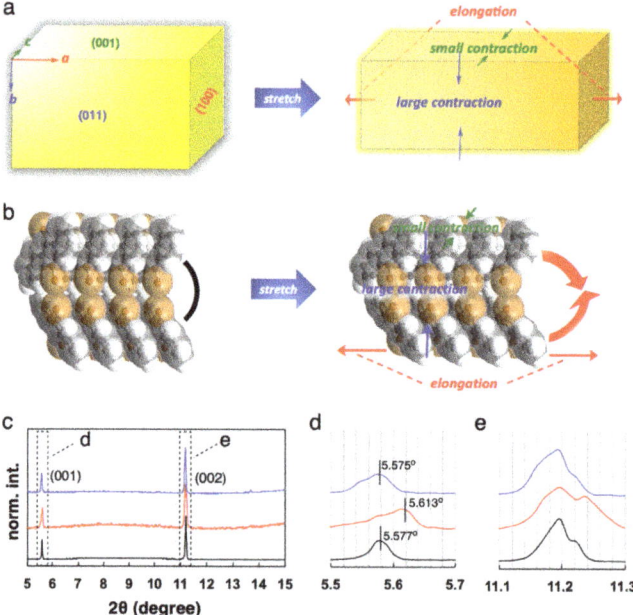

**Figure 10.** (**a**) Anisotropic Poisson's effect. Schematic illustration based on the results of macroscopic observation of the elastic crystal of 9,10-dibromoanthracene (**6**) crystal. (**b**) Illustration of the simulation calculation results. Simulation of crystal structure under stretching deformation. (**c**–**e**) 1D X-ray diffraction patterns of original (black), elongated (red), and relaxed (blue) elastic 9,10-dibromoanthracene crystals. Reproduced from [21] with permission of Wiley-VCH Verlag GmbH & Co. KGaA, Weinheim, copyright 2020.

## 5. Unique Properties of Elastic Organic Crystals: Mechanically Induced Shaping into Perfect Crystal Fibers

Crystals of π-conjugated molecules have attracted much attention owing to their potential applications in organic optoelectronic devices [58–61]. The macroscopic shapes of the materials are important for various organic device applications and depend on bottom-up fabrication processes [58–61]. Consequently, it is difficult to control their shapes. Top-down shaping is a practical method for crystal shape control (i.e., forming small fine crystals from larger crystals), but it is only feasible for flexible materials. Typical flexible materials, such as polymers, are of great interest because their flexibility allows various shapes to be formed by facile mechanical shaping. However, unlike flexible polymer materials, applying stress to common less flexible organic crystals generally causes them to disintegrate into powders and cracked small crystals.

Thermal- or photochemical-stimulus-triggered splitting deformations of organic crystals into small or fine crystals (e.g., salient effect) are very interested in crystal engineering [62–64]. However, these deformations randomly occur at crystal defects and are thus not suitable for top-down-controlled crystal shaping method. Laser fabrication of microcrystals into nanoparticles is a common top-down shaping approach. However, this method can only produce nanoscale crystals. Mechanical shaping is one of the ideal process for fabricating organic devices, but the brittleness of organic crystals makes it difficult to produce the exact shapes that are required for use in organic devices. However, if large-scale organic crystals (i.e., greater than micrometer scale) could be endowed with elastic bending flexibility and toughness, the bulk crystals could be easily processed into various fine shapes, such as fibers and films, by mechanical shaping.

Mechanically induced shaping of organic single crystals is one of an undeveloped area of research [17,18]. The author has described the mechanical splitting of the elastic organic crystal of **1**, which shows fibril lamella crystal morphology (Figure 3), and a facile shaping method for centimeter-scale of elastic organic crystals into various fine crystalline fibers (~50 μm thickness, ~150 μm width, and ~25 mm length) (Figure 11a) [17,18]. The produced fibers maintained their original crystal structure and optical properties (i.e., quantum efficiency and elastic flexibility). Thus, these long, fine and flexible photoluminescent organic crystal fibers would show potential for optoelectronic applications. Moreover, crystalline films (2D crystal) could be fabricated using the Scotch tape method like a synthesis of graphene. Elastic organic crystals based on fibril lamella structure (Figure 3) provide a new approach for fabrication of crystal fibers and films (top-down synthesis of supramolecular one-dimensional and two-dimensional materials). In addition, mechanically induced shaping of elastic organic crystals with similar crystal morphologies to **1** and other elastic crystals (1,4-bis(2-thienyl)-2,3,5,6-tetrafluorobenzene **2** and cupper acetylacetonate) which also show fibril lamella crystal morphologies can be performed.

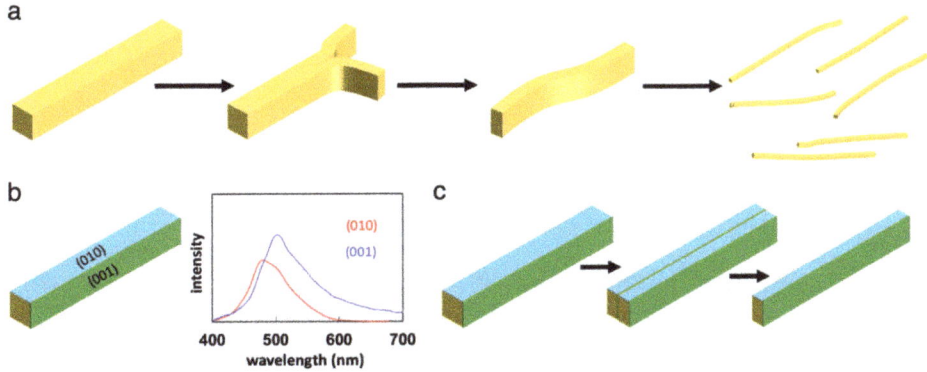

**Figure 11.** *Cont.*

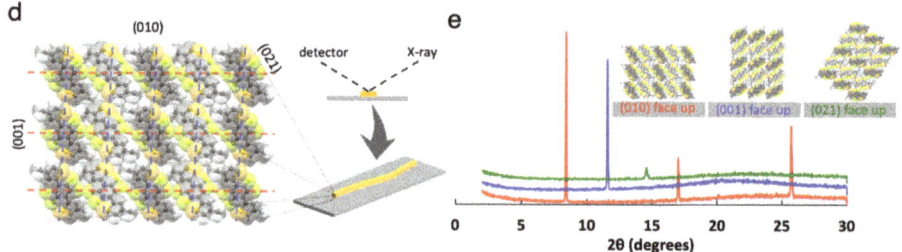

**Figure 11.** (a) Schematic illustration for mechanically induced shaping of the elastic organic crystals into the fibers. (b) Directional specific photoluminescence. PL spectra of 1,4-bis(4-methylthien-2-yl)-2,3,5,6-tetrafluorobenzene (**1**) crystal at (010) and (001). (c) Schematic illustration for mechanically induced shaping of the elastic organic crystals. The mechanically induced shaping of the centimeter-scale of crystal appeared new green-colored photoluminescent (001) face in the cross section. (d) Crystal structure at (100) face and illustration of lamella spaces. (e) 1D XRD patterns of the crystal fiber. Reproduced from [18] with permission of Royal Society of Chemistry, copyright 2019.

Interestingly, the centimeter-scale of organic crystal of **1** showed directional PL. Sky blue PL ($\lambda$ = 481 nm) of the (010) face and green PL ($\lambda$ = 502 nm) of the (001) face were observed because the (010) and (001) faces of the crystal have different directional molecular alignment orientation (Figure 11b). More interestingly, the (010) and (001) faces of the crystal before shaping showed sky blue and green PL, respectively (Figure 11c). A green PL line was observed when the crystal of **1** was cut along its length using a knife (Figure 11c). The shaped crystal showed green PL of the (001) face in the cross-section (Figure 11c).

Spectroscopic measurements and 1D X-ray analysis have been performed to understand the crystal structural features before and after mechanically induced shaping of the crystals. From absorption and PL spectral measurements, it was found that there was no remarkable change in the optical properties of the crystal and the fiber. However, grinding caused a change in the spectrum [37]. In addition, a comparison of the obtained fiber and the crystal before shape processing by XRD measurement showed that there was no change in the structural pattern. From this result, the physical properties and structure were maintained after machining. When the crystal fiber (Figure 11d) changed its direction on the substrate and a powder XRD pattern was recorded, a pattern derived from each lamella interval was observed (Figure 11e). In other words, a perfect crystal fiber in which the original crystal properties were transferred by crystal processing was obtained.

The phenomenon whereby an organic crystal mechanically breaks into a specific shape has been observed [62–65]. However, there has been no research on processing such an organic crystal to obtain a single crystal with a desired shape. Therefore, it can be considered that a very new phenomenon has been revealed from the aspect of crystal engineering. In addition, from the viewpoint of the fiber materials, it is noteworthy that a completely new material was produced. Heretofore, there have been several examples of low molecular weight fiber growth. Because it is more difficult to adjust low molecular weight fiber material than high molecular weight fiber, the former fiber has attracted attention. However, it is difficult to grow a crystal without branching, and there is no example of a fiber that can be considered to be a perfect crystal. Therefore, I believe that fibers obtained by processing organic crystals are expected to become new fiber materials.

## 6. Unique Properties of Elastic Organic Crystals: Mechanically Induced Photoluminescence Change

The most interesting character of the light-emitting organic crystals is the deformation-induced photophysical property changes (called mechanochromism and mechanofluorochromism) [19–21]. The molecular packing and intermolecular interactions in light-emitting crystals are perturbed or changed by external mechanical forces (e.g., shearing, grinding, tension, and hydrostatic pressure), which can cause a change in the light-emission of the crystals. There have been few reports of reversible color and PL changes under application and release of mechanical stress (or pressure) [66–68]. Consequently, flexible and tough crystals of π-conjugated molecules are candidate materials for reversible mechanochemical sensors (Figure 12).

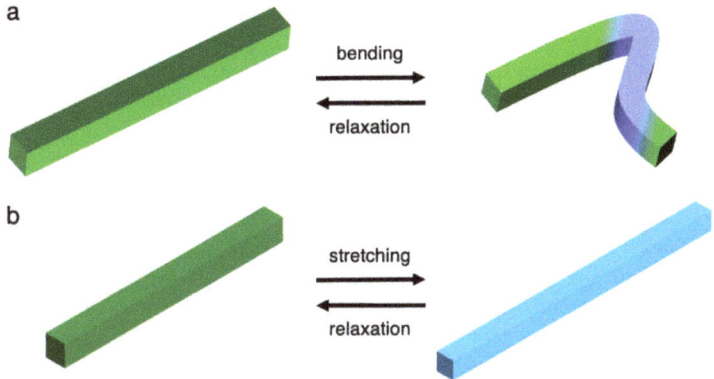

**Figure 12.** (a) Schematic illustration of elastically bending deformation-induced photoluminescence change. (b) Schematic illustration of elastically stretching deformation-induced photoluminescence change.

The author has also investigated the crystal of readily available 4,7-dibromo-2,1,3-benzothiadiazole **5** [19]. The centimeter-scale of needle-shaped single crystal of **5** bent under applied stress and rapidly reverted to its original crystal shape upon relaxation. Moreover, the crystal showed greenish blue PL, while under application of bending stress near its elastic bending limit (about 30°) the PL color changed to sky blue. The unique mechanical and light-emitting properties of the crystal referred to as mechanofluorochromism are based on the mechanical bending–relaxation. The change in the PL is probably due to a change of the center-to-center separation length in the face-to-face slip-stacked molecular packing, similar to the packing change in the crystal of 9,10-dibromoanthracene (**6**) (Figure 10) [21]. The molecular packing of **5** shifts from stable to metastable under application of mechanical bending stress, resulting in a blue-shifted PL band. Relaxation of the bent crystal of **5** allows for the recovery of the stable crystal packing.

To estimate the change in the structure and physical properties due to bending in detail, it is necessary to focus on a very small region of the crystal. For example, μ-focused X-ray analysis using a synchrotron can analyze the deformed part of the crystal in detail (Figure 13). In addition, the changes outside and inside the bent crystal can be observed [55]. Thus, it is possible to confirm the changes in the physical properties due to deformation of the crystal by performing μ-focused spectroscopic analysis of the outside and inside of the bent crystal.

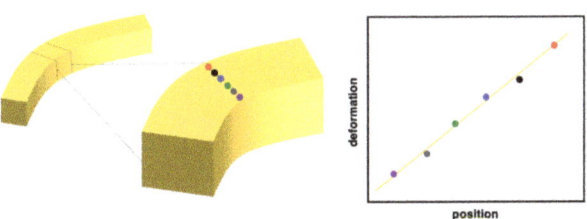

**Figure 13.** Schematic illustration of μ-focused X-ray analysis.

The stress-induced changes in the photophysical properties, that is, the deformation-induced PL changes of the crystal, have been successfully observed by spatially resolved µ-PL (µPL) measurement (Figure 14a) [21]. The PL spectra of the original crystal without stress are shown in Figure 14c,e, and the PL maximum wavelength (λ) was 500 nm. When bending stress was applied to the crystal, elongation and contraction simultaneously occur at both the outside and inside position sides of the curvature. On the other hand, no strain occurred at the center position. PL measurement of the microscopic areas—inside, center, and outside positions—of the bent crystal was performed using a spatially resolved µ-photoluminescent measurement system, µPL (Figure 14b–g). The PL spectra of the inside, center, and outside positions of the straight crystal revealed that λ = 500 nm (Figure 14b,c). However, a red-shift (λ = 506 nm) and a blue-shift (λ = 497 nm) of the PL bands, respectively, compared with that of the center position (λ = 505 nm) were clearly observed under bending strain of ε = 1.3% (Figure 14d,e). These observations suggested that an elongation deformation-induced red-shift and a contraction deformation-induced blue-shift indeed occurred at the outside and inside positions of the bent crystal, respectively. The bending strain induces slipping of the π-plane of face-to-face slip-stacked structure to increase the J-aggregate character at the outside of the crystal and H-aggregate character at the inside of the crystal, which would lead to the red-shift and blue-shift, respectively. In bending–relaxation, PL spectrum changes similar to the outside of the bent crystals were observed by crystal elongation (Figure 14f). Upon elongation along the *a* axis with $\varepsilon_a$ = 1.0%, a clear red-shift of the PL band (λ = 505 nm) was observed (Figure 14g, red line). The PL band returned to the original crystal shape when the elongation stress was released (Figure 14g, blue line). A larger red-shift (λ = 511 nm) was also observed upon elongation along the *a* axis with $\varepsilon_a$ = 2.0% (Figure 14g, green line). The spectral changes were reversible due to elastically changes in the crystal structure. This deformation-induced PL change probably resulted from the molecular sliding in the molecular packing.

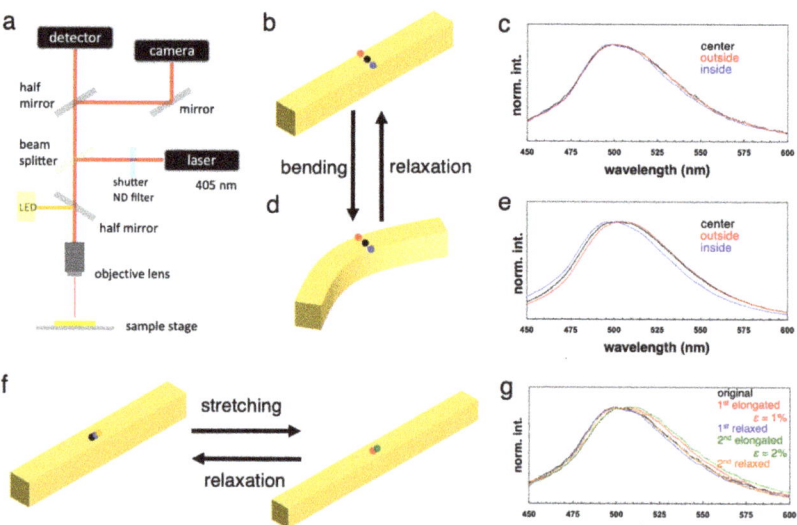

**Figure 14.** (**a**) Schematic illustration of specially resolved µ-photoluminescent measurement system. (**b**) Illustration of straight crystal. (**c**) PL spectra of the straight 9,10-dibromoanthracene crystal at outside, center, and inside. (**d**) Illustration of bent crystal. (**e**) PL spectra of the bent 9,10-dibromoanthracene crystal at outside, center, and inside. (**f**) Illustration of crystal stretching. (**g**) Reversible change of PL band of the 9,10-dibromoanthracene crystal. Reproduced from [21] with permission of Wiley-VCH Verlag GmbH & Co. KGaA, Weinheim, copyright 2020.

Twisted elastic organic crystals are thought to be able to give chirality to the crystal structure. The measurement of circular dichroism (CD), circular PL dichroism (CPL), and second harmonic generation (SHG) for the mechanically deformed elastic organic crystal will enable an investigation regarding the creation of chiral organic crystals.

## 7. Unique Properties of Elastic Organic Crystals: Flexible Optical Waveguide

The optical waveguide (OWG) has been investigated as a tool for capturing light and transmitting the light, but there is no flexible and highly efficient material because of the trade-off relationship of each material. The characteristics of the materials that have been applied as OWGs are summarized in Table 1 [69]. The refractive index, which is important for this tool, is one of the most important parameters for optical materials. Essentially, the refractive index values are large for high density materials. Inorganic crystals have less flexibility and tunability of optical properties, but often exhibit sufficient characteristics for OWGs [69]. In polymer chemistry, attention is often paid to the sizes of the atomic and molecular units to control or improve the refractive index. For π-conjugated polymer fiber, a flexible fiber can be prepared by the electrospray method or a similar method [70]. There has been no report containing a flexible waveguide using a π-conjugated polymer fiber, but because there are many defects inside and outside the material, in addition to low performance, the little phenomenon is observed [70]. In addition, organic-inorganic hybrids such as fluorescent lanthanoid metal-organic frameworks and a mixture of fluorescent lanthanoid coordination polymers and other polymers are known as new waveguiding materials [71]. Such new crystal and amorphous polymer materials show potentials for high-performance that differ from the summary in Table 1. Therefore, it can be considered that an organic crystal with a close-packed structure of organic molecules has a very high refractive index and thereby is a suitable material. In addition, most of the reported optical waveguides in organic materials chemistry are organic crystals because they have the advantage that there are few molecular defects, such as amorphous domains, inside and outside the material [69]. However, as mentioned above, since many organic crystals are inflexible, flexible OWGs cannot be made.

**Table 1.** Comparison of inorganic crystals, polymer solids, and organic crystals in terms of photoluminescent waveguide applications.

| Materials | Inorganic Crystals | Polymer Solids | Organic Crystals |
|---|---|---|---|
| Preparation methods | Vapor-phase deposition | Template polymerization, Electro-spinning | Vapor-phase deposition, Self-assembly |
| Crystallinity | Single crystal | Amorphous, Less-crystalline | Single crystal |
| Surface defects | Less of defects | Unavoidable defects | Less of defects |
| Interactions | Covalent bond, Ionic bond | Weak interactions [1] | Weak interactions [1] |
| Photo-stability | Relative stable | Decomposition | Bleaching |
| Optical properties | Excellent but not easy tunable | Tunable [2] | Tunable [3] |
| Refractive index, $n$ | >2 | 1.5–2.0 | >1.5 |
| Flexibilities | Less flexible | Flexible (viscoelastic) or brittle | No flexibility |
| Waveguide types | Passive and active | Passive and active | Active |

[1] van der Waals force, π-π interactions, hydrogen bonds; [2] by modification of the substituents; [3] by design and synthesis of molecules and crystal structure.

The waveguide characteristics differ depending on the shape and molecular orientation of the luminescent organic crystals (Figure 15a,b). To investigate this, it is necessary to focus on a very small area of the crystal, irradiate the laser, and detect the light emitted from the edge of the organic crystal. For example, the needle-shaped crystal of 2,5-dimethoxybenzene-1,4-dicarbardehyde show waveguide ability (Figure 15c) [72]. To measure the waveguide performance of the organic crystal, the loss coefficient of the waveguides was determined by measuring the spatially resolved PL spectra through local excitation of the crystal. When excited by a focused laser at different positions along its length, the green colored emission was observed from the edge and both ends of the crystal, irrespective of the laser-focused position. This is a typical characteristic of OWG. The spectral intensity and band profiles of the laser light excitation did not substantially change with the position along the crystal. However, the PL intensity at the end of the crystal almost exponentially decreased with increasing distance between excited site and emitting tip (Figure 15d). The optical loss coefficient, $\alpha$, by single exponential fitting ($I_{in}/I_{out} = A\exp(-\alpha X)$ where $X$ is the distance between the excited site and the emitting tip were calculated by recorded PL intensities at the excited site ($I_{in}$) and emitting tip ($I_{out}$). The $\alpha$ value was thus calculated to be 0.00120 dB/μm (Figure 15e). It should be noted that the $\alpha$ value is relatively high for this reported material. The $\alpha$ value and red-shifted feature are mainly affected by reabsorption during propagation of the light along the crystal. In the short wavelength region, the PL band overlaps with the absorption band in the solid state and the excitation spectrum of the crystal, resulting in a high $\alpha$ value. In the long wavelength region, the PL and absorption spectra of the crystal are well separated, which is beneficial for propagating light in the crystal, resulting in a low loss coefficient. This result gave a red-shift of the band and a very narrow full width at half-maximum (FWHM) (Figure 15d).

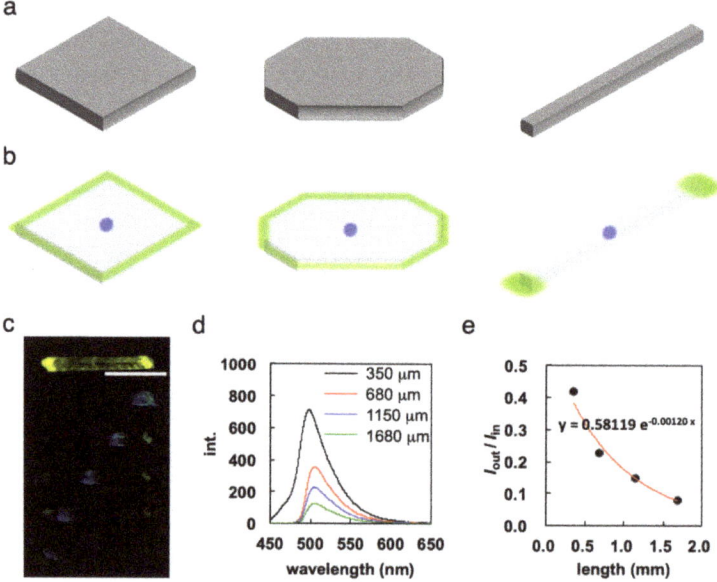

**Figure 15.** (**a**) Schematic illustration of crystal shapes. (**b**) Optical waveguiding of the crystals. Blue spot: laser excitation. Green region: waveguided photoluminescence. (**c**) Example of optical waveguiding for the crystal of 2,5-Dimethoxybenzene-1,4-dicarbardehyde crystal. Microscopy images under ambient light and collected upon laser 405 nm excitation of the identical crystal at different positions. Scale bar: 1.0 mm. (**d**) PL spectra at tip of a single crystal. (**e**) $I_{out}/I_{in}$ values on length. Reproduced from [72] with permission of Wiley-VCH Verlag GmbH & Co. KGaA, Weinheim, copyright 2019.

The organic crystals exhibited unique OWG characteristics (Figure 16) [20]. The refractive index are very important factors for the performance and a small amount of surface defects is negative factor against the OWG [69–77]. Organic crystals of light-emitting π-conjugated molecules are suitable OWG materials compared with flexible light-emitting π-conjugated polymer fibers, but their crystals are generally less flexible. Thus, light-emitting elastic organic crystals are considered to be suitable or new for high performance and flexible OWG materials (Table 1). By excitation of the straight original crystal of **3** with a focused laser (405 nm) at different positions, reddish orange colored emission was detected from the end of the straight crystal of **3**, irrespective of the excitation position (Figure 16a). This is a typical result of OWG materials [69]. The elastic strain ($\varepsilon$) value for the crystal of **3** was calculated to be 1.9% (Figure 16b). The PL spectra of the straight and bent crystals at the illuminated position (black dotted line) and end of the crystal (solid line) are shown in Figure 16c,d. The PL band at the end of the crystal showed a peak at 597 nm with a narrower FWHM (34 nm) than at the illuminated position (573 nm, FWHM = 56 nm). The spectral profile did not substantially change with the illumination position, although the PL intensity at the end very slightly decreased with increasing distance (Figure 16c,d). The PL intensities at the illuminated position ($I_{in}$) and end of the crystal ($I_{out}$) were measured to calculate the optical loss coefficient $\alpha$ values by measuring the spatially resolved PL spectra and fitting by a single exponential curve. The $\alpha$ values were calculated to be 0.043 (the straight crystal) and 0.047 dB/mm (bent crystals), respectively (Figure 16e). It is noteworthy that OWG applications with various physical properties have been reported for various light-emitting elastic organic crystals by Zhang and co-workers [75–77]. These are good showcases for the great potential of such crystals. The author expects that more diverse applications will be developed in the future.

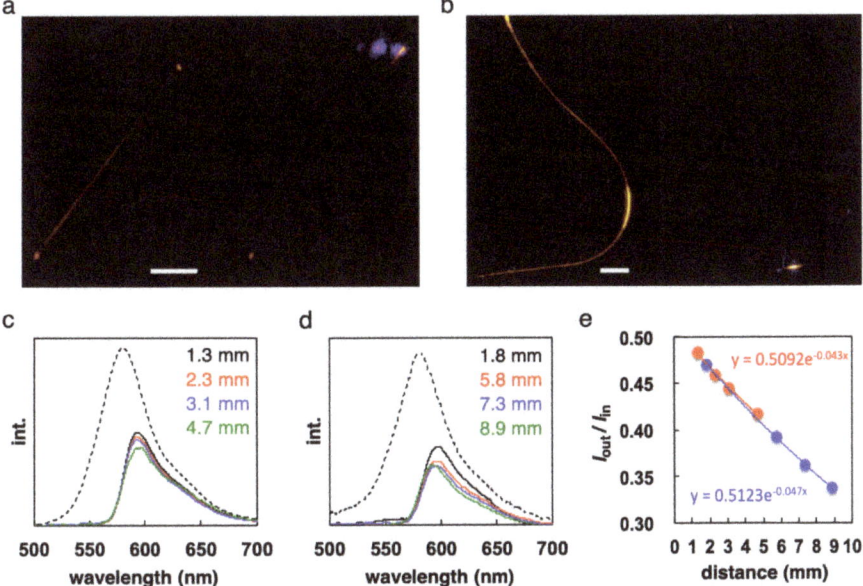

**Figure 16.** (a) Optical waveguiding for the straight elastic organic crystal of **3**. *left*: under UV irradiation. *right*: laser excitation. (b) Optical waveguiding for the bent elastic organic crystal of **3**. *left*: under UV irradiation. *right*: laser excitation. (c) PL spectra at the end of a straight crystal. Dotted line: PL spectrum at the illuminated position. (d) PL spectra at the end of a bent single crystal. Dotted line: PL spectrum at the illuminated position. (e) Relative $I_{out}/I_{in}$ values as a function of distance from the illuminated position to the end. Red line: Straight crystal. Blue line: Bent crystal. Reproduced from [20] with permission of Wiley-VCH Verlag GmbH & Co. KGaA, Weinheim, copyright 2018.

## 8. Conclusions

Here, the author described the deformation-induced structures and changes to the physical properties of organic crystals that behave with enthalpy elasticity. The structural features and material possibilities of flexible and functional organic crystals, especially elastic organic crystals of $\pi$-conjugated molecules, have been discussed. Recently, flexible organic crystals have become one of the active areas in crystal engineering. Therefore, various challenges have been proposed to discover these crystals and methods for their investigation. A design of the organic molecules for such materials (supramolecular synthon) and its applications (e.g., deformation-induced photoluminescence changes, conductivity, optical waveguides, and organic devices) would be accepted. I think that more detailed information of regarding the structure and deformation-induced changes in the physical properties remain. The promotion of mechanical property measurement methods, including nanoindentation [78], calculation simulation such as energy frameworks [79], and other effective crystal investigation methods will provide more information in this area. In the future, more accurate analysis and the application of new ideas and technological innovation will be required. In addition, flexible and functional organic crystals have not been utilized yet functioned as flexible organic devices. Verifying the combination of flexible organic crystals with various substrates will also be an important issue.

**Funding:** This research received no external funding.

**Acknowledgments:** S.H. acknowledges a KAKENHI (Grant-in-Aid for Scientific Research B: No. 18H02052, Grant-in-Aid for Scientific Research on Innovative Areas 'π-figuration': No. 17H05171, 'coordination asymmetry': No. 19H04604 and 'soft crystal': No. 20H04684) of the Japan Society for the Promotion of Science (JSPS). This work was performed under the Cooperative Research Program of "Network Joint Research Center for Materials and Devices".

**Conflicts of Interest:** The author declares no conflict of interest.

## References

1. Saha, S.; Mishra, M.K.; Reddy, C.M.; Desiraju, G.R. From molecules to interactions to crystal engineering: Mechanical properties of organic solids. *Acc. Chem. Res.* **2018**, *51*, 2957–2967. [CrossRef] [PubMed]
2. Desiraju, G.R. Crystal engineering: From molecule to crystal. *J. Am. Chem. Soc.* **2013**, *135*, 9952–9967. [CrossRef] [PubMed]
3. Reddy, C.M.; Kirchner, M.T.; Gundakaram, R.C.; Padmanabhan, K.A.; Desiraju, G.R. Isostructurality, polymorphism and mechanical properties of some hexahalogenated benzenes: The nature of halogen···halogen interactions. *Chem. Eur. J.* **2006**, *12*, 2222–2234. [CrossRef] [PubMed]
4. Panda, M.K.; Ghosh, S.; Yasuda, N.; Moriwaki, T.; Mukherjee, G.D.; Reddy, C.M.; Naumov, P. Spatially resolved analysis of short-range structure perturbations in a plastically bent molecular crystal. *Nat. Chem.* **2015**, *7*, 65–72. [CrossRef]
5. Reddy, C.M.; Gundakaram, R.C.; Basavoju, S.; Kirchner, M.T.; Padmanabhan, K.A.; Desiraju, G.R. Structural basis for bending of organic crystals. *Chem. Commun.* **2005**, *31*, 3945–3947. [CrossRef]
6. Reddy, C.M.; Padmanabhan, K.A.; Desiraju, G.R. Structure-property correlations in bending and brittle organic crystals. *Cryst. Growth Des.* **2006**, *6*, 2720–2731. [CrossRef]
7. Nguyen, T.T.; Arkhipov, S.G.; Rychkov, D.A. Simple crystallographic model for anomalous plasticity of L-Leucinium hydrogen maleate crystals. *Mater. Today Proc.* **2020**, *25*, 412–415. [CrossRef]
8. Takamizawa, S.; Takasaki, Y. Shape-Memory Effect in an Organosuperelastic Crystal. *Chem. Sci.* **2016**, *7*, 1527–1534. [CrossRef]
9. Takamizawa, S. Dynamic gas-inclusion in a single crystal. *Angew. Chem. Int. Ed.* **2015**, *54*, 7033–7036. [CrossRef]
10. Takamizawa, S.; Miyamoto, Y. Superelastic organic crystals. *Angew. Chem. Int. Ed.* **2014**, *53*, 6970–6973. [CrossRef]
11. DeRosa, C.; Auriemma, F. From entropic to enthalpic elasticity: Novel thermoplastic elastomers from syndiotactic propylene–ethylene copolymers. *Adv. Mater.* **2005**, *17*, 1503–1507. [CrossRef]
12. DeRosa, C.; Auriemma, F. Structure and physical properties of syndiotactic polypropylene: A highly crystalline thermoplastic elastomer. *Prog. Polym. Sci.* **2006**, *31*, 1145–1237. [CrossRef]
13. Ghosh, S.; Reddy, C.M. Elastic and bendable caffeine cocrystals: Implications for the design of flexible organic materials. *Angew. Chem. Int. Ed.* **2012**, *51*, 10319–10323. [CrossRef] [PubMed]

14. Ghosh, S.; Mishra, M.K.; Kadambi, S.B.; Ramamurty, U.; Desiraju, G.R. Designing elastic organic crystals: Highly flexible polyhalogenated N-benzylideneanilines. *Angew. Chem. Int. Ed.* **2015**, *54*, 2674–2678. [CrossRef]
15. Hayashi, S.; Koizumi, T. Elastic organic crystal of a fluorescent π-conjugated molecule. *Angew. Chem. Int. Ed.* **2016**, *55*, 2701–2704. [CrossRef]
16. Hayashi, S.; Asano, A.; Kamiya, N.; Yokomori, Y.; Maeda, T.; Koizumi, T. Fluorescent organic single crystals with elastic bending flexibility: 1,4-bis(thien-2-yl)-2,3,5,6-tetrafluorobenzene derivatives. *Sci. Rep.* **2017**, *7*, 9453. [CrossRef]
17. Hayashi, S.; Koizumi, T. Mechanically induced shaping of organic single crystals: Facile fabrication of fluorescent and elastic crystal fibers. *Chem. Eur. J.* **2018**, *24*, 8507–8512. [CrossRef]
18. Hayashi, S.; Koizumi, T. Direction-specific fluorescence of an engineered organic crystal and the appearance of a new face caused by mechanically induced shaping. *CrystEngComm* **2019**, *21*, 5990–5996. [CrossRef]
19. Hayashi, S.; Koizumi, T.; Kamiya, N. Elastic bending flexibility of fluorescent organic single crystal: New aspects of commonly used building block "4,7-dibromo-2,1,3-benzothiadiazole". *Cryst. Growth Des.* **2017**, *17*, 6158–6162. [CrossRef]
20. Hayashi, S.; Yamamoto, S.; Takeuchi, D.; Ie, Y.; Takagi, K. Creating elastic organic crystals of π-conjugated molecules with flexible optical waveguide and bending mechanofluorochromism. *Angew. Chem. Int. Ed.* **2018**, *57*, 17002–17008. [CrossRef]
21. Hayashi, S.; Ishiwari, F.; Fukushima, T.; Mikage, S.; Imamura, Y.; Tashiro, M.; Katouda, M. Anisotropic Poisson's effect and deformation-induced fluorescence change of elastic 9,10-dibromoanthracene single crystals. *Angew. Chem. Int. Ed.* **2020**, *59*, 16195–16201. [CrossRef] [PubMed]
22. Hayashi, S. Elastic organic crystals of π-conjugated molecules: Anisotropic densely packed supramolecular 3D polymers exhibit mechanical flexibility and shape tunability. *Polym. J.* **2019**, *51*, 813–823. [CrossRef]
23. Hayashi, S. Flexible and densely packed π-figuration system: Creating elastic organic crystals of π-conjugated molecules. *J. Synth. Org. Chem. Jpn.* **2020**, *58*, 962–970. [CrossRef]
24. Hayashi, S. Elastic Organic Crystals. *J. Synth. Org. Chem. Jpn.* **2020**, *58*, 985.
25. Saito, M.; Shinokubo, H.; Sakurai, H. Figuration of bowl-shaped π-conjugated molecules: Properties and function. *Mater. Chem. Front.* **2018**, *2*, 635–661. [CrossRef]
26. Hayashi, S.; Inagi, S.; Fuchigami, T. Synthesis of 9-substituted fluorene copolymers via chemical and electrochemical polymer reaction and their optoelectronic properties. *Macromolecules* **2009**, *42*, 3755–3760. [CrossRef]
27. Ajayaghosh, A.; Praveen, V.K. π-Organogels of self-assembled p-phenylenevinylenes: Soft materials with distinct size, shape, and functions. *Acc. Chem. Res.* **2007**, *40*, 644–656. [CrossRef] [PubMed]
28. Holzwarth, J.; Amsharov, K.Y.; Sharapa, D.I.; Reger, D.; Roschyna, K.; Lungerich, D.; Jux, N.; Hauke, F.; Clark, T.; Hirsch, A. Highly regioselective alkylation of hexabenzocoronenes: Fundamental insights into the covalent chemistry of graphene. *Angew. Chem. Int. Ed.* **2017**, *56*, 12184–12190. [CrossRef]
29. Omachi, H.; Segawa, Y.; Itami, K. Synthesis of cyclophenylenes and related carbon nanorings: A step toward the controlled sysnthesis of carbon nanotubes. *Acc. Chem. Res.* **2012**, *45*, 1378–1389. [CrossRef]
30. Hayashi, S.; Asano, A.; Koizumi, T. Modification of pyridine-based conjugated polymer films via Lewis acid: Halochromism, characterization and macroscopic gradation patterning. *Polym. Chem.* **2011**, *2*, 2764–2766. [CrossRef]
31. Hayashi, S.; Koizumi, T. From benzodiazaborole-based compound to donor-acceptor polymer via electropolymerization. *Polym. Chem.* **2012**, *3*, 613–616. [CrossRef]
32. Hayashi, S.; Yamamoto, S.; Koizumi, T. Effects of molecular weight on the optical and electrochemical properties of EDOT-based π-conjugated polymers. *Sci. Rep.* **2017**, *7*, 1027. [CrossRef]
33. Hayashi, S.; Hirai, R.; Yamamoto, S.; Koizumi, T. A simple route to unsymmetric cyano-substituted oligo(p-phenylene-vinylene)s. *Chem. Lett.* **2018**, *47*, 1003–1005. [CrossRef]
34. Hayashi, S.; Takigami, A.; Koizumi, T. Solvent control over supramolecular gel formation and fluorescence for a highly crystalline π-conjugated polymer. *Chem. Asian J.* **2018**, *13*, 2014–2018. [CrossRef] [PubMed]
35. Hayashi, S.; Yamamoto, S.; Nishi, K.; Asano, A.; Koizumi, T. Synthesis of network polymer emitters: Tunable detection of chemicals by the geometric design. *Polym. J.* **2019**, *51*, 1055–1061. [CrossRef]

36. Hayashi, S.; Sakamoto, M.; Ishiwari, F.; Fukushima, T.; Yamamoto, S.; Koizumi, T. A versatile scaffold for facile synthesis of fluorescent cyano-substituted stilbenes. *Tetrahedron* **2019**, *75*, 1079–1084. [CrossRef]
37. Hayashi, S. Highly crystalline and efficient red-emissive π-conjugated polymer film: Tuning of macrostructure for light-emitting properties. *Mater. Adv.* **2020**, *1*, 632–638. [CrossRef]
38. Hayashi, S.; Nishioka, N.; Nishiyama, H.; Koizumi, T. π-Conjugated alternating copolymer based on the 3,5-dinitro-9-fluorenone for electron-acceptor type materials. *Synth. Met.* **2012**, *162*, 1485–1489. [CrossRef]
39. Facchetti, A.; Yoon, M.-H.; Stern, C.L.; Katz, H.E.; Marks, T.J. Building blocks for n-type organic electronics: Regiochemically modulated inversion of majority carrier sign in perfluoroarene-modified polythiophene semiconductors. *Angew. Chem. Int. Ed.* **2003**, *42*, 3900–3903. [CrossRef]
40. Reese, C.; Bao, Z. Organic single crystals: Tools for the exploration of charge transport phenomena in organic materials. *J. Mater. Chem.* **2006**, *16*, 329–333. [CrossRef]
41. Inada, Y.; Yamao, T.; Inada, M.; Itami, T.; Hotta, S. Giant organic single-crystals of a thiophene/phenylene co-oligomer toward device applications. *Synth. Met.* **2011**, *161*, 1869–1877. [CrossRef]
42. Kim, T.-D.; Lee, K.-S. D-π-A conjugated molecules for optoelectronic applications. *Macromol. Rapid. Commun.* **2015**, *36*, 943–958. [CrossRef] [PubMed]
43. Desiraju, G.R. Supramolecular synthon in crystal engineering—A new organic synthesis. *Angew. Chem. Int. Ed.* **1995**, *34*, 2311–2327. [CrossRef]
44. Ghosh, S.; Mishra, M.K.; Ganguly, S.; Desiraju, G.R. Dual stress and thermally driven mechanical properties of the same organic crystal: 2,6-dichlorobenzylidene-4-fluoro-3-nitroaniline. *J. Am. Chem. Soc.* **2015**, *137*, 9912–9921. [CrossRef] [PubMed]
45. Crouch, D.J.; Skabara, P.J.; Heeney, M.; McCulloch, I.; Coles, S.J.; Hursthouse, M.B. Hexyl-substituted oligothiophenes with a central tetrafluorophenylene unit: Crystal engineering of planar structures for p-type organic semiconductors. *Chem. Commun.* **2005**, 1465–1467. [CrossRef]
46. Crouch, D.J.; Skabara, P.J.; Lohr, J.E.; McDouall, J.J.W.; Heeney, M.; McCulloch, I.; Sparrowe, D.; Shkunov, M.; Coles, S.J.; Horton, P.N.; et al. Thiophene and selenophene copolymers incorporating fluorinated phenylene units in the main chain: Synthesis, characterization, and application in organic field-effect transistors. *Chem. Mater.* **2005**, *17*, 6567–6578. [CrossRef]
47. Ahmed, E.; Karothu, D.P.; Naumov, P. Crystal adaptronics: Mechanically reconfigurable elastic and superelastic molecular crystals. *Angew. Chem. Int. Ed.* **2018**, *57*, 8837–8846. [CrossRef]
48. Mishra, M.K.; Sun, C.C. Conformation directed interaction anisotropy leading to distinct bending behaviors of two ROY polymorphs. *Cryst. Growth Des.* **2020**, *20*, 4764–4769. [CrossRef]
49. Wang, K.; Mishra, M.K.; Sun, C.C. Exceptionally elastic single-component pharmaceutical crystals. *Chem. Mater.* **2019**, *31*, 1794–1799. [CrossRef]
50. Mishra, M.K.; Mishra, K.; Asif, S.A.S.; Manimunda, P. Structural analysis of elastically bent organic crystals using *in situ* indentation and micro-raman spectroscopy. *Chem. Commun.* **2017**, *53*, 13035–13038. [CrossRef]
51. Horstman, E.M.; Keswani, R.K.; Frey, B.A.; Rzeczycki, P.M.; LaLone, V.; Bertke, J.A.; Kenis, P.J.A.; Rosania, G.R. Elasticity in macrophage-synthesized biocrystals. *Angew. Chem. Int. Ed.* **2017**, *56*, 1815–1819. [CrossRef] [PubMed]
52. Gupta, P.; Karothu, D.P.; Ahmed, E.; Naumov, P.; Nath, N.K. Thermally twistable, photobendable, elastically deformable, and self-healable soft crystals. *Angew. Chem. Int. Ed.* **2018**, *57*, 8498–8502. [CrossRef] [PubMed]
53. Commins, P.; Karothu, D.P.; Naumov, P. Is a bent crystal still a single crystal? *Angew. Chem. Int. Ed.* **2019**, *58*, 10052–10060. [CrossRef] [PubMed]
54. Ono, T.; Sugimoto, M.; Hisaeda, Y. Multicomponent molecular puzzles for photofunction design: Emission color variation in Lewis acid–base pair crystals coupled with guest-to-host charge transfer excitation. *J. Am. Chem. Soc.* **2015**, *137*, 9519–9522. [CrossRef]
55. Worthy, A.; Grosjean, A.; Pfrunder, M.C.; Xu, Y.; Yan, C.; Edwards, G.; Clegg, J.K.; McMurtrie, J.C. Atomic resolution of structural changes in elastic crystals of copper(II) acetylacetonate. *Nat. Chem.* **2018**, *10*, 65–69. [CrossRef]
56. Hayashi, S. Facile investigation of reversible nanostructure changes in flexible crystals. *Res. Sq.* **2020**. [CrossRef]

57. Kenny, E.P.; Jacko, A.C.; Powell, B.J. Mechanomagnetics in elastic crystals: Insights from [Cu(acac)$_2$]. *Angew. Chem. Int. Ed.* **2019**, *58*, 15082–15088. [CrossRef]
58. Wang, C.; Dong, H.; Hu, W.; Liu, Y.; Zhu, D. Semiconducting π-conjugated systems in field-effect transistors: A material odyssey of organic electronics. *Chem. Rev.* **2012**, *112*, 2208–2267. [CrossRef]
59. Gidron, O.; Bendikov, M. α-Oligofurans: An emerging class of conjugated oligomers for organic electronics. *Angew. Chem. Int. Ed.* **2014**, *53*, 2546–2555. [CrossRef]
60. Kobaisi, M.A.; Bhosale, S.V.; Latham, K.; Raynor, A.M.; Bhosale, S.V. Functional naphthalene diimides: Synthesis, properties, and applications. *Chem. Rev.* **2016**, *116*, 11685–11796. [CrossRef]
61. Uegaki, K.; Nakabayashi, K.; Yamamoto, S.; Koizumi, T.; Hayashi, S. Donor–acceptor random regioregular π-conjugated copolymers based on poly(3-hexylthiophene) with unsymmetrical monothienoisoindigo Units. *RSC Adv.* **2020**, *10*, 19034–19040. [CrossRef]
62. Kakkar, S.; Bhattacharya, B.; Reddy, C.M.; Ghosh, S. Tuning mechanical behaviour by controlling the structure of a series of theophylline co-crystals. *CrystEngComm* **2018**, *20*, 1101–1109. [CrossRef]
63. Seki, T.; Mashimo, T.; Ito, H. Anisotropic strain release in a thermosalient crystal: Correlation between the microscopic orientation of molecular rearrangements and the macroscopic mechanical motion. *Chem. Sci.* **2019**, *10*, 4185–4191. [CrossRef] [PubMed]
64. Nath, N.K.; Panda, M.K.; Sahoo, S.C.; Naumov, P. Thermally induced and photoinduced mechanical effects in molecular single crystals—A revival. *CrystEngComm* **2014**, *16*, 1850–1858. [CrossRef]
65. Sahoo, S.C.; Sinha, S.B.; Kiran, M.S.; Ramamurty, U.; Dericioglu, A.F.; Reddy, C.M.; Naumov, P. Kinematic and mechanical profile of the self-actuation of thermosalient crystal twins of 1,2,4,5-tetrabromobenzene: A molecular crystalline analogue of a bimetallic strip. *J. Am. Chem. Soc.* **2013**, *135*, 13843–13850. [CrossRef] [PubMed]
66. Nagura, K.; Saito, S.; Yusa, H.; Yamawaki, H.; Fujihisa, H.; Sato, H.; Shimoikeda, Y.; Yamaguchi, S. Distinct responses to mechanical grinding and hydrostatic pressure in luminescent chromism of tetrathiazolylthiophene. *J. Am. Chem. Soc.* **2013**, *135*, 10322–10325. [CrossRef]
67. Ono, T.; Tsukiyama, Y.; Taema, A.; Sato, H.; Kiyooka, H.; Yamaguchi, Y.; Nagahashi, A.; Nishiyama, M.; Akahama, Y.; Ozawa, Y.; et al. Piezofluorochromism in charge-transfer inclusion crystals: The influence of high pressure versus mechanical grinding. *ChemPhotoChem* **2018**, *2*, 416–420. [CrossRef]
68. Zhao, J.; Chi, Z.; Zhang, Y.; Mao, Z.; Yang, Z.; Ubba, E.; Chi, Z. Recent progress in the mechanofluorochromism of cyanoethylene derivatives with aggregation-induced emission. *J. Mater. Chem. C* **2018**, *6*, 6327–6353. [CrossRef]
69. Zhang, C.; Zhao, Y.S.; Yao, J. Optical waveguides at micro/nanoscale based on functional small organic molecules. *Phys. Chem. Chem. Phys.* **2011**, *13*, 9060–9073. [CrossRef]
70. Di Benedetto, F.; Camposeo, A.; Pagliara, S.; Mele, E.; Persano, L.; Stabile, R.; Cingolani, R.; Pisignano, D. Patterning of light-emitting conjugated polymer nanofibres. *Nat. Nanotechnol.* **2008**, *3*, 614. [CrossRef]
71. Chen, X.; Sun, T.; Wang, F. Lanthanide-based luminescent materials for waveguide and lasing. *Chem. Asian J.* **2020**, *15*, 1–33. [CrossRef]
72. Hayashi, S.; Koizumi, T.; Kamiya, N. 2,5-Dimethoxybenzene-1,4-dicarboxaldehyde: An emissive organic crystal and highly efficient fluorescent waveguide. *ChemPlusChem* **2019**, *84*, 247–251. [CrossRef] [PubMed]
73. Li, Y.; Ma, Z.; Li, A.; Xu, W.; Wang, Y.; Jiang, H.; Wang, K.; Zhao, Y.; Jia, X. A single crystal with multiple functions of optical waveguide, aggregation-induced emission, and mechanochromism. *ACS Appl. Mater. Interfaces* **2017**, *9*, 8910–8918. [CrossRef] [PubMed]
74. Guo, Z.-H.; Lei, T.; Jin, Z.-X.; Wang, J.-Y.; Pei, J. T-Shaped donor–acceptor molecules for low-loss red-emission optical waveguide. *Org. Lett.* **2013**, *15*, 3530–3533. [CrossRef]
75. Liu, B.; Di, Q.; Liu, W.; Wang, C.; Wang, Y.; Zhang, H. Red-emissive organic crystals of a single-benzene molecule: Elastically bendable and flexible optical waveguide. *J. Phys. Chem. Lett.* **2019**, *10*, 1437–1442. [CrossRef] [PubMed]
76. Huang, R.; Wang, C.; Wang, Y.; Zhang, H. Elastic self-doping organic single crystals exhibiting flexible optical waveguide and amplified spontaneous emission. *Adv. Mater.* **2018**, *30*, 1800814. [CrossRef] [PubMed]
77. Liu, H.; Lu, Z.; Zhang, Z.; Wang, Y.; Zhang, H. Highly elastic organic crystals for flexible optical waveguide. *Angew. Chem. Int. Ed.* **2018**, *57*, 8448–8452. [CrossRef] [PubMed]

78. Varughese, S.; Kiran, M.S.R.N.; Ramamurty, U.; Desiraju, G.R. Nanoindentation in crystal engineering: Quantifying mechanical properties of molecular crystals. *Angew. Chem. Int. Ed.* **2013**, *52*, 2701–2712. [CrossRef]
79. Turner, M.J.; Thomas, S.P.; Shi, M.W.; Jayatilaka, D.; Spackman, M.A. Energy frameworks: Insights into interaction anisotropy and the mechanical properties of molecular crystals. *Chem. Commun.* **2015**, *51*, 3735–3738. [CrossRef]

**Publisher's Note:** MDPI stays neutral with regard to jurisdictional claims in published maps and institutional affiliations.

© 2020 by the author. Licensee MDPI, Basel, Switzerland. This article is an open access article distributed under the terms and conditions of the Creative Commons Attribution (CC BY) license (http://creativecommons.org/licenses/by/4.0/).

*Review*

# Resonance in Chirogenesis and Photochirogenesis: Colloidal Polymers Meet Chiral Optofluidics

Michiya Fujiki

Division of Materials Science, Graduate School of Science and Technology, Nara Institute of Science and Technology, 8916-5 Takayama, Ikoma, Nara 630-0192, Japan; fujikim@ms.naist.jp

**Citation:** Fujiki, M. Resonance in Chirogenesis and Photochirogenesis: Colloidal Polymers Meet Chiral Optofluidics. *Symmetry* **2021**, *13*, 199. https://doi.org/10.3390/sym13020199

Academic Editors: Rui Tamura and Ivan Fernandez-Corbaton
Received: 15 November 2020
Accepted: 18 January 2021
Published: 26 January 2021

**Publisher's Note:** MDPI stays neutral with regard to jurisdictional claims in published maps and institutional affiliations.

**Copyright:** © 2021 by the author. Licensee MDPI, Basel, Switzerland. This article is an open access article distributed under the terms and conditions of the Creative Commons Attribution (CC BY) license (https://creativecommons.org/licenses/by/4.0/).

**Abstract:** Metastable colloids made of crystalline and/or non-crystalline matters render abilities of photonic resonators susceptible to chiral chemical and circularly polarized light sources. By assuming that μm-size colloids and co-colloids consisting of π- and/or σ-conjugated polymers dispersed into an optofluidic medium are artificial models of open-flow, non-equilibrium coacervates, we showcase experimentally resonance effects in chirogenesis and photochirogenesis, revealed by gigantic boosted chiroptical signals as circular dichroism (CD), optical rotation dispersion, circularly polarized luminescence (CPL), and CPL excitation (CPLE) spectral datasets. The resonance in chirogenesis occurs at very specific refractive indices (RIs) of the surrounding medium. The chirogenesis is susceptible to the nature of the optically active optofluidic medium. Moreover, upon an excitation-wavelength-dependent circularly polarized (CP) light source, a fully controlled absolute photochirogenesis, which includes all chiroptical generation, inversion, erase, switching, and short-/long-lived memories, is possible when the colloidal non-photochromic and photochromic polymers are dispersed in an achiral optofluidic medium with a tuned RI. The hand of the CP light source is not a determining factor for the product chirality. These results are associated with my experience concerning amphiphilic polymerizable colloids, in which, four decades ago, allowed proposing a perspective that colloids are connectable to light, polymers, helix, coacervates, and panspermia hypotheses, nuclear physics, biology, radioisotopes, homochirality question, first life, and cosmology.

**Keywords:** symmetry breaking; biomolecular handedness; circular dichroism; circularly polarized luminescence; non-equilibrium; colloid; absolute asymmetric synthesis; conjugated polymer; dissipative structure

## 1. Introduction—Historical Backgrounds, Knowledge, Then, and Now

*1.1. What Is the Life?—Open System, Negative Entropy, Non-Equilibrium Thermodynamics*

Humankind has been thinking about life for a long time; where did life come from? Where will life go? Does life exist on earth only? So far, many scientists have postulated several scenarios throughout the primordial eons to address these curious questions. Regarding the first life on earth, the origin of biomolecular handedness remains an unanswered question in the scientific community [1–41].

In 1944, Erwin Schrödinger, one of the leading quantum physicists in his day, wrote an essay book titled *What is Life? With Mind and Matter and Autobiographical Sketches* [1]. Entropy inevitably increases only in a closed system-like universe as the consequence of the second law of thermodynamics. Living matters, thus, cannot elude conventional physics laws, but should involve other physics laws. To avoid a thermodynamic equilibrium by a spontaneous decay, living matters have to acquire and maintain negative entropy in an open system. Because living organisms exist as metastable states, metabolism that is a biological term of maintaining negative entropy is needed through all processes of eating, drinking, breathing, and, in case of plants, assimilating. To my knowledge, the two key concepts are rarely stated in most chemistry textbooks and not lectured in schools and colleges, although Schrödinger was the first to invoke an importance of these concepts.

The chemistry textbooks and lectures teach us positive entropy, close system, enthalpy-driven chemical reaction, and Arrhenius plot, but do not involve quantum tunneling, resonance, hierarchy from elemental particles, atoms, and molecules, four fundamental forces in nature, parity violation in weak nuclear force, the origin of homochirality, and the origin of life.

In 2002, Soffer et al. stated mathematically that non-equilibrium thermodynamics and equilibrium thermodynamics, respectively, correspond to the excited-state and the ground-state solutions of a Schrödinger wave equation [39]. In 2017, by referring to Schrödinger's notes, Ornes described the core concept, concerning how non-equilibrium thermodynamics are connected to the mystery of life and other fields [40]. Non-equilibrium biological systems play key roles in gene transcription, diffusion in cytoplasm, transporting substances between cells with molecular motors, and cell signaling. Moreover, non-equilibrium phenomena are commonly recognized in the fields of fluid dynamics, plasma physics, meteorology, and astrophysics.

Likewise, an issue of left–right preference in the biomolecular handedness in the ground and excited states should involve these key concepts. When an equal probability of a simple chiral molecule, such as L- and D-alanine is considered, two possible states exist, corresponding to an increase in entropy. Living organisms only comprising optically active ingredients on Earth are possible to exist under open-flow conditions of energy and chemical resources because the metastable life requires low-entropy foods with chirality, and harvests solar/thermal energy [1,10,17]. If life existed on earth in the past, one of the greatest mysteries is whether stereogenic centers and/or stereogenic bonds in Hadean and Precambrian eons are identical to those of the on-going life on earth or are the extraterrestrial origin [22–38]. A conclusive answer to the question appears difficult because any fossil records of biomolecules and biopolymers associated with their handedness were decomposed. The snowball earth hypothesis, however, claims that prokaryotes inhabited the Precambrian eon for ~2 billion years [27,28]. Researchers in archaeological bacteriology have accepted that lighter $^{12}C$-containing stuffs are enriched in living organisms for their entire lifetimes. Indeed, the fact that $^{13}C$ in $^{13}C$-/$^{12}C$-isotopic ratio is significantly depleted would be evidence that the methanogenic microbes existed in the Archaean eon 3.5–3.8 billion years ago, associated with evolution of geodynamo of Earth, although Earth was born 4.54 billion years ago already [33–36].

Characterization of isotopic ratios and L-D ratios is powerful to discuss the origin of amino acids and carboxylic acids on Earth whether it is extraterrestrial origin or terrestrial origin. In 1987, Epstein et al. invoked that amino acids and carboxylic acids extracted from the Murchison meteorite were extraterrestrial in origin from isotopic ratios of D in D/H and $^{15}N$ in $^{15}N/^{14}N$ [37]. In 1997, Engel and Macko supported the extraterrestrial origin of amino acids and carboxylic acids in the meteorite from the analysis of the isotopic ratios and L-D ratios of alanine and glutamic acids [38].

Regarding the origin of biological homochirality associated with the first life on Earth, several scenarios are postulated. For example, the possible scenarios are classified to: (i) extraterrestrial and terrestrial origins; (ii) the by-chance and the necessity mechanisms; (iii) physical force origins and chemical substance origins. Importantly, all chemical processes and reactions should obey the physics laws regardless of extraterrestrial and terrestrial origins and chance-and-necessity mechanisms. The only exception would be that these physics laws, and chemistry, cannot apply to black holes.

*1.2. Physical Advantage Factors for Left–Right Asymmetry*

Goldanskii et al. comprehensively reviewed that 13 physical advantage factors called $g^*$ can cause a left–right preference to address the homochirality question at molecular and polymeric levels [42–44]. The $g^*$-values defined in physics may be a half of the Kuhn's anisotropy or Kuhn's dissymmetry ratio, $g$-values, defined in chemistry. The capabilities of chiral physical forces can access primitive molecules, followed by polymerization processes, which enable the homochiral world on terrestrial and extraterrestrial conditions. The global advantage factors are universal, deterministic for the biomolecular handedness, while the local advantage factors are mirror-symmetric, hence, provide the by-chance mechanism. Among the advantage factors, the $g^*$ of circularly polarized (CP) light source is considerably high on the order of $10^{-4}$–$10^{-2}$.

On the other hand, the $g^*$ for handed $\beta^-$-electron is extremely small, only $10^{-9}$–$10^{-11}$, while the values of $g^*$ for: (i) static magnetic field with linearly polarized light, known as the Faraday effect; (ii) Coriolis force (rotation) with static electric filed and static magnetic field; and (iii) Coriolis force with static magnetic field and gravitational field are on the order of $10^{-4}$. Coriolis force with gravitational force is a mirror symmetric chiral force, corresponding to the vectoral hydrodynamic flowing, often called swirling flow and vortex flow. Handed weak neutral current provides the smallest $g^*$ on the order of $10^{-17}$. Static magnetic field, static electric filed (so-called Stark effect), and gravitational field alone do not induce the left–right preference. CP-light, Coriolis force, static electric filed, static magnetic field, and gravitational field are parity-conserving mirror-symmetrical physical forces, while a weak neutral current and $\beta^-$-electron are parity-violating handed physical sources.

In 1953, Frank treated, mathematically, the first seminal bifurcation model account for a spontaneous mirror symmetry breaking in terms of evolution of biological homochirality [45]. For visibility and clarity, Goldanskii et al. illustrated two bifurcation models connecting to double-well and single-well potential curves [42–44]; a mirror-symmetric bifurcation model with a mirror-symmetric double-well potential curve and a non-mirror-symmetric bifurcation model with a non-mirror-symmetric double-well potential curve. Based on these bifurcation models associated with the $g^*$ values discussed above, Goldanskii et al. showed two scenarios, allowing us to well recognize both the bi-chance and necessity mechanisms for the spontaneous mirror symmetry breaking. Knowing the fluctuation behaviors around these bi-furcation points with those $g^*$ values, namely, a phase transition characteristics from the single-well to the double-well or from the double-well to the single-well is the key to rationally design the left–right preference, followed by the homochirality systems [7,8,46].

1.2.1. Circularly Polarized Light Source

Historically, in 1874, LeBel, and in 1894, van't Hoff, proposed independently a possibility of absolute asymmetric synthesis (AAS) using $r$- and $l$-CP light as chiral physical source [46–49]. A half century after their predictions, in 1929, Kuhn and Broun experimentally realized their conjectures as photodestruction mode AAS [46]. Their pioneering work prompted allowed many researchers to investigate AAS for a century because expensive chiral chemical substances are no longer needed [47–50]. However, most researchers have long believed that $l$-CP light produces left-hand (or right-hand) molecules preferentially and vice versa because the product chirality is determined solely by the hand of CP light. The AAC with $r$- and $l$-CP light covers absolute asymmetric photosynthesis, photodestruction, and photoresolution modes [47–49]. Generating left–right preference endowed with CP-light was coined photochirogenesis by Yoshihisa Inoue as the key concept of CP light–matter interactions in 1996 [7,8,51].

In the 1970s, Calvin et al. reported an anomaly of an excitation wavelength dependent CP light-driven photosynthesis mode AAS recognized as a switching product chirality of [8]-helicene in homogeneous toluene solution [52]. In 2014, Meinert et al. reported the wavelength-dependent CP light-driven photodestruction mode AAS revealing switching chirality when *rac*-alanine film was decomposed upon irradiation of two vacuum-UV light sources (184 and 200 nm) [53]. A recent development of wavelength-dependent CP light driven photoresolution mode AAS employed by us will be given in Section 4.5 [54–56].

1.2.2. Static Electric Field

According to recent works, with static electric fields, so-called Stark effects, in the absence of magnetic field, Auzinsh et al. reported chirogenesis as emission modes on the order of $g$ = 0.20 (10% for circularity) at $^7D_{3/2}$ state and $g$ = 0.15 (7.5% for circularity) at $^9D_{3/2}$ state of parity-violating, paramagnetic Cs vapor placed between two mirrored electrodes [57]. On the other hand, Datta et al. indicated an occurrence of lowering molecular symmetry of achiral diamagnetic aromatic molecules by a computer simulation. Coronene, a 6-fold highly symmetric molecule, undergoes $D_{6h} \rightarrow C_2$ distortions via vibrational instability upon application of sufficiently intense static external electric field [58]. The electric field induced structural distortion is understood as resulting from excess charge accumulation of planar rings circumvented by symmetry lowering. Contrarily, tribenzopyrene, a contorted polyaromatic compound due to a repulsion by two pairs of syn H-atoms, undergoes, spontaneously, $C_{2v} \rightarrow C_2$ distortions in the absence of the electric field. Coronene, tribenzopyrene, and other fused aromatics are assumed to be the representative polycyclic aromatic hydrocarbons, so-called PAHs, which are abundant in the universe [59]. Because $C_2$-symmetric molecules are chiral, a naive question remains to be elucidated experimentally, whether $C_2$-coronene and $C_2$-tribenzopyrene exist as racemic mixtures or single enantiomers in connection with the homochirality question in the universe. Collision-free, gas-phase circularly polarized luminescence (CPL) spectroscopy may allow us to provide an answer to this question.

### 1.2.3. Static Magnetic Field

With the origin of biological chirality in mind, Pasteur was the first to attempt a chemical process called enantioselection, showing whether left-handed or right-handed molecules are produced under magnetic field, but failed because the magnetic field is a pseudo-vector that cannot be coupled with molecular chirality [60]. In 1994, Zadel et al. published an astonishing paper titled *Enantioselective Reactions in a Static Magnetic Field* [61]. Several research teams attempted to reproduce these results immediately, but failed [62,63], because, prior to the experiments, in the solutions, under 2.1 Tesla of static magnetic field (NMR instrument) in the paper, all reaction solutions were in advance design and manipulated [64]. Static magnetic field on the order of 2 Tesla does not induce any detectable enantioselective reaction in a fluidic solution under non-restricted Brownian molecular motions. Uniform static magnetic fields do not induce molecular chirality because of the time-odd, axial vector [60,65]

Naaman and Wadeck reviewed theory and experiments of chiral-induced spin selectivity (CISS), where ordered films of chiral molecules on surfaces can act as electron spin filters [66]. By applying the idea of CISS, in 2018 and 2019, Naaman and his international team succeeded in enantiospecific crystallization of three L-/D-amino acids and thiolated L-/D-alanine-based helical oligomers at the magnetized surface with a gold-coated ferromagnet cobalt film [67,68]. Enantioselectivity is determined by north-up and south-up geometry of the magnet and the nature of chiral substances. An attractive force between electrons in the substrate and in the molecules is on the order of several tens of kJ mol$^{-1}$ when a molecule surface distance is 0.1–0.2 nm [68]. Brownian motion of floppy molecules is considerably restricted at the molecule surface interface, followed by a great suppression of the molecular motions during crystallization at the surface. Since a magnetic field is a short-distance force compared to the electric field, a proximity effect of the molecule at the magnetized surface should be considered.

In 1955, Akabori proposed the polyglycine hypothesis—foreproteine as the origin of protein homochirality on Earth [22]. The hypothesis involves three steps: (i) formation of aminoacetonitrile from formaldehyde, ammonia, and hydrogen cyanide, which were ubiquitous in primordial Earth; (ii) polymerization of aminoacetonitrile at a solid surface, such as a Kaolinite that is a two-dimensional aluminosilicate $Al_4Si_4O_4(OH)_4$, followed by hydrolysis to yield polyglycine and ammonia; (iii) introduction of side chains to polyglycine. The experimental results showed that: (i) methylene groups of polyglycine are adsorbed on Kaolinite; and (ii) glycyl residues are converted to serine and threonine residues reacted with formaldehyde and acetaldehyde, respectively. Although achiral glycine has two equal C-H bonds, the two C-H bonds at each residue of helical oligoglycine and polyglycine are no longer equal. The two C-H bonds at each glycine residue of the helix may reveal different reactivity toward chemical species. Akabori conjectured that the origin of all L-selectivity in amino acids of proteins results from the by-chance mechanism because there is no chirality in Kaolinite. The hypothesis, followed by experiments, were reported just before the groundbreaking fever in 1956–1958 that the parity in the β-decay process of $^{60}$Co is violated, though a left–right equality of enantiomers was a concrete common sense in chemistry in those days [69–76].

Knowing the magnetic minerals and the first geodynamo on Earth is an important clue how Earth's core, atmosphere, biomolecular chirality, and life evolved [33–36]. In 2015, Tarduno et al. claimed that, by analyzing inside of zircon crystals, which were indestructible for nearly 4.4 billion years, in Western Australia, Earth's magnetic field had evolved already from more than 4 billion years ago [33]. Likewise, from the oldest rocks in South Africa, Earth's magnetic evolved around 3.5 billion years ago [34]. Moreover, a recent work indicates that, from magnetic minerals in ancient Greenlandic rocks, Earth's magnetic field arose at least 3.7 billion years ago [35,36], enabling to magnetically shield primitive life and chiral substances evolved on Earth.

The CISS theory [66], along with the experimental results [67,68] proven by Naaman et al., encourages to test whether: (i) helical oligoglycine and polyglycine are adsorbed vertically at the magnetized surface; (ii) the two C-H bonds at each residue reveal a different chemical reactivity toward several chemical species in the absence and presence of UV-light around 190–220 nm; and (iii) enantioselectivity is determined solely by north-up and south-up geometry.

In 2019, Stevenson and Davis proposed a possibility of magnetophoresis that radical species of D- and L-substances are separable and sortable under an ultrastrong magnetic field gradient of $>4.4 \times 10^9$ Tesla [77]. Such an ultrastrong magnetic field and gradient can no longer obey the standard Maxwell equations. Paramagnetic chiral molecular species in the universe may be passing through in the vicinity of magnetar, which are newborn neutron stars spinning very fast that generate the ultrastrong magnetic field. Kouveliotou et al. think that more than 100 million magnetars are wandering through the interstellar universe and trillions of organic paramagnetic materials are imbedded in cold molecular clouds [78]. The ultrastrong magnetic field causes vacuum birefringence, photon splitting, scattering suppression, and distortion of atoms [78]. Although magnetic field gradients are widely utilized in NMR imaging, so-called MRI, the feasibility of the ultrastrong magnetic field gradient is challenging.

1.2.4. Hydrodynamic Flowing as a Model of Coriolis Force with Gravitational Force

When artificial molecular chromophores and luminophores are dispersed in the fluidic medium, hydrodynamic swirling flowing clockwise (CW) and counterclockwise (CCW) are experimentally testable models to validate the origin of the gravitational field's left–right preference [79–83]. The vectoral hydrodynamic flowing is often called as swirling flow and/or vortex flow. The unidirectional hydrodynamic flowing in the CW or CCW direction imparts the left–right preference at the molecular and supramolecular levels. The left–right preference in the northern hemisphere may be opposite to that in the southern hemisphere due to parity-conserving physical force on Earth.

In 1993, Ohno et al. reported chirogenesis of circular dichroism (CD)-active J-aggregates from achiral free-base porphyrin derivative in acidic water solution by mechanically swirling in CW and CCW directions [79]. They observed two clear couplet-like CD bands around 410–450 nm due to Soret band and 480–500 nm due to Q-band. The couplet-like CD band profiles were inverted by choosing the CW or CCW direction during the propagation of J-aggregate. Very weak non-couplet CD band around 480 nm, however, was recognized under a stagnant condition. Obviously, the hydrodynamic flowing appears responsible for the chirogenesis, but a statistical analysis was not performed yet. It is unclear whether this event occurs using the specific porphyrin derivative at a specific laboratory.

In 2001, Ribó et al. confirmed, by the statistical analysis from nearly a hundred of independent experiments that macroscopic swirling forces in the CW and CCW directions indeed generate homochiral aggregates from water soluble achiral porphyrin derivatives that slightly differ from the porphyrin above [80]. The hydrodynamic flowing indeed acted as a trigger of the spontaneous homochiral *l*-or-*r* aggregation [81].

Although water-soluble porphyrin derivatives carrying multiple phenyl rings at meso-positions are directly connected to biomolecular substances, the idea of hydrodynamic flowing is applicable to chain-like synthetic polymers and several chromophore/luminophore during their association processes that feel the handed force of the hydrodynamic flowing, regardless of the northern and the southern hemispheres on Earth and other exoplanets. The author assumes that, since unsubstituted free-base porphyrin framework is very floppy, these porphyrin derivatives substituted with tetraaryl groups at meso-positions may adopt dynamically twisting two structures, between left and right, in a fluidic solution at ambient temperature.

In line with the bifurcation scenario, in 2011, Okano et al. succeeded in chirogenesis from achiral rhodamine B as a red luminescent molecular probe doped to a water-soluble sol-gel polymer, revealed by CPL spectroscopy [82]. Based on a statistic analysis, the sign in CPL signals at 580 nm was controlled by the CW/CCW motions of hydrodynamic flowing in a cuvette. The critical temperature ($T_c$) for gelation was controlled by tuning the concentration of the polymer in aqueous solution. The polymer solution was in a sol state with a low viscosity, while stirring above the $T_c$, while the solution below the $T_c$ spontaneously underwent a non-flowing gel state in a very high viscosity.

In 2018, by mimicking submarine rock micropores at primordial Earth, Liu and collaborators designed a sophisticated microfluidic chirogenesis system [83]. The fluidic system was composed of ten pairs of inclined microchambers to efficiently generate pairs of CW and CCW microvortices using achiral 3-fold molecular symmetric aromatic molecules that possessed supramolecular gelation capability. By a statistical analysis based on 56 totally independent experiments, they ascertained that the microfluidic chirogenesis system works very efficiently by simulating a behavior of hydrodynamic flowing in the microchamber. They re-confirmed that the microfluidic system is viable for a very rapid chirogenesis within 1 ms of the water-soluble porphyrin derivative. These results should shed light on the origin of the biomolecular homochirality; how oceanic vortices play a critical role in protein folding and self-organization in primordial Earth.

### 1.2.5. Static Magnetic Field with Polarized Light

Static magnetic field with linearly polarized light and static magnetic field with unpolarized light, which are called magneto-optical effects, including Faraday and polar Kerr effects, are possible to be chiral physical sources, enable to induce mirror symmetrical, left–right imbalance in the ground and photoexcited states. The left–right chiroptical preference from achiral substances in the ground and photoexcited states is determined by north-up and south-up magnetic fields to propagation of incident light, called magnetic circular dichroism (MCD) [84–87] and magnetic circularly polarized luminescence (MCPL) [88–91], respectively. The phenomena are closely connected to Zeeman splitting of degenerate molecular orbitals. Recently, Imai, Fujiki, and coworkers verified mirror symmetrical MCPL characteristics from diamagnetic achiral organic and racemic lanthanide luminophores. These luminophores radiate *l*- and *r*-CP light upon unpolarized light at north-up and south-up Faraday geometry [92–94]. However, it is obscure whether mirror symmetrical MCPL arises from Zeeman splitting of degenerate molecular orbitals in the ground states or in the photoexcited states because a comparison with the corresponding MCD spectra is not elucidated yet. Nevertheless, in analogy with the first report of Raman scattering in 1928 [95], the MCPL characteristics teach that the handed light emission under external magnetic field is becoming a new type of secondary radiation sources, *l*- and *r*-CP light, in the absence of any chemical chirality, when achiral luminophores are, by-chance, placed on magnetized substances (even on ubiquitous), but a weak geomagnetic field of

~$3 \times 10^{-5}$ T. In that case, the magnetic fields should satisfy north-up and south-up Faraday geometries. A magnetic field-caused CP light source from achiral luminophores may be ubiquitous in the universe, and was therefore the source of circularly polarized radiation as seeds of molecular chirality in the past (and now all over universe).

Rikken and Raupach designed a static magnetic field-driven photoresolution molecular system using racemic $K_3Cr^{III}$(oxalate)$_3$ complex in water upon irradiation of Ti:sapphire laser at 696 nm [96]. They demonstrated an occurrence of mirror-symmetrical photoresolution from the complex, in which a preferential chirality is determined by north-up and south-up Faraday geometries. Recently, Sharma calculated (based on the MCD effect) that unpolarized ultraviolet sunlight, so-called UV-C region, combined with atmospheric paramagnetic oxygen diluted with carbon dioxide can possibly trigger an initial enantiomeric excess (ee) to determine the biomolecular handedness under the geomagnetic force in Archaean Earth [97]. Notably, when the UV-C light destructs several biomolecular constituents, a preferential chirality between purine and pyrimidine nucleosides depends on a partial pressure of carbon dioxide. However, a preference in the product chirality of amino acids is always L-form. Thus, a static magnetic field with unpolarized light becomes one candidate for efficient photodestruction, and photoresolution modes in racemic amino acids and nucleosides in water disregard of terrestrial and interstellar conditions.

### 1.2.6. Longitudinally Polarized Left-Handed $\beta$-Electron and Right-Handed $\beta$-Positron

In 1956, Lee and Yang theoretically indicated a possibility of parity-violation in certain nuclear reactions. In 1957, Wu et al. experimentally confirmed the parity-violation in the reaction $^{60}Co \rightarrow ^{60}Ni + e^- +$ anti-$\nu_e$ using $\beta^-$-decay of radioisotope $^{60}Co$. Likewise, a parity-violation in $\beta^+$-decay process that the $^{58}Co \rightarrow ^{58}Fe + e^+ + \nu_e$ was confirmed, while positron and anti-$\nu_e$ were antimatters of electron and $\nu_e$, respectively [69–76]. The parity-violation relies on a weak charged current of the radioisotopes. In 1958, Goldhaber et al. determined that helicity of massless electron neutrino is left-handed from the resonance scattering experiment of $\gamma$-ray at 960 keV that an excited nucleus $^{63}Sm^{152}$ (1− state) spontaneously radiates a handed $\gamma$-ray to relax a ground state nucleus $^{62}Sm^{152}$ (0+ state) [74].

In 1959, this discovery prompted Vester and Ulbricht to propose so-called Vester–Ulbricht (V–U) hypothesis: parity-violating, spin-polarized left-handed $\beta$-electron causes circularly polarized bremsstrahlung that preferentially decomposes D-amino acids and left-handed DNA precursors, thus remaining *l*-amino acids and right-handed DNA on Earth [98,99]. In 1984, Bonner reported the failure of the V–U hypothesis—that several racemic amino acids cause radioracemization when polarized $\beta$-electron Bremsstrahlen from radioisotopes $^{90}$Sr-$^{90}$Y, $^{14}$C, and $^{32}$P is applied [100].

In 1982 and 1984, Zitzewitz et al. utilized low-energy, right-handed $\beta$-positron in place of left-handed $\beta$-electron, and reported a weak asymmetry of leucine while cystine and tryptophan had little V–U effects [101,102]. In 2014, Dreiling and Gay confirmed successfully that the V–U hypothesis is valid, based on the detection of a preferential L-D decomposition of racemic 3-bromocamphor [103]. In 2018, Dreiling et al. confirmed volatile ester derivatives of DL-amino acids and DL-sugar molecules with low-energy $\beta$-electron and $\beta$-positron may be suited to verify the V–U hypothesis.

### 1.2.7. Handed Weak Neutral Current Mediated by $Z^0$ Boson

The $\beta^-$- and $\beta^+$-decay processes are mediated by massive $W^-$ boson (80.4 Gev) and $W^+$ (80.4 Gev) boson, respectively. On the other hand, parity-violating weak neutral current mediated by massive $Z^0$ boson (91.2 Gev) is ultimately a handed chiral physical bias for sub-atoms, atoms, molecules, supramolecules, polymers, colloids, crystals, inorganics, possibly our life, and upward disregard of terrestrial and extraterrestrial origins. According to the electroweak theory, massive three $W^-$, $W^+$, $Z^0$ bosons and massless photon (boson, light) can be unified to the same family above ~$10^{15}$ eV [104]. Upon rapid cooling of our universe, the parity-violating weak force and parity-conserving electromagnetic force are separated at ~$10^{15}$ eV by the bifurcation mechanism.

Since the discovery in 1956–1957 of the parity-violating $\beta^-$- and $\beta^+$-decay processes [69–76], people assumed that the handed weak neutral current is responsible for deterministic factor of L-amino acid and D-ribose [3,6,105]. Parity-violation hypothesis in atoms, so-called atomic parity-violation, is established theoretically and experimentally, based on absorption and photoluminescence spectroscopy of collision-free atomic vapors [106–112]. As for the question: "Is parity violated in atoms?"—most atomic, elemental particle, and cosmological physicists say, no question, it is common sense. On the other hand, to the similar question of "Is parity violated in close-shell chiral molecules?"—most chemists believe it is skeptical, no question, nonsense, a big burden in one's research carrier and honor.

Up until now, many scientists in chemistry and molecular physics argued theoretically a possibility of parity-violation in molecules, in relation to the origin of biomolecular handedness on Earth, often called molecular parity-violation hypothesis. Parity-violation energy difference (PVED) of L- and D-molecules predicted theoretically is ultrasmall on the order of $10^{-8}$–$10^{-14}$ kcal mol$^{-1}$ or $10^{-9}$–$10^{-15}$% $ee$. In 1983, Mason and Tranter showed that, from calculation of the PVED with Slater-type-orbitals (STO) with 6-31G basis set, L-alanine and L-polypeptide in $\alpha$-helix and $\beta$-sheet are energetically stabilized compared to the corresponding D-alanine and D-polypeptide, due to parity violating weak nuclear current [3,113]. However, the PVED is ultrasmall on the order of $10^{-23}$ kcal mol$^{-1}$. Most chemists, therefore, think that the molecular parity-violation hypothesis is doubtful and skeptical. Even if the hypothesis is true, it is experimentally impossible to prove by ordinary spectroscopy and enantioseparation column chromatography.

Most theories are treating ideal isolated molecules, including simple amino acids, carbohydrates, hypothetical, and realistic molecules, including heavier atoms in a vacuum at absolute temperature of zero-Kelvin. Additionally, the molecular parity-violation theorists always require experimentalists, who directly detect the tiny PVED for a pair of left- and right-molecules and well-defined oligomers in a collision-free condition, such as a reduced pressure, using precision high-resolution spectroscopy. Theorists think that solutions, solids, aggregates, gels, and polymers should be excluded in testing the molecular parity-violation hypothesis. Most theories focused on the PVED at the electronic, vibrational, and ro-vibrational ground states [3,113–117]. Experimentalists attempted to detect the PVED as absorption mode, using several sophisticated high-resolution spectrometers [118]. In 2003, Berger calculated the PVED of formyl aldehyde, HFC = O, in the electronically excited states that are suited to experimentally test the PVED hypothesis because of an enhanced PVED amplitude from a well-defined photoexcited parity [119].

In an analogy of Kasha's rule in photochemistry, radioisotopes $^{60}$Co and $^{58}$Co as excited states of the corresponding stable isotopes $^{60}$Ni and $^{58}$Fe, respectively, allowed to unveil the parity-violation in their $\beta^{\pm}$-decay radiation processes. Similarly, detecting atomic parity violation as emission mode from atomic vapors is more obvious, rather than absorption mode [109–112]. These spontaneous radiation processes to the ground state, connecting to parity, commonly arise from the far-from, non-equilibrium excited state.

The $\beta^{\pm}$-decay process, atomic parity violation as emission mode, and Berger's theory stimulated the author and coworkers to measure spontaneous radiation processes from $S_1$ state of nearly sixty non-rigid rotamers as achiral and/or racemic luminophores homogeneously dissolved in achiral solvents (not optofluidic solvents) in the photoexcited states to detect difference as emission modes between L-D molecular state upon excitation of unpolarized light [120,121]. All of these rotamers showed (−) sign CPL signals in the UV-visible region, suggesting temporal generation of energetically inequal, non-racemic structures under conditions of far-from, open-flow, non-equilibrium $S_1$-state, realized only upon a continuous irradiation of incoherent unpolarized light. Unpolarized light is an equal mixture of left- and right-CP light. For comparison, unsubstituted achiral fused aromatic luminophores that adopt rigid planar structures did not show CPL signals. Furthermore, D- and L-camphor, an enantiomeric pair of rigid chiral XYC = O dialkyl ketone luminophores, showed ideal mirror-image CPL spectra. From these comprehensive results, the author and coworkers invoked that the parity of non-rigid molecular rotamers in the photoexcited is violated, followed by radiating (−) sign CPL signals without any exception [120,121].

Currently, experimentally testing the molecular parity-violation hypothesis remains a hot topic and very challenging. Collision-free, gas-phase CPL spectroscopy may be a useful tool to provide a definitive answer to this long-standing question. Although this topic is not a major issue in this paper, the author, his coworkers, his students, his researchers, and his technical staff are aware of the facts that non-ideal mirror-image CPL and CD spectral characteristics can be often seen in our colloidal polymers dispersed in optofluidic media with tuned RIs. Some of the readers may be aware of the non-ideal mirror-image CPL and CD spectral characteristics in this paper. The author conjectures that these non-ideal mirror-image chiroptical characteristics from the colloidal polymers in tuned optofluidic media could be connected to certain mechanisms, in an enhancement in weak neutral currents at molecular and polymer levels. Those who are interested in this topic should read related parity-related reviews, monographs, and original papers, shown above.

### 1.3. Extraterrestrial Origins of Life and Chirality—Panspermia Hypotheses

Apart from the physical aspects in the previous sections, the fascinating, awesome extraterrestrial scenarios might be several panspermia hypotheses [122,123]. Panspermia means *seeds everywhere* in a Greek word. The panspermia hypothesis assumes that seeds of life ubiquitously exist in the past and, even now, all over universe, allowing to spread from one to another spaces and locations. The hypotheses may invoke us that seeds of chirality are ubiquitous in the past and, even now, all over universe.

In the late 19th century, Louis Pasteur, William Thomson (Lord Kelvin), and Hermann von Helmholtz conjectured extraterrestrial origin of life (and biological chirality). In 1903, Svante Arrhenius, Swedish physicist and chemist, proposed the radio-panspermia hypothesis; certain seeds of life with sizes of 200–300 nm traveled by radiative pressure of solar light and stellar light, landed on Earth [122]. In the 1960s–1980s, Hoyle and Wickramasinghe hypothesized cometary panspermia based on detection of astronomical radiation and reflectance in the range of 2.5–4.0 μm (~3000–3800 cm$^{-1}$) from dust coma of Comet Halley and Comet 67P/Churyumov-Gerasimenko [123]. The cometary panspermia, including comets, interstellar dusts, and asteroids, can carry several viruses and microbials, which landed on Earth from the primordial era (and even now) [123]. The scenarios led to propose further lithopanspermia (often referred to interstellar panspermia) and ballistic panspermia hypotheses (often referred to interplanetary panspermia). In lithopanspermia, impact-expelled rocks and meteorites from a surface of planet acts as spaceships to spread biological materials and microbials from one solar system to another solar system, while, in ballistic panspermia, the same events occur within the same solar system.

In 2013, Kawaguchi, Yamagishi, and coworkers hypothesized massapanspermia that microbial aggregates are transferrable between planets [124]. In 2020, they confirmed that aggregates of *Deinococcus radiodurans* (a representative radioresistant bacteria) can survive under the very severe space environment for 2–8 years, enabling the interplanetary travel that delivers the seeds of life and related chemical stuffs for many years [125]. The new hypotheses might encourage astrobiologists and astrochemists to pursue the possibility of the extraterrestrial life and related chemical substances, including liquid water and chiral molecules conducted by sophisticated spacecrafts, orbiting telescopes, and high-resolution spectrograph on Earth, followed by recent triumphs in astroscience.

For example, astroscientists confirmed plumes of water vapor and are now confident about the existence of water ocean on Europa (that is one of Jupiter's moons), with help from auroral emissions of hydrogen and oxygen by the Hubble space telescope (HST, National Aeronautics and Space Administration: NASA), observation by the Galileo spacecraft (NASA), and direct detection of water vapor using near-IR spectrograph at the 10-m Keck Observatory (Hawaii) [126]. The Rosetta space probe (European Space Agency: ESA) detects prebiotic constituents and water on Comet 67P/Churyumov-Gerasimenko, one of the comets in the Jupiter family [127,128]. The Cassini spacecraft (NASA), the Saturn probe launched in 1997, discovers water under the surface of Enkelados, which is one of the Saturnian moons, with help from the quadrupole gravity measurement system [129]. JPL (Jet Propulsion Laboratory at NASA) reported that Titan, which is the largest Saturnian moon, possesses water comprising inorganic salts [130]. The Kepler space telescope (NASA) discovered over 2600 exoplanets; among them, numbers of Earth-like planets are hidden in habitable zones [131,132]. In 2016, astronomers discovered interstellar chiral propylene oxide generating in *SagittariusB2* of the Milky Way Galaxy by analyzing the unique absorption bands in the range of mm-wavelength of Parkes radio telescope at Australia Observatory [133]. A preference between L- and D-propylene oxides remains obscure because the telescope is not a polarized microwave spectrometer. In 2019, an international team from Japan and USA showed, for the first time, evidence of extraterrestrial origin $^{13}$C-enriched sugars from carbonaceous chondrites in two meteorites (NWA801 and Murchison) by means of gas chromatography–mass spectrometry (GC–MS) [134].

In December 2014, aiming to verify the panspermia hypothesis associated with the origin of life and molecular chirality on Earth, the Hayabusa-2 spacecraft (Japan Aerospace Exploration Agency: JAXA (Chofu, Tokyo, Japan)) launched to land on the surface of 'Ryugu', which is one of the carbon-rich asteroids. In December 2020, on the way to the next target (1998KY26, a rapidly spinning asteroid), the spacecraft delivering a small capsule, called Urashima's treasure chest, containing organic and gaseous ingredients for life, had just returned to Earth [135]. Importantly, the substances captured from the sub-surface of Ryugu did not contain any contaminations existing on Earth. Japanese researchers are analyzing all of the ingredients. The team anticipates evidence of extraterrestrial-origin molecular substances associated with preferential chirality if the substances are chiral.

*1.4. Terrestrial Origins of Life and Chirality*

In 1953 and 1959, Miller and Urey reported production of nearly forty amino acids including glycine, $\alpha$-alanine, $\beta$-alanine, aspartic acid, and so on, by applying electric discharge under reducing atmosphere containing $CH_4$, $NH_3$, $H_2O$, and $H_2$ [136,137]. The system is regarded as an open-flow, recycling chemical reaction system that is experimentally testable at ordinary lab level to mimic primitive Earth. Although the product chirality was not characterized at those days, the electric discharge was thought be, at least for the author, an unpolarized high-energy electromagnetic force that produces racemic amino acids. In 2013, Bada reviewed his insights of prebiotic chemistry under the volcanic condition [138].

In 2017, a Japanese team reported an astonishing result that thundering is generating $\gamma$-ray, called atmospheric photonuclear reaction [139]. The thundering acts as a naturally occurring accelerator, which might ubiquitously generate in primordial Earth, and is regarded as a realistic electric discharge in the Miller–Urey experiments [136,137]. The naturally occurring $\gamma$-ray is responsible for several photonuclear reactions. When $\gamma$-ray collides to the atmospheric $^{14}$N molecule, radioactive $^{13}$N ($t_{1/2}$ 598s) and neutron ($t_{1/2}$ ~890s) will be generated. The unstable $^{13}$N causing $\beta^+$-decay reaction produces stable $^{13}$C, positron, and neutrino. The neutron with $^{14}$N produces radioactive $^{14}$C ($t_{1/2}$ 5700±30yr). The unstable radioactive $^{15}$N ($t_{1/2}$ 122s) by capturing the neutron re-generates $\gamma$-rays. The Japanese team detected $\gamma$-rays at 0.511MeV regenerated by electron-positron annihilation. These scenarios infer that all electron, positron, neutron, and radioactive atoms may become parity-violating handed physical sources.

A non-naturally occurring radioisotope $^{60}$Co, which is synthesized from stable isotope $^{59}$Co by capturing one neutron only in a nuclear reactor, cannot contribute to the left–right imbalance on primordial Earth. However, naturally occurring, long-lifetime radioisotopes $^{235}$U, $^{238}$U, $^{232}$Th, $^{87}$Rb, and $^{40}$K, embedded to continental crust, mantle, oceanic crust, and sediment of Earth contribute in principle to the left–right imbalance through the $\beta^-$-decay channels radiating handed elemental particles and radiating $\gamma$-rays [140–142].

*1.5. Extraterrestrial Origins of Circularly Polarized Radiation Sources*

On the other hand, scientists have long argued that several kinds of circularly polarized electromagnetic sources, including $\gamma$-ray, X-ray, and vacuum-UV in the universe, are responsible for the biomolecular handedness [4–9,13–16]. Alternatively, a subtle imbalance of L- and D-amino acids as tiny chiral chemical seeds significantly catalyze asymmetric chemical reactions to yield carbohydrates with a high ee, leading to the handed biomolecular world [48]; a nearly racemic substance in $10^{-5}$% ee attained a nearly ~100% ee when the Soai reaction was regarded as a testable model of the chemical evolution of chirality in prebiotic eons [143,144].

When these substances are homogeneously dissolved in solutions, a probability of CP light (photon chirality) that reacts with racemic and prochiral biomolecular is, at most, one-chance with quantum efficiency of 1.0 by assuming the Beer–Lambert law. We assume that non-rigid, efficient, photonic resonators adaptable to photon chirality is possibly suited for recycling CP light–matter interactions, such as whispering gallery mode (WGM).

*1.6. Coacervate Hypothesis*

In the 1920s–1930s, Oparin [23] and Haldane [145] proposed the coacervate hypothesis, which is a prototype of living cells during the chemical evolution of life. The coacervate refers to sphere shape, cell-wall free colloidal droplets, from 1 μm and 100 μm in diameter, surrounded by fluidic water. The colloids comprising organic substances were postulated to spontaneously lead to a metabolism after a prolonged time. The hypothesis relies on a scenario of spontaneous self-organization arising from non-covalent interactions, such as hydrogen bonds, electrostatic, van der Waals, dipole–dipole, and so on. The hypothesis was mostly abandoned because the closed coacervate systems in an equilibrium state did not evolve living organisms.

In 2020, Lu and Spruijt demonstrated experimentally and theoretically how immiscible multicomponent droplets consisting of oppositely charged polyelectrolytes in the presence of several additives undergo a spontaneous formation of ca. 1–10 μm size multi-phase coacervates, with help from interfacial tensions and concentration of salts [146]. The multiphase coacervates dispersed in water containing salts spontaneously propagate to hierarchically self-organize cellular structures with time. The coacervates hypothesis is not an old-fashioned concept. A remodeled coacervates hypothesis sheds light on a new insight of how non-equilibrium colloidal particles in fluidic solutions propagate spatiotemporally [146]. In-situ observations of spatiotemporal propagation behaviors associated with a preferred handedness in chirogenesis of coacervates, using sophisticated circular dichroism (CD) and circularly polarized luminescence (CPL) microscopy, are challenging [147–152].

*1.7. Colloids*

Colloid science has a long history [153]. Colloids involve micelles, vesicles, shape-controlled polymers, molecular and macromolecular aggregates, emulsion of polymers, core-shell nanoparticles, spherulite, microgel, and possibly, coacervate. In 1919, le Chatelier hypothesized that the size of colloids ranges from 1 nm to 1000 nm [154]. In the 1920s, *makromolekül* (macromolecule) was categorized to one of the colloids because the makromolekül hypothesis proposed by Staudinger was controversial [155]. Although the concept was established in the 1930s, charged colloids made of macromolecules and/or polymers did not appear to be well-recognized in those days. It is obscure whether Oparin and Haldane recognized in those days that the coacervate hypothesis was intimately connected to colloid science, macromolecular/polymer science, and supramolecular science. In 1952, Terayama proposed polyion complexes in water of oppositely charged water-soluble polyelectrolytes, nowadays, widely known as the colloid titration method [156]. Non-charged co-colloidal polymers, however, did not appear to be well-recognized until recently. It is still unclear how co-colloidal chiral polymers propagate with time.

*1.8. Ostwald Ripening and Viedma Ripening*

In 1990, Kondepudi et al. found that, from saturated solution of $C_{3v}$-symmetric achiral sodium chlorate (NaClO$_3$) in water, L- or D-crystal in nearly 100% ee is dominantly generated in a beaker solely by mechanically stirring. Contrarily, no preference in the L-D crystal was obtained under stagnant condition [157]. The homochiral L- or D-world in the beaker is determined by the by-chance mechanism. Martin et al. indicated that damages in the first-generation crystals, led by the stirring, can produce numbers of the second-generation crystals in smaller size, which are responsible for the production of L- or D-crystal in a vessel [158]. The initial chirality in the first-generation crystals act as the seed of chirality in the second-generation crystals during mechanical stirring, leading to autocatalytic homochiral crystallization. It is noted that hydrodynamic convection flow as mimic model of Coriolis force was not responsible for the L-or-D crystallization [158].

These studies stimulated another autocatalytic homochiral crystallization, known as the Viedma ripening. In 2005, Viedma observed attrition-enhanced de-racemization phenomenon in a mixture of solid/liquid of enantiomorphous crystals using NaClO$_3$ [159]. In 2008, Viedma and coworkers developed this approach to evolution of homochiral crystal D- or L-aspartic acid in the presence of salicylaldehyde as a catalyst at elevated temperatures (90–160 °C) [160]. Enantiomerically pure aspartic acid that is rapidly racemized upon heating is assumed to be a realistic amino acid existing on the primordial Earth.

Possible mechanisms underlining should involve four steps [161]: (i) Ostwald ripening with an initial chirality with help from the gliding to introduce crystal damage/defects and non-racemic additives; (ii) enantiospecific aggregation to a larger aggregation with the same chirality; (iii) a breakage of the aggregates; and (iv) racemization. Ostwald ripening with chirality is a dynamic process of chirality crystal growth and dissolution due to the crystal-size dependent solubility. A larger size chiral crystal has a lower solubility due to a smaller specific surface area, conversely, a smaller size chiral crystal has a higher solubility due to a large specific surface area.

In 2007, Viedma tested the parity-violation hypothesis of whether handed crystals (L or D) using NaClO$_3$ and NaBrO$_3$ are generated [162]. Heavy Br atom was expected to boost the parity-violation effect obeying 5–6 power dependency of the weak neutral current based on a large spin-orbit coupling. Systematic experimental results of mirror symmetry breaking on a macroscopic level, however, did not support the parity-violation hypothesis. The results were ascribed to unidentified cryptochiral impurity existing in environmental conditions. The crystallization protocol is, possibly, very susceptible to external impurities as chiral seeds, rather than the weak neutral current effect.

These studies infer to us that, by controlling the dynamic process between chirality and non-chirality at the solid–liquid interface, an alternative heterogeneous chirogenic system comprising colloid and its dispersion solvent is possible, when proper colloidal polymers with chiral liquefied media are chosen.

*1.9. Optofluidics Connecting to Colloids and Circularly Polarized Light*

In 1986, Qian et al. reported the first laser emission from free-falling droplets doped with red-color dyes upon excitation of a 532 nm laser based on WGM mode, or morphological dependent resonance mode, whereas rhodamine 590 and rhodamine 640 were used as dyes [163]. In 2006 to 2010, the optofluidics, which is analogous to the corresponding solid-state optical devices, was coined for a conceptional fusion of integrated optics and microfluidics [164–167]. Microfluidics is an essential science of precisely controlling and manipulating fluidic medium in the range of ten-to-hundred μm in diameter. One can design unique μm-scale liquid-based optical devices with a greater flexibility; for example, it is easy to: (i) tailor several optical properties of the fluidic medium, such as refractive index (RI), wettability, and viscosity; (ii) construct optically smooth colloid–liquid interfaces; and (iii) confine massless light into an optical resonator. Moreover, Whitesides, Psaltis, and coworkers demonstrated that ultralow threshold dual-color lasing from rhodamine 560 and rhodamine 640 in 20–40 μm size droplets of benzyl alcohol ($n_D$ = 1.54) is possible when dispersing in $C_7F_{15}OC_2H_5$, which is a low RI fluorinated solvent with $n_D = 1.29$ [168]. Note that $n_D$ means an average RI value at 589.0 and 589.6 nm of doublet D-lines arising from vapor of atomic sodium. The lasing droplet device is realized based on the WGM scheme [163,168,169]. The idea of optofluidics was recently realized as Varioptic® from Corning® (Corning, NY, USA), which is a commercial product of variable focus liquid lens. The Fabry-Pérot resonator in a planar microfluidic geometry using Bragg-type grating reflectors shows a resonantly enhanced transmission with a sharp peak at a well-tuned RI in a fluidic medium [170].

In a light-harvesting plant system, stroma in chloroplasts is adaptable to any alterations, such as osmotic pressure, $Mg^{2+}/K^+$ ions, sunlight intensity, temperature, and so on [171–174]. The stroma may work as naturally occurring, self-tuning optofluidic medium. The underlying mechanism teaches that, to maintain the non-CP and CP light-driven photophysical and biological properties, the chirally assorted macro-colloids surrounding optofluidic stroma appear key [171–174]. Likewise, the adaptability and flexibility to any alterations in chemical and physical biases are crucial in the chemical evolution and propagation of life.

The noticeable advantages facilitate anyone to fabricate the low-reflection-loss chromophoric and/or luminophoric colloidal molecules and polymers with high RIs by surrounding a fluidic medium with a lower RI to resonantly boost chiroptical spectral responses in the ground and the photoexcited states. Based on the optofluidics, the colloidal polymers in the ground and photoexcited states as a function of RI of the surrounding solvents are rarely studied. We showcase several optofluidic effects of colloidal polymers in the ground and photoexcited states by tuning the RI of the solvents in the following.

### 1.10. Microdroplets and Aerosols—Prebiotic Chemical Reactors

Atmospheric μm-sized aerosols and μm-sized droplets called microdroplets in water offer unique chemical and photochemical reactors by concentrating molecules, which serve several key precursors, such as oligopeptides and ribonucleosides in the prebiotic Earth [175–177]. The μm-sized aerosols and μm-sized microdroplets are regarded as realistic models testable of the coacervate hypothesis on a laboratory scale.

Recently, researchers became aware of the fact that various chemical reactions and processes in the microdroplets are significantly accelerated by several orders of the magnitudes, compared with the corresponding reactions and processes in the bulk phase [178]. These reactions and processes confined in the microdroplets differ from the conventional chemistry in homogeneous solutions. The reactions involve protein unfolding [178], helix formation [179], phosphorylation in ribonucleosides [176,177], production of nanostructures [180], autoreduction [181], and so on. Although the oil/water, air/water, and silica/water interfaces are used as the platforms, the cause of the acceleration mechanisms in these microdroplets is as a matter of debate.

To address the issue, in 2018, Zare et al. designed—to experimentally and computationally study the distribution and photoluminescence (PL) polarization anisotropy of rhodamine 6G as a PL probe doped into microdroplet oils (3M™ Fluorinate® FC-40—a mixture of two perfluoroalkyl amines, $n_d = 1.29$) dispersed in water ($n_d = 1.33$) [182]. They found that, when a radius of the microdroplet increases and a concentration of rhodamine 6G decreases, the density of rhodamine 6G is significantly higher on the surface than in the center of the microdroplets and that the ratio of the surface density to that of the center grows. Moreover, PL polarization anisotropy on the surface of the microdroplets is significantly large, indicating that rhodamine 6G is well-aligned on the surface. The reduced entropy affects the significant change in the free-energy for the reactions [182].

The concept of optofluidics is applicable to photoexcited microdroplets and aerosols that are regarded as open-flow, non-equilibrium, volatile photochemical resonators. Relative RIs of organic microdroplets in water and aerosols in air are $n_d$(droplet)/$n_d$(water) = 1.4–1.6/1.33 and $n_d$(water solution)/$n_d$(air) $\geq$ 1.33/1.00, respectively. When prochiral substances in the photoexcited microdroplets and aerosols are concentrated on their surfaces, one can anticipate WGM-driven accelerated photochemical reactions in absolute asymmetric synthesis (AAS) or absolute photochirogenesis (APC) with a CP light source in the absence of an external magnetic field and unpolarized light in the presence of an external magnetic field.

*1.11. Our Hypothesis for Chirogenesis and Photochirogenesis*

As mentioned above, the multiple concepts of panspermia and coacervate hypotheses, $g^*$ value, bi-furcation model, circularly polarized (CP)-light driven photochirogenesis, optofluidics, and WGM stimulated us to investigate open-flow, non-equilibrium colloidal systems composing of artificial chain-like polymeric chromophores and luminophores dispersed in optofluidic biomolecular liquefied chiral terpenes, as chiral foods and beverages. Moreover, we explored a possibility of AAS or APC of the artificial coacervate as a suspension state in achiral organic solvents upon excitation of a wavelength-dependent CP-light source as a model of cosmos-origin chiral electromagnetic forces. These external chirality-driven, open-flow systems allowed to mimic the tempo-spatial transcription of molecular chirality and light chirality in the cell-wall free colloids on Earth at ambient temperatures in the absence and presence of CP light sources.

The open-flow, dissipative structures connected to molecular motions is a curious subject in the present special issue [183]. This is beneficial because chirogenesis and photochirogenesis from optically inactive colloids of $\pi$-/$\sigma$-conjugated polymers in the ground state and in the photoexcited state are easy to quantitatively characterize by means of CD, optical rotation dispersion (ORD), CPL, and CPL excitation (CPLE) spectroscopy. The $\pi$-/$\sigma$-conjugated polymers are excellent chromophores and luminophores as probes in place of amino acids and sugars.

Our knowledge and understandings of the $\pi$-/$\sigma$-conjugated polymers at ordinary laboratories on the northern hemisphere of the spinning Earth could shed light on rethinking about possible scenarios of the first life associated with L-amino acids and D-sugars, regardless of terrestrial and extraterrestrial origins.

*1.12. Resonance Effects from Colloidal Systems in Optofluidic Media in the Ground and Photoexcited States*

In the following sections, we highlight the importance of optofluidics for achieving an efficient chirogenesis and an efficient photochirogenesis using several $\pi$-/$\sigma$-conjugated polymers in UV-visible-near IR (NIR) regions. Particularly, a fine tuning of the RI value of the surrounding liquids allows us to resonantly boost chiroptical signals and photochemical reactions within a μm-size colloidal polymers. To our knowledge, such comprehensive, systematic studies, showcasing marked resonance effects of chirogenesis and photochirogenesis in the realms of chiroptical spectroscopy, asymmetric chemistry, synthetic chemistry, polymer chemistry, colloid chemistry, photochemistry, photophysics, and materials chemistry, have not been reported yet.

All of the spectral analyses of colloidal substances in the ground and photoexcited states were conducted by practical chiroptical approaches using JASCO J-725/J-820/CPL-200 spectrometers equipped with commercial and custom-made accessories. First, a classical ORD spectroscopy shows us, exactly, a left–right imbalance of chiral chromophores in the ground state as differences in relative RIs and/or light speeds between left- and right-circular polarized light as a function of wavelength in a vacuum. Next, CD spectroscopy allows detecting a left–right imbalance of chiral chromophores in the ground state as a difference in absorption (attenuation) between left- and right-circular polarized light as a function of wavelength in a vacuum. Moreover, CPL spectroscopy can detect a left–right imbalance of chiral luminophores in the photoexcited state as a difference in radiation probability between the left- and right-CP light emission processes when unpolarized light is excited at a specific wavelength of the chromophore as a function of wavelength in a vacuum [184]. In an analogy to achiral photoluminescence excitation (PLE) spectroscopy, CPLE spectroscopy provides differences in absorption band profiles in the ground state responsible for the specific CPL signals in the photoexcited state solely by monitoring, at a specific CPL band, as a function of wavelength in a vacuum. Although CPLE spectroscopy is rare in the realm of chiroptical spectroscopy, we often apply this technique to clarify which unpolarized absorption and CD bands are responsible for CPL/PL bands, particularly, dual/multiple CPL bands with opposite chiroptical signs [185]. Importantly, CPL, CPLE, PL, and PLE spectroscopy cannot apply to non-luminescent and ultraweak luminescent chromophores. More importantly, achiral and 50:50 racemic mixture of chromophores in the ground state and luminophores in the photoexcited state reveal zero-signals because of any differences between left- and right-ORD, between left- and right-CD, between left- and right-CPL, and between left- and right-CPLE signals. Notably, we should pay attention to artifact signals when anisotropic specimens, oriented films, solid substrate with periodicity acting as additional linear polarizers are employed [186]. It is beneficial that, because μm-size colloidal particles are randomly dispersed in fluidic medium that is similar to optically anisotropic molecules in homogeneous solutions, the chiroptical artifact is not a serious issue, even if an individual colloid is optically anisotropic.

## 2. Colloid-Induced Chiroptical Enhancement and Aggregation-Induced Emission

Firstly, we briefly touch upon two histories in the communities of artificial chiral chemistry, solution chemistry, and colloid/aggregation chemistry, including colloid-induced chiroptical enhancement (CIE) and aggregation-induced emission (AIE).

A photophysical transition from nearly non-emissive molecules dissolved in homogeneous solutions to highly photoluminescent colloidal particles dispersed in heterogeneous solutions was established as a well-known phenomenon in the last two decades [187,188]. In 2001, Tang et al. found an anomaly in the enhancement of quantum yield from aggregates made of 1-methyl-1,2,3,4,5-pentaphenylsilole [189]. The silole derivative revealed an abrupt enhancement in the quantum yield ($\phi$) of PL by solely adding water from a weak $\phi \approx 0.06\%$ in ethanol to high $\phi \approx 90\%$ in water-ethanol cosolvents. The aggregation process is responsible for the enhanced quantum yield. The phenomenon was coined aggregation-induced emission (AIE). Soon, the origin of the AIE phenomenon was explained by restricted intramolecular rotations of multiple C–C bonds of phenyl groups at peripheral position of the silole ring [190]. Either increasing solvent viscosity or decreasing solution temperature efficiently enables suppressing the rotational freedom at a molecular level in the aggregates. This idea was viable for several aggregates made of polyacetylenes substituted with floppy 1,1,2,3,4,5-hexaphenylsilole (HPS) in acetone-water cosolvents. In 2012, the aggregate of silole derivative carrying two long D-sugar-based tails in $n$-hexane-dichloromethane cosolvents showed the first AIE-circular dichroism (AIE-CD) in the ground state and AIE-circularly polarized luminescence (AIE-CPL) in the photoexcited state [188,191]. The Kuhn's dissymmetry ratio in the CPL spectrum ($g_{CPL}$) value [188] reached $\approx -0.32$ at 500 nm, in which the degree of circular polarization was $-0.16$. Similarly, the aggregate of tetraphenylethylene (TPE) substituted with two L-valines in a mixture of dichloroethane and methanol revealed AIE-CPL of $g_{CPL} \approx -5 \times 10^{-3}$. High-resolution images of scanning electron microscope and transmission electron microscope (TEM) indicated that one-dimensional helical fibers of the TPE aggregate are responsible for the AIE-CPL and AIE-CD [188,192].

We should touch on major differences between AIE-CD/AIE-CPL and CIE-ORD/CIE-CD/CIE-CPL/CIE-CPLE because both phenomena are related to significant amplification of chiroptical signals in the ground and photoexcited states.

First, AIE-chiroptical systems rely on significant suppressions of rapid intramolecular rotations among single bonds based on very low rotational/twisting barrier heights, which are responsible for thermally excited, non-radiation processes in the photoexcited and ground states. To achieve the suppressions at a molecular level, adding poor- or non-solvent-like water to homogeneous solutions, e.g., tetrahydrofuran (THF) and acetone as good solvents, of the weakly emissive substances is the key. The poor-/non-solvent and good solvents are completely miscible in any volume fractions. Actually, the building blocks of AIE, such as TPE and HPS, are highly emissive in the solid states, owing to great suppression of molecular motions and vibrations, where they dissolved in good solvents, are weakly emissive due to rapid rotational/twisting mobilities. In my view, controlling RI values of a mixed co-solvents of poor-/non-solvent and good solvent does not appear to be a major factor. However, the $n_D$ value might alter in response to volume fractions of THF–water cosolvents because a lower RI water ($n_D$ = 1.333) and a higher RI THF ($n_D$ = 1.407) are mixed together. Thus, we assume that AIE-chiroptical signals of the aggregates are possible, to maximize at specific volume fractions, namely, specific $n_D$, of good-poor co-solvents, to satisfy a resonance condition.

Contrarily, CIE chiroptical systems are largely affected by relative RI values between colloidal polymers and the surrounding fluidic media. To rationally realize the resonance condition in chiroptical signals, we often use semi-flexible and rod-like chain-like chromophoric polymers with a narrower molecular weight distribution. These polymers are inherently highly emissive, even in dilute homogeneous solutions, and in solid films. The reason for the semi-flexible and rod-like polymers is to restrict rotational freedoms along huge numbers of single bond sets. These polymers ensure spontaneous association of well-ordered packing structure during a controlled slow addition of a poorer solvent to the polymer solution, monitored by the naked eye. This approach belongs to a fractionation technique, to isolate the desired molecular weight of polymer from a broad molecular weight distribution of synthetic polymers, during a two-phase separation phenomenon. In our case, fine tuning the RI value of poor and good cosolvents, however, is the key, because polymers in the colloids often involve the surrounding co-solvents and usually exist as a swollen state. Thus, we have to experimentally determine the best volume fraction, namely, RI value, of poor and good cosolvents by measuring CD spectral shape with CD signal magnitudes of the colloidal systems. However, more importantly, when a specific resonance condition in the colloidal system is established with help from the RI-tuned cosolvents, the tiny chiroptical signals are enhanceable by huge times of reflection at the colloid–liquid interfaces due to confinement of traveling light in line with WGM, as mentioned in the following section.

### 3. Open-Flow, Non-Equilibrium Coacervate Hypothesis Meets Optofluidics

Our early assumption had to be remodeled when we came across optofluidics—that is an influential combined concept, at least for us, between liquid-based microoptics and soft-matter based optical devices. A traveling light, massless photon energy source, enables to efficiently confine polarized light in the interior of a droplet in a fluidic medium. The confinement causes a recurring polarized light–matter interaction within the droplet and/or at the droplet–liquid interfaces. An extension from optofluidics to chiral optofluidics might be applicable to chirogenesis and photochirogenesis from (at least) 1 to 100 µm-sized colloidal particles, with a higher RI, by dispersing in a fluidic medium with a lower RI, although their RIs have to be tuned precisely. The medium is not restricted to fluidic media. Likewise, solidified media, such as glassy polymers and inorganic solids, are possible to use.

Linearly polarized light is a superposition of left (*l*-) and right (*r*-) circularly polarized (CP) light. The *l*- and *r*-CP-light carry angular momenta of integer spins ($\pm\hbar$), respectively [193–196]. Previously, Ghosh and Fischer [193] and Silverman et al. [194] simulated that the optical rotation from chiral liquids is boosted by several orders of magnitude when a coupled geometry of multiple prisms is filled with optically active biomolecular liquids, such as limonene, carvone, and a mixture of camphorquinone and methanol. The concept of AAS with light, one of the curious topics in the present special issue of *Symmetry* [197], is classified to: (i) absolute asymmetric photosynthesis; (ii) photodestruction; and (iii) photoresolution [47–49].

According to a theory led by Mortensen et al., the slow light of colloidal particles, by filling with a liquid medium, could enhance the CD signal at the edges of the optical band gap [195,196]. This theory is verified by enhanced ORD signals in a fluidic medium with a tuned RI. ORD spectroscopy allows us to detect differences in light speeds between *l*- and

$r$-CP light as a function of incident light. In an analogy of the ORD signal enhancement, one can assume that CD spectrum as a function of wavelength of incident $l$- and $r$-CP light is possible to detect enhanced differences in light speeds between $l$- and $r$-CP light when optically active colloids are dispersed into a proper liquid medium by solely tuning RI.

These seminal theories of optofluidics and our CIE chiroptical phenomena prompted us to focus on the chiral optofluidic medium suited to μm-sized colloids, consisting of luminous σ-/π-conjugated polymers. With this mind, we designed several optically active fluidic media containing limonene, α-pinene, and other biomolecular solvents [54–57,198–212]. Moreover, we expected an efficient APC when the CP-light source, with sense, was employed as a function of excitation wavelength [54–56].

Herein, we highlight a series of CIE-CD, CIE-ORD, CIE-CPL, and CIE-CPLE experiments from the colloids of π-/σ-conjugated polymers and co-colloids consisting of both π- and σ-conjugated polymers, in terms of artificial models of open-flow, non-equilibrium coacervates in the ground state and the photoexcited state. These colloidal polymers are readily dispersed in a tuned optofluidic medium by inspiring the possible scenarios of the homochirality questions on Earth. By controlling RI of the surrounding medium, the μm-size colloidal π-/σ-conjugated polymers in the optofluidic medium resonantly reveal enhanced characteristics of chirogenesis and photochirogenesis [198–212]. The chirogenesis is susceptible to chiral optofluidic medium. An ability of APC from achiral colloidal polymers in achiral optofluidic medium is possible by choosing tandem factors of wavelengths and a sense of CP light sources [54–56], which are purely massless chiral photons carrying integer spins of $\pm \hbar$ [54–56,213–215].

## 4. CIE-CD in the Ground State and CIE-CPL in the Photoexcited States
*4.1. Steady-State CD and CPL Spectroscopies*

Steady-state CPL spectral datasets tell us information of the short-lived chiral species in the photoexcited state, while steady-state CD and ORD spectral datasets indicate long-lived chiral species in the ground state. Here we applied Kasha's rule and Jablonski diagram [216,217] to the nature of optically active luminophores and chromophores (Figure 1). The short-lived species ($S_1$, $S_2$ ... ) upon excitation of unpolarized light are generated on the timescale of $\sim 10^{-15}$ s based on the Franck–Condon principle [218]. The short-lived species, followed by non-radiative relaxation processes with vibrational modes, relax spontaneously to the lowest vibronic state ($S_1$ with $v' = 0$), such as the metastable photoexcited state at the time scales of $10^{-11}$–$10^{-12}$ s. Finally, a spontaneous relaxation occurs from the low-entropy handed $S_1$-state ($v' = 0$) to the high-entropy, non-handed $S_0$-state ($v = 0, 1, 2 \ldots$), proven by CPL signals with a lifetime of $10^{-9}$–$10^{-6}$ s.

Absolute temperature of 300 K equals to 0.60 kcal mol$^{-1}$ (0.0259 eV and 208 cm$^{-1}$). When luminescent molecules and polymers are photoexcited at 400 nm in a vacuum, the 400 nm light energy corresponds to 35,970 K (3.10 eV, 71.5 kcal mol$^{-1}$, 25,000 cm$^{-1}$). CPL/CPLE and CD/ORD spectroscopy, thus, allow to dictate different chiral information in the photoexcited and ground states, respectively. Although CPL signals dictate a temporal short-lived chiral species at 35,970 K, CD and ORD spectral data provide information of long-lived chiral species at 300 K. When one can ensure the identity chiral species in the ground and photoexcited states, enantiomeric pairs of rigid luminophores and rigid chromophores are needed to restrict by intra-/intermolecular rotations at 35,970 K. Optically active luminescent colloids are candidates to design WGM-based chiroptical resonators between a pair of (+)- and (−)-CIE-CPL signals.

**Figure 1.** A brief concept of the modified Jablonski diagram, Kasha's rule, and Franck–Condon principle [212,216–218] of chiral σ-/π-conjugated chromophores with/without rotational freedom in the absence of optical resonator, whispering gallery mode (WGM), and optofluidic effects. Single-headed arrow stands for restricted rotation clockwise (CW) or counter-clockwise (CCW); double-headed arrow indicates a free rotation CW and CCW. Modified from an original article [199] and a book chapter [212].

This idea predicts that, although a non-rigid chiral luminophore arising from substantial rotational freedom in the ground state does not reveal detectable CD and weak CD signals known as cryptochirality, the luminophores does not emit clearly CPL signal, and emits, apparently, no detectable CPL with the corresponding unpolarized PL due to an equal probability of *left*- and *right*-hand chiral structures in the photoexcited state. However, if a non-rigid chiral luminophore produces certain chiral colloids induced by the restricted rotational freedom, one can detect the spatiotemporal chiral colloid as CIE-CPL signal as well as CIE-CD and CIE-ORD signals. One can detect steady-state CIE-CPL signals because the lifetime of the spatiotemporal species is on the order of ≈1–10 ns.

One may question whether the chiral optofluidics is valid for CIE-CD and CIE-CPL beyond an extension of AIE-CPL and AIE-CD. To prove the idea that the concept is generally applicable to a wide range of colloidal system suspensions in the fluidic media, we had to elucidate whether the chiroptical amplitudes in the resulting CD and CPL signals from several colloidal polymers were resonantly enhanced at specific RI values of the surrounded media. However, we had to consider serious concerns of the suspension heterogeneous systems. This is because anisotropic films and the vortex flowing from the chromophores and luminophores often cause unfavorable chiroptical artifacts and apparent sign inversion, as well as alteration in the absolute magnitudes in the CD and CPL signals [186]. To practically solve these artifacts, the use of optically anisotropic colloids dispersed in isotropic fluidic medium is one candidate. This approach is routinely applied to artifact-free CD and CPL spectral measurements by dissolving optically anisotropic molecules by dissolving in isotropic solution, although inherent anisotropic molecular information is lost. This is similar to routinely obtaining a solution of $^1$H- and $^{13}$C NMR spectra to measure highly anisotropic molecules in homogeneous solutions by rotating sample tubes and solid-state $^{13}$C NMR spectrum of an anisotropic solid sample, by rapidly rotating sample tubes with a magic angle (54.7°) to cancel anisotropic terms.

Therefore, extremum wavelengths, chiroptical amplitudes, and signs in CPL/CPLE/CD/ORD signals, as well as those in PL excitation (PLE) and PL signals are not significantly affected by Rayleigh–Mie scatterings in proportion to $\lambda^{-4}$ ($\lambda$: wavelength of light in a vacuum). The scatterings often cause an unfavorable increment and apparent spectral shift embedded into the background UV-visible spectrum. Instrumental knowledge is crucial to characterize inherent chiroptical signals from optically active colloids in the ground and photoexcited states. It is noted that the RI value alters as a function of wavelength. ORD spectroscopy can measure the difference in RI between left *(l)*-

and right *(r)*-CP light. Unpolarized light is a superposition of *r*-CP and *l*-CP light. Fine tuning RIs of colloidal particles and the surrounding medium is key when the idea of the chiral optofluidics is applied to CIE-driven chiroptical enhanced signals. We showcase several examples in the following sections.

*4.2. Gigantic Enhanced CD, ORD, and CPL from Colloidal Optically Active Helical Poly-silanes—Importance of Controlled RI in Optofluidic Medium*

Herein, we showcase, for the first time, a clear example that a precisely controlled RI of the surrounding medium, which is the essential concept of optofluidic, is critical to boost CIE-CD, CIE-ORD, and CIE-CPL signals for colloidal particles of several helical polysilanes. We tested poly{*p*-(*S*)-2-methylbutoxypheneyl-*n*-propylsilane} (**1S**) [219], poly(*n*-decyl-(*S*)-3-methylpentylsilane) (**2S**) [220] poly(*n*-decyl-(*S*)-2-methylbutylsilane) (**3S**) [221], poly(*n*-dodecyl-(*S*)-2-methylbutylsilane) (**4S**) [21,199,221], and poly(*n*-dodecyl-(*R*)-2-methylbutylsilane) (**4R**) [21,199] (Figure 2). Poly(*n*-hexyl-(*S*)-2-methylbutylsilane) (**5S**) [208–210,222,223] and poly(*n*-hexyl-(*R*)-2-methylbutylsilane) (**5R**) [208–210,222,223] were used as helical scaffoldings of producing chiral co-colloidal particles with several optically inactive π-conjugated polymers.

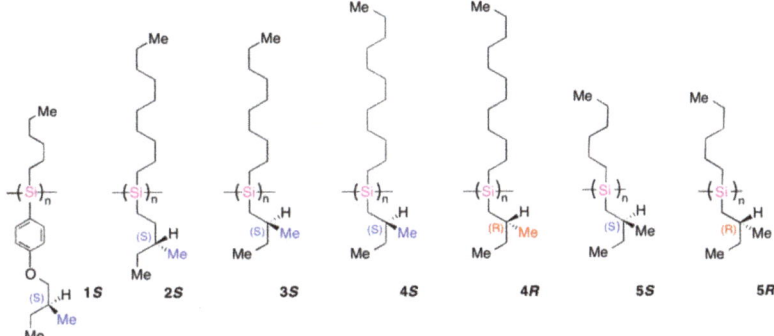

**Figure 2.** Chemical structures of semi-flexible helical alkylarylpolysilane (**1S**) and rod-like helical dialkyl polysilanes (**2–5**) carrying enantiomerically pure chiral substituents.

First, we highlighted the colloidal **3S** as a function of RI value of the surrounding cosolvents, by assuming that RI value of the colloidal **3S** is $n_D \approx 1.7$ [199]. The CIE-CD, CIE-ORD, and CIE-CPL spectra of the **3S** are shown in Figure 3a–c. From Figure 3a, the **3S** clearly shows negative exciton couplet CD bands at the lowest Siσ–Siσ* band at 323 nm. The Kuhn's dissymmetry ratio in the CD spectrum ($g_{CD}$) attain −0.31 at 325 nm and +0.33 at 313 nm, respectively. These $g_{CD}$ values are comparable to ≈16% left (*l*)-circular polarization and ≈17% right (*r*)-circular polarization, respectively, while pure *l*- and *r*-circular polarizations are $g_{CD} = \pm 2.0$, respectively [224].

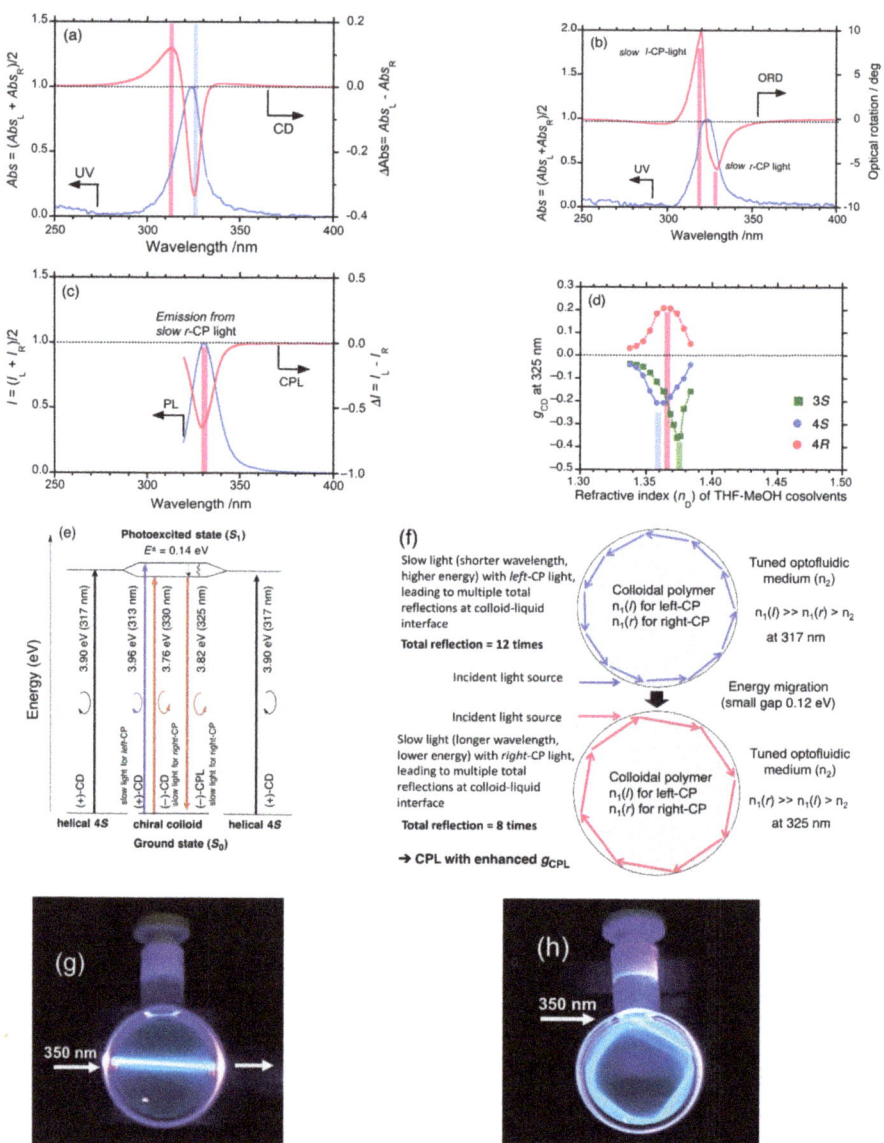

**Figure 3.** (**a**) Colloid-induced enhancement circular dichroism (CIE-CD) and UV spectra; (**b**) CIE optical rotation dispersion (CIE-ORD) and UV spectra; and (**c**) CIE circularly polarized luminescence (CIE-CPL) and PL spectra of the **3S** colloids; (**d**) The Kuhn's dissymmetry ratio in the CD spectrum ($g_{CD}$) of the colloidal **3S**, **4S**, and **4R** as a function of refractive index (RI) in methanol-tetrahydrofuran (THF) cosolvents. The CIE-CD and CIE-CPL spectra are normalized by Kuhn's dissymmetry ratio in the ground and the photoexcited states [212,224,225]; (**e**) Schematic Jablonski diagram of the **3S**; (**f**) a possible explanation for the CIE-CPL by confining left (*l*)- and right (*r*)-circularly polarized (CP) light in the **3S** in optofluidic medium with a tuned RI [199]. Modified from an original article [199] and a book chapter [212]; (**g**) incident UV light at ~0° of the cuvette surface travels a very dilute CHCl$_3$ solution of stilbene 420 upon excitation of unpolarized light at 350 nm of cylindrical quartz cuvette (22 mm in diameter, 1 mm in thickness of cell wall, 10 mm in cuvette optical path); (**h**) incident light at ~50° of the cuvette surface infinitely travels the stilbene solution, undergoing to WGM by four-time reflections per one-cycle at the CHCl$_3$-quartz interfaces.

As shown in Figure 3b, the colloidal **3S** clearly shows negative bisignate ORD spectra at the 323 nm Siσ–Siσ* transition. The traveling light speed of *r*-CP at 330 nm significantly slows down compared to the traveling speed of *l*-CP light at 320 nm. Likewise, a traveling speed of *l*-CP light at 320 nm slows down relative to the traveling speed of *r*-CP light at 320 nm. Therefore, the relative RI value between *r*- and *l*-CP light largely relies on wavelengths of incident light. A higher RI value results in slower CP-light along with shortening of wavelength because the wavenumber is unchanged, regardless of RI, by the conservation law of light energy.

The **3S** showed clearly single (−)-CPL signal at the 330 nm PL band associated with a small Stokes' shift of 466 cm$^{-1}$, as depicted in Figure 3c. This small Stokes' shift indicates a very minimal structural reorganization owing to a very restricted rotational motion in the colloidal **3S** in the photoexcited state. The magnitude of $g_{CPL}$ attained −0.65 at 330 nm, corresponding to ≈33% *r*-circular polarization, whereas the ideal *l*- and *r*-circular polarization in the photoexcited state occurs at $g_{CPL}$ = ±2.0, respectively [225].

Both CIE-CD and CIE-CPL characteristics of the **3S** are nearly similar to those of the **4S** and **4R**. Notably, the CIE-CD and CIE-CPL characteristics of the **3S**, **4S**, and **4R** boost at the very specific $n_D$ of mixed solvents of methanol and tetrahydrofuran (THF) (Figure 3d). The $g_{CD}$ value of the **3S** is resonantly boosted at $n_D$ = 1.374, which is a specific volume fraction of the cosolvent. Likewise, the $g_{CD}$ values of the **4S** and **4R** are resonantly boosted at $n_D$ = 1.359 and $n_D$ = 1.365, respectively. The $n_D$ dependent chiroptical resonance effects are an evidence of the chiral optofluidics, where noticeable differences in a traveling light speed between *l*- and *r*-CP in the μm-scale colloids in the ground state is evinced by tuning $n_D$ of a liquid medium. The resonance effects are responsible for CIE-CD signals, followed by CIE-CPL signals in the photoexcited state.

From Figure 3a–c, a modified Jablonski diagram with chirality of the **3S** associated with the Kasha's rule and exciton coupling theory is schematically illustrated in Figure 3e [199]. With the aid of the chiroptical resonator for selective *l*- or *r*-CP light, a large difference in the RI between *l*- or *r*-CP light is crucial.

In Figure 3f, we provide a plausible explanation for the CIE-CPL occurring in an optical resonator as a recurring WGM mode endowed with an efficiently confining *l*- and *r*-CPL in the **3S** with a high RI surrounded by optofluidic medium with a lower RI [199]. The optically active spherical colloids should work as chiroptical resonator for the respective *l*- and *r*-CP light. As a result, efficient separation between *l*- and *r*-CP light as radiation from the colloids is realized. Internal total reflection modes (Brewster angle and recurring numbers) at the colloid–liquid interface should be kept in mind.

Firstly, unpolarized light at 317 nm (3.90 eV), which is an equal mixture of *l*- and *r*-CP light, excites simultaneously the colloidal **3S** with a high $n_2$ surrounded by the liquid with a lower RI. The RI value of *l*-CP light, $n_1(l)$, at 317 nm is higher than that of *r*-CP light, $n_1(r)$, at 317 nm. The respective $n_1(l)$, $n_1(r)$, and $n_2$ are assumed to 1.8, 1.6, and 1.4. By Snell's law, the respective critical angles of refraction ($\theta_c$) for *l*- and *r*-CP light are evaluated to 51° and 61°. The difference in the $\theta_c$ angles of *l*- and *r*-CP light, thus, efficiently acts as chiroptical filter between *l*- and *r*-CP light at the colloid–liquid interface, such as a quarter-wave plate toward unpolarized light.

For clarity and visibility of the WGM concept, we demonstrate how an incident light travels interior of the μm-size colloidal resonator using two macroscopic models consisting of a fused quartz cylindrical cuvette with optically smooth surfaces (22 mm in diameter, 1 mm cell wall in thickness) filled with a very dilute CHCl$_3$ solution of stilbene 420, which is a sky-blue-color luminophore, upon excitation of unpolarized light at 350 nm. The cm-size cylindrical cuvette is a cross-section model of a μm-size colloidal sphere shape. The macroscopic demonstration is a zoom-in model of light–matter interaction occurring in a μm-scale resonator to persuade non-specialists.

From Figure 3g, the incident UV light at ~0° of the cuvette surface to excite stilbene 420 travels straightly and emits from the opposite side of the cuvette. In this case, only one-time interaction of light with the luminophore is possible. On the other hand, it is obvious from Figure 3h that, when an incident light at a very specific angle of ~50° of the cuvette surface, the light reflects four times per one-cycle at the $CHCl_3$-quartz interface and undergoes into the WGM with the traveling light circulation mode. This recycling enables the incident light to infinitely interact with the luminophore, if attenuation by the absorbing light is neglected. One can expect that endless light–matter interactions are possible, thus, causing significantly boosted chiroptical signals in the ground and photoexcited states.

When open-flow, μm-size colloidal polymers with a smooth surface adaptable to the external physical and chemical biases are surrounded by an optofluidic medium, subtle chiral physical forces and chiral chemical sources are enhanceable by huge numbers of recycling times ($N$) in the optofluidic colloidal polymers, in line with the scheme of WGM. This means that a subtle left–right imbalance interior of the colloid, e.g., $10^{-6}$% $ee$, is boostable by $10^{-6} \times N$. This scenario is contrast to a conventional stereochemistry, photochemistry, and photophysics in homogeneous solution systems, in which the light–matter interaction occurs one chance ($N$ = 1) only. When aggregation-induced emission systems chiroptically fulfil the resonance condition, one can anticipate the significant enhancements in CPL and CD signals.

The recycling light should occur even at the μm-size colloidal polymers that fulfil the resonant condition with relative RI values, between the polymers and fluidic medium. However, practically, the colloidal shape may not be an ideal sphere with an optically smooth surface, and an irregular shape with an optically rough surface diminishes this booster effect. The concept of WGM at the colloid–liquid interface, satisfying a resonance condition, along with specific RI values is applicable to boost chiral light–matter, light–chiral matter, and chiral light-chiral matter interactions.

To design the light recycling at the colloid–liquid interface, fine tuning RI at the specific wavelengths of the liquid medium is essential. It should be noted that RI values vary as a function of wavelength in a vacuum. This unique idea is valid for any optically opaque system, of chromophores and luminophores, in a heterogeneous medium, but is not applicable to optically transparent, homogeneous solutions. If the RIs between the colloid and surrounding medium are, by chance identical, any refraction, any reflection, and any scattering at the colloid–liquid interface do not occur. One can say that the colloidal mixture is optically transparent. The representative example is optically transparent pellet of poly(4-methyl-1-penetene) (TPX®, Mitsui Chemicals (Tokyo, Japan)). This transparency arises from the nearly identical RIs of crystalline and non-crystalline phases of TPX®.

The traveling light speeds of $l$- and $r$-CP in the colloidal polysilanes at 317 nm slow down to $1.67 \times 10^8$ ms$^{-1}$ and $1.88 \times 10^8$ ms$^{-1}$, respectively. However, the wavelengths of the incident $l$- and $r$-CP light in a vacuum shorten to 176 and 198 nm in the colloid, respectively. As a result, the incident 317 nm $l$-CP light results in a greatly shortened $l$-CP light of 176 nm. Similarly, the incident 317 nm $r$-CP light becomes a slightly shortened light of 198 nm. If the 176 nm slow-down $l$-CP light was employed, multiple total internal reflections occurred efficiently (for example, 12 times) in the colloids compared to that of the slow-down 198 nm $r$-CP light. Increasing the number of total internal reflection (TIR) of CP light at the colloid–liquid interface meets the greater opportunity of chiral light–matter interactions. The recurring multiple TIR process at the colloid–liquid interface is an important step.

The slow-down 317 nm $l$-CP light relaxes to the 330 nm $r$-CP light along with a change in CD sign within the colloids. The faster 317 nm $r$-CP light migrates to the 330 nm $r$-CP light without significant change in the CD sign. Owing to a great suppression of the 325 nm $l$-CP light, the slow-down 330 nm $r$-CP light emits dominantly from the $S_1$ state with an apparent minimal Stokes' shift. This explanation is verified by the ORD spectrum in the fluidic medium with the tuned $n_D$ (Figure 3c). The ORD spectrum, as a function of the incident wavelength of $l$- and $r$-CP light, can detect differences in traveling light speed between $l$- and $r$-CP in an optically active colloidal particle.

## 4.3. Controlled Chirogenesis from Optically Inactive Helical Polysilanes Endowed with Limonene Chirality

Optically active, colloidal helical polysilanes are instantly generated by solely adding poor solvent to a homogeneous solution of optically active helical polysilanes carrying chiral substituents [219]. This approach is, however, inevitably needed to use expensive chiral biomolecular and artificial sources and multiple synthetic steps when we intend to introduce chiral substituents to Si–Si backbone [219–223]. Herein, we showcase a facile, inexpensive, environmentally-friendly approach to instantly generate CIE-CPL and CIE-CD signals of colloidal polysilanes from optically inactive helical polysilanes in a minute (Figure 4) [200,226]. Herein, the authors used artificial chiral alcohols and biomolecular chiral solvents for chirogenesis of colloidal σ- and π-conjugated polymers (Figure 5) [198–207,212].

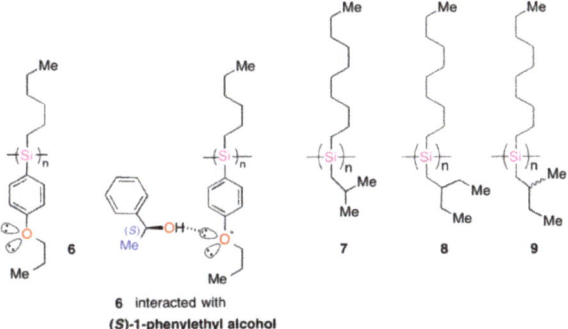

**Figure 4.** Chemical structures of CD-silent helical alkylarylpolysilane (**6**) with (S)-1-phenylethyl alcohol and dialkyl polysilanes (**7**–**9**).

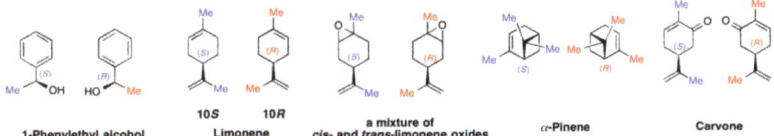

**Figure 5.** Chemical structures of artificial chiral alcohols and biomolecular chiral solvents used for chirogenesis of colloidal polymers.

Previously, we reported that optically inactive colloidal helical **6** exhibits couplet-like CIE-CD by transferring molecular chirality of (S)- and (R)-1-phenylethyl alcohols and other chiral alkyl alcohols when these chiral alcohols were used as solvent quantities [226]. It should be noted that **6** homogeneously dissolved in the alcohol solutions, adopting a CD-silent racemic helical state arising from a dynamic twisting between the left- and right-helices in a low energy barrier height in a double-well. By adding a poor solvent (methanol), CD-silent **5** generates CIE-CD effect from the colloidal **6** arising from mirror symmetry breaking (MSB). In this case, the resulting CD sign is determined solely by the chirality of alcohols used. It was driven by solvent molecules. The handed chiral CH/O interaction is responsible for the CIE-CD (Figure 4) [226]. We, at that time, however, were not aware of the RI dependency of the chiral alcoholic solvents because any ideas of optofluidics, optical resonator, confinement of chiral light, and chiral optofluidics were lacking.

Our early studies prompted further experimentally tests on whether three CD-silent, rod-like, helical dialkyl polysilanes (**7**,**8**,**9**) provide CIE-CD (**7**,**8**,**9**) and CIE-CPL (**8**,**9**) in the presence of inexpensive (S)-limonene (**10S**) and (R)-limonene (**10R**) [200]. We assumed that non-covalent attractive intermolecular interactions, such as chiral CH/π, London dispersion, and van der Waals forces were possible to emerge CIE-CD and CIE-CPL signals. A precisely controlled RI of limonene-containing tersolvent was critical as well.

The CIE-CD and the corresponding UV spectra of the colloidal **7** and **8** are shown in Figure 6a,b. In the **7**, dispersed in **10R**-based tersolvent, the $g_{CD}$ amplitudes at the bisignate CD spectra boosted to +0.022 at 327 nm and −0.031 at 309 nm, respectively [200]. As expected, the **7** in **10S**-based tersolvent gave the mirror-symmetric $g_{CD}$ amplitudes with −0.021 at 327 nm and −0.033 at 309 nm, respectively [200]. The CIE-CD values of the **7** are comparable to those of the **9**, but those in the **8** decreased by the magnitude of one-third. The bisigned $g_{CD}$ values of the **8** in **10R**-based tersolvent were −0.007 at 331 nm and +0.010 at 317 nm, respectively. Conversely, the **8** in **10S**-based tersolvent gave the mirror-symmetrical $g_{CD}$ values of +0.005 at 331 nm and −0.007 at 317 nm, respectively. The degree of circular polarization of the **7**, **8**, and **9** was, however, not intense on the order of 1.0–1.5%. The absolute magnitudes of the CIE-CD values of the **7**, **8**, and **9** are greatly dependent of $M_w$. A representative $M_w$ dependent $g_{CD}$ values from the **7** is given in Figure 6c. The specific **7** with $M_w = 2.7 \times 10^4$ afforded the greatest $g_{CD}$ value; **8** and **9** had similar $M_w$-dependency of the CIE-CD effects [200].

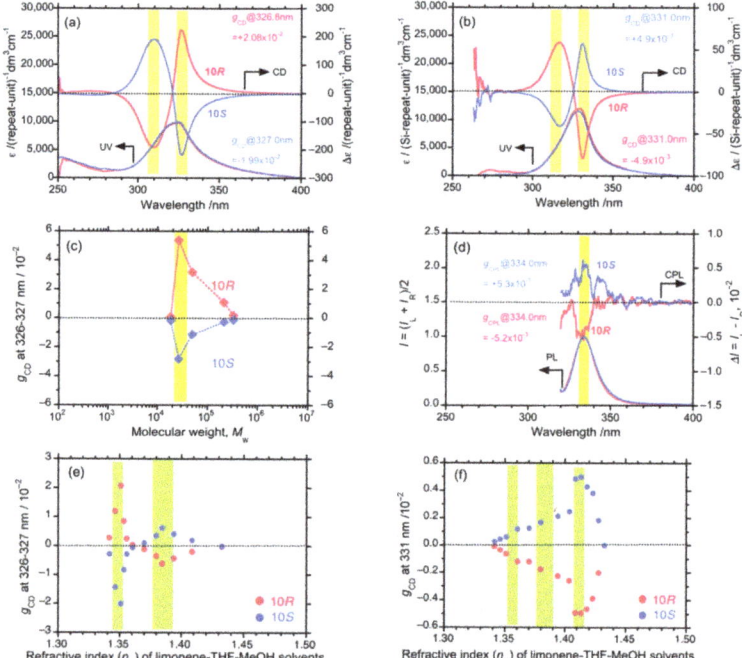

**Figure 6.** The CIE-CD and UV spectra from the colloidal: (**a**) **7** and (**b**) **8**. (**c**) The $g_{CD}$ of the **7** as a function of weight average molecular weight of polymer ($M_w$) of the **7**. (**d**) CIE-CPL and PL spectra of the **8**. The $g_{CD}$ of the: (**e**) **7**; and (**f**) **8** as a function of refractive index at Na-D line ($n_D$) in the limonene containing tersolvent. Modified from an original article [200] and the book chapter [212].

Among the **7**, **8**, and **9** showing weaker $g_{CD}$ values, only the **8** and **9** revealed weaker CIE-CPL associated with the absolute value of $g_{CPL}$, $|g_{CPL}| \approx 0.005$ (Figure 6d) [200]. The **7** did not emit any PL due to unresolved reasons, thereby, revealed no detectable CPL signals. From Figure 6a,b, the CD sign of the **7** is opposite of that of the **8** when the same limonene chirality was applied. However, the CD sign of the **7** inverted at the specific $n_D = 1.35$ of the **10R**- and **10S**-based tersolvents (Figure 6e). The unchanged CD sign of the **8** showed an abrupt increment at $n_D = 1.41$ (Figure 6f). Thus, subtle structural alterations in achiral alkyl side chains of helical polysilanes significantly affect the amplitudes and chiroptical sign of CD and CPL signals.

### 4.4. Chirogenesis from Achiral π-Conjugated Polymers Endowed with Limonene Chirality

From three CD-silent helical **7**, **8**, and **9** in homogeneous solution, we knew the CIE-CPL effects from the colloidal **8** and **9** and CIE-CD effect from the **7**, **8**, and **9** in the presence of **10S** and **10R** [200]. The results prompted to further investigate CIE-CD and CIE-CPL effects from achiral π-conjugated luminescent polymers in the presence of **10S** and **10R** and (S)- and (R)-pinene (Figure 5) [198,201–212]. Herein, the first successful limonene chirality transfer allowed for realizing CIE-CD and CIE-CPL effects from achiral poly [(9,9-dioctylfluorenyl-2,7-diyl)-*alt*-bithiophene] (**11**) [198] among several chain-like π-conjugated polymers (Figure 7) and hyperbranched poly(dioctylfluorene) [206]. The CIE-CD and CIE-CPL characteristics of **12** [203], **13** [202], **14** [204], **15** [198], **16** [198], **17** [205], **18** [205], **19** [201], and the hyperbranched polyfluorene [206] are nearly similar to those of **15**. Moreover, the CIE-CD and CIE-CPL characteristics, including $g_{CD}$ and $g_{CPL}$ values in these colloidal polymers, were resonantly boosted at very specific RIs, associated with specific volume fractions in the limonene-based tersolvents.

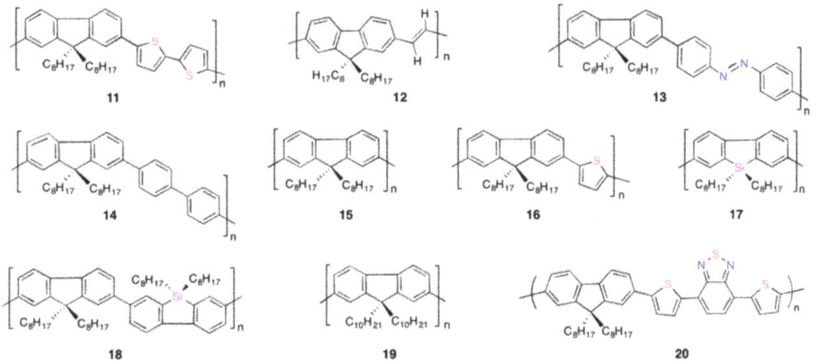

**Figure 7.** Chemical structures of optically inactive, achiral, and CD-silent π-conjugated polymers.

The UV-visible and PL spectra of **11** in homogeneous CHCl$_3$ solution are given in Figure 8a. The PL spectrum is accompanied with three well-resolved vibronic signals at 500, 534, and 578 nm with a spacing of ≈1350 cm$^{-1}$ [198]. Contrarily, the corresponding UV-visible spectrum is structureless broad bands at ≈457 nm (major) and 478 nm (a shoulder) [198]. This indicates that **11** in the photoexcited state adopts a highly ordered structure as a low entropy state. Conversely, **11** in the ground state adopts a randomly fluctuated disordered structure as a high entropy state. The UV-visible and PL spectral characteristics in the homogeneous solution are representative of random coiled luminescent chromophoric polymers in dilute solutions [220–223]. The rotational freedom with an equal probability in twisting between *left* and *right* in C–C, Si–Si, and Si–C bonds in the ground state results in the apparently broadened UV-visible absorption bands, leading to no detectable CD signals. The barrier heights of these single bonds are as small as 1–2 kcal mol$^{-1}$ from our calculations [198,201,203,220].

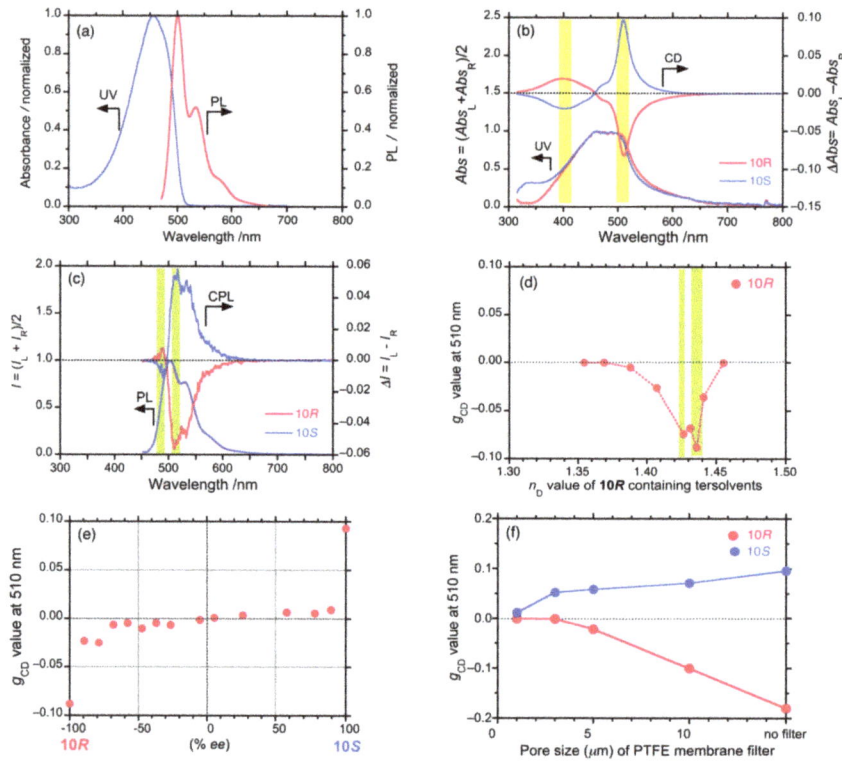

**Figure 8.** (a) UV-visible and PL spectra of **11** in a homogeneous CHCl$_3$ solution; (b) CIE-CD and UV spectra of the unfiltered colloidal **11** suspension in limonene-CHCl$_3$-methanol tersolvent; (c) CIE-CPL and PL spectra of the **11** in limonene-CHCl$_3$-methanol tersolvent; (d) the $g_{CD}$ value at 510 nm as the first Cotton band as a function of the $n_D$ in the tersolvent; (e) The $g_{CD}$ value of the unfiltered **11** as a function of the limonene enantiomeric excess (ee) value; (f) The $g_{CD}$ value of the filtered **14** as a function of the colloid size. Modified from an original article [198] and the book chapter [212].

The colloidal **11** in the limonene-based tersolvents reveals nearly mirror-symmetrical CIE-CD and CIE-CPL spectra by the lack of rotational freedom in the ground and photoexcited states (Figure 8b,c) [198]. The bisignate $g_{CD}$ values are −0.085 at 510 nm and +0.042 at 394 nm (for **10R**), respectively, and +0.114 at 510 nm and −0.041 at 394 nm (for **10S**), respectively. Concurrently, the bisignate $g_{CPL}$ values are evaluated to +0.012 at 489 nm and −0.058 at 511 nm (**10R**), respectively, and −0.010 at 489 nm and +0.056 at 511 nm (**10S**), respectively. In the **10R**-based tersolvent, the 511 nm (−)-CPL signal originates from the 510 nm-(−)-CD signal, while the 489 nm (+)-CPL band arises from the 394 nm (+)-CD signal. Likewise, the **10S**-based tersolvent induces the opposite to those of the **11** led by the **10R**-based tersolvent [88].

The $g_{CD}$ value of the **11** is resonantly boosted at $n_D$ = 1.44 of the limonene-containing tersolvent (Figure 8d). Notably, the $g_{CD}$ value varies extraordinary in response to the ee value of limonene, exhibiting the so-called, negative cooperative effect (Figure 8e). The negative cooperativity indicates that enantiomerically pure limonene chirality is needed to attain the highest $g_{CD}$ value. Moreover, we are aware that the $g_{CD}$ value increases as the colloidal size increases (Figure 8f) [198]. A larger size of the colloidal particles fulfils morphology-dependent resonance condition in the WGM-origin chiroptical resonators known as cavity quality, indicated by theories [227,228] and experiments [229,230].

## 4.5. Fully Controlled Absolute Photochirogenesis from Achiral π-Conjugated Polymers Endowed with Excitation Wavelength Dependent Circularly Polarized Light Chirality

The colloids exhibiting significantly boosted CIE-CD and CIE-CPL signals are instantly generated at room temperature with help from limonene with (S)- and (R)-chirality from optically inactive σ-/π-conjugated polymers. The $g_{CD}$ and $g_{CPL}$ signals are resonantly boosted at the specific RI of the optofluidic media. These results encourage us to further investigate a possibility of AAS with light or APC from two optically inactive colloidal **13** [55] and **11** [54] by dispersing to achiral solvents with a very specific RI endowed with CP light chirality (Figure 9). Recently, we demonstrated that the controlled APC with CP light (CAPC) from achiral colloidal polymethacrylate carrying azobenzene **21** in a tuned achiral cosolvent is possible by CP light, with wavelength and sense as purely chiral physical force [56] (Figure 8). We had confidence that, in the colloidal systems of **11**, **13**, and **21**, a hand of CP light is not a deterministic factor for their product chirality and that both a hand and an excitation wavelength of CP light are subject for the CAPC with CP light.

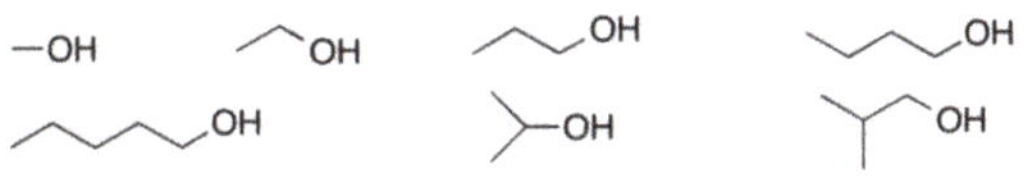

**Figure 9.** Chemical structures of alcohols used for *l*- and *r*-CP light controlled absolute photochirogenesis (CAPC) experiments for the colloidal **11** and **13**.

### 4.5.1. Achiral π-Conjugated Polymer Containing Azobenzene Unit as a Backbone Endowed with Circularly Polarized Light Chirality

In this section, we demonstrate that CAPC operated in the RI-tuned optofluidic medium is efficiently realized when wavelength-controlled *r*- and *l*-CP light is employed as chiral electromagnetic force. The CAPC conducted by achiral optofluidic medium enables all chiroptical polarization (chirogenesis), depolarization (racemization or chiroptical erase), inversion (anti-chirogenesis), retention (memory), and switching in the μm-size colloidal **11**, **13**, and **21** (see, later Section 4.5.3), as evidenced by their CIE-CD and CIE-CPL spectral characteristics.

As mentioned in Section 1.2.1, LeBel and 1894, van't Hoff proposed AAS with *r*- and *l*-CP light. Kuhn and Broun proved their idea as a photodestruction mode of AAS [46]. Their works prompted researchers to catalyze AAS study for a century because expensive chiral chemical substances are no longer needed [47–49,231–233]. Possibly, most researchers might concur that *l*-CP light produces *left*-hand (or *right*-hand) molecules preferentially, or vice versa, because the product chirality is determined solely by the hand of CP light. The author was one of the researchers.

In Figure 10a,b, the CIE-CD, CIE-ORD, and UV spectra of the colloidal **13** are displayed when *r*- and *l*-CP light source is excited at 436 nm [54]. The bisignate $g_{CD}$ values at the first and second Cotton bands are −0.025 at 500 nm and +0.021 at 367 nm for *r*-CP light, respectively [54]. Conversely, they are +0.025 at 509 nm and −0.027 at 364 nm for *l*-CP light, respectively. The **13** with negative-couplet CIE-CD upon excitation of *r*-CP light at 436 nm results in nearly zero CIE-CD signals upon excitation of *l*-CP light at 436 nm for 5–10 min (Figure 10c). A prolonged excitation of *l*-CP light for 51 min led to an ideal, mirror-symmetric, positive-couplet CIE-CD signal (Figure 10c). An alternative excitation between *r*- and *l*-CP light sources enables chiroptical inversion and switching between the positive- and negative-couplet CIE-CD signals [54]. It should be noted that **13** and unsubstituted azobenzene dissolved in dilute homogeneous $CHCl_3$ solutions are weakly emissive. The **13**, however, does not reveal any detectable CIE-CPL and PL spectra due to aggregation-caused quenching mechanism [187].

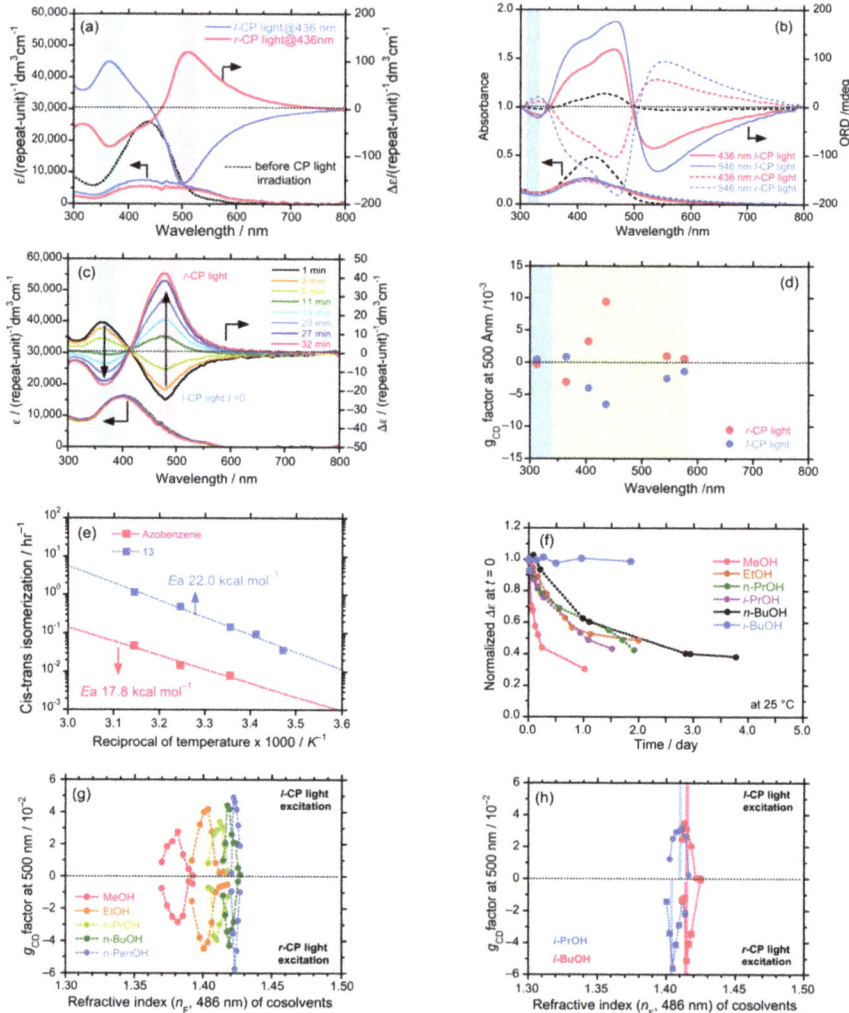

**Figure 10.** (a) CIE-CD and UV spectra of the colloidal **13** upon excitation with *r*- and *l*-CP light at 436 nm; (b) CIE-CD of the **13** initially generated by *l*-CP light at 436 nm, followed by *l*-CP light excitation at 436 nm; (c) The Arrhenius plots of the **13** and unsubstituted azobenzene during *cis-trans* thermal isomerization; (d) Alcohol-dependent thermal chiroptical stability of the CP-light source induced the colloidal *cis*-**13** at 25 °C. The $g_{CD}$ values as a function of the $n_F$ values of (e) non-branched alcohol- $CHCl_3$ cosolvents and (f) isoalcohol-$CHCl_3$ cosolvents. Modified from an original article [54] and the book chapter [212].

We are aware of an anomaly of an apparent chiroptical inversion for the product chirality of **13** when excited at 313 nm with the same *l*-CP light source; the positive-sign, weak CD signals excited at 313 nm only, while the negative-sign, intense CD signals upon excited at 365 nm, 405 nm, 436 nm, 549 nm, and 577 nm (Figure 10d). Likewise, under the same *r*-CP light excitation, the product chirality at 313 nm is opposite to that at 365 nm, 405 nm, 436 nm, 549 nm, and 577 nm.

The Arrhenius plots of the colloidal **13** and unsubstituted azobenzene during *cis-trans* thermal isomerization indicates that the activation energy ($E_a$) from *cis*-**13** to *trans*-**13** in a $CHCl_3$-methanol cosolvent is $E_a \approx 22$ kcal mol$^{-1}$ [54,121], which is slightly higher than that of the unsubstituted azobenzene of $E_a \approx 18$ kcal mol$^{-1}$ (Figure 10e). The higher $E_a$ value of the **13** contributes to the long-term chiroptical retention memory at ambient temperature. Likewise, from the Eyring plot that indicates the activation enthalpy ($\Delta H^{\ddagger}$)–activation entropy ($\Delta S^{\ddagger}$) relationship, the thermally excited *cis-trans* isomerization of azobenzene moieties in the **13** should obey the rotation mechanism rather than the inversion one [54,234]. The rotation mechanism arises from restricted rotational freedom of azobenzene moieties in the colloids. Thus, the chirality of *r*- and *l*-CP light can efficiently drive all of the chiroptical modes; generation, inversion, erase, switching, and short-term/long-term memory.

The thermal chiroptical stability of the **13** largely depends on the structure of the alcoholic solvents used (Figure 9). The $g_{CD}$ value of the **13** in methanol gives the shortest lifetime of 4–5 h. The $g_{CD}$ values in other non-branched alcohols and isopropanol diminishes in two days (Figure 10f). Notably, the $g_{CD}$ value in isobutanol remains unchanged for at least two days. The proper choice of alcohols allows for the efficient generation and desired retention times of the chiroptical properties.

In the **13** systems, we re-emphasize that the precisely controlled RI in the alcohol-CHCl$_3$ cosolvents is key for the CIE-CD signals disregard of the non-branched and branched alcohols (Figure 10g,h). Using either *r*- or *l*-CP light as a chiral electromagnetic force, the absolute $g_{CD}$ values, $|g_{CD}|$, of the **13** are resonantly enhanced at a specific $n_F$ value—that is, RI at 486.1 nm; $n_F$ = 1.382 for methanol, 1.404 for ethanol, 1.410–1.412 for *n*-propanol, 1.418 for *n*-butanol, 1.426 for *n*-pentanol, 1.405–1.411 for isopropanol, and 1.415 for isobutanol.

4.5.2. Achiral Luminescent π-Conjugated Polymer Endowed with Excitation Wavelength Dependent Circularly Polarized Light Chirality

The knowledge of the non-emissive colloidal **13** with azobenzene moiety led us to design the CAPC of the colloidal polymer made of highly luminescent **11** lacking photochromic moieties upon excitation of *r*- and *l*-CP light sources at six different wavelengths (313 nm, 365 nm, 405 nm, 436 nm, 549 nm, and 577 nm). The CAPC behaviors are readily characterized by the CIE-CPL and CIE-CD spectra.

Figure 11a,b displays the CIE-CD and UV spectra of the colloidal **11** by exciting the two different *l*-CP light sources at 546 and 365 nm and by the two different *r*-CP light sources at 546 and 365 nm, respectively. Unexpectedly, it is obvious that the positive-couplet CD spectrum endowed with the 546 nm *l*-CP light source inverts to negative-couplet CD of the 365 nm *l*-CP light source. Likewise, the negative-couplet CD spectrum induced by the 546 nm *r*-CP light source becomes positive-couplet CD when the 365 nm *r*-CP light source is applied. An excitation of CP light at 313 and 405 nm shows a similar tendency as the 365 nm excitation, while an excitation of CP light at 436 nm and 577 nm reveals a similar tendency as the 546 nm excitation. Upon excitation of the same *l*-CP (or *r*-CP) light, the choice of shorter (UV) and longer (visible) wavelengths of CP light causes an inversion of chiroptical sign of the **11**. Thus, we conclude that, the hand of CP-light, whether *left* or *right*, is not a deterministic factor for the CIE-CD sign of the **11**.

Figure 11c displays a comparison between the CIE-CPL and PL spectra of the **11** upon excitation of the 546 nm *l*- and *r*-CP light sources. A positive couplet-like CPL spectrum led by the 546 nm *r*-CP light source and negative-couplet-like CPL spectrum induced by the 546 nm *l*-CP light source is obvious. The absolute magnitude of $g_{CPL}$, $|g_{CPL}|$, is approximately 10$^{-3}$. Upon excitation of *r*-CP light, the colloidal **11** reveals weak (+)-CPL at 570 nm arising from the 540 nm (+)-CD band, while the colloids showed a weak (−)-CPL signal at 518 nm, originating from the 380 nm (−)-broad CD signal. A tandem controlling hand and photoexcited wavelength of the CP light source allowed for the production of CPL-functioned **11** with $\phi$ = 8% and $|g_{CPL}|$ = (2 − 4) × 10$^{-3}$ at 540 nm. The (+)-CIE-CD of the **11** generated by the 546 nm *r*-CP light for 30 nm irradiation completely inverted to negative-couplets solely by the 54 nm *l*-CP light for a prolonged irradiation of 120 nm, as shown in Figure 11d. In other word, angular momentum ($\pm\hbar$) of massless photon chirality enables to non-photochromic colloidal substances, resulting in induction and inversion of chemical chirality from achiral substances at ambient temperatures. Evidently, by optofluidically tuned RI of the achiral cosolvents, the CP light-driven CIE-CD signals of the **11** resonantly boosted at $n_D$ = 1.412 disregard of *r*- and *l*-CP light as excitation sources (Figure 11e).

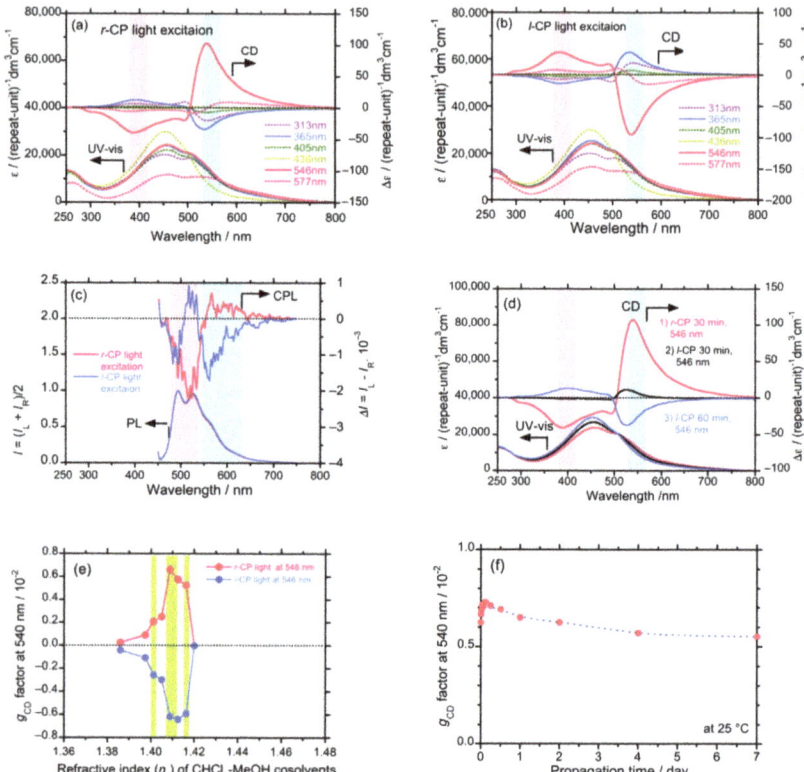

**Figure 11.** CIE-CD and UV spectra of the colloidal **11** in a CHCl$_3$/MeOH cosolvent (2.1/0.9 (volume/volume) (v/v)) with $n_D$ = 1.41 led by: (**a**) *r*-CP; and (**b**) *l*-CP light sources excited at six wavelengths (313 nm, 365 nm, 405 nm, 436 nm, 546 nm, and 577 nm); (**c**) CIE-CPL and PL spectra excited at 400 nm of the **11** led by the 546 nm *r*-CP and *l*-CP light sources for 60 min; (**d**) chiroptical erase (racemization) and inversion (anti-chirogenesis) in CIE-CD associated with change in UV-visible spectra of the **11** conducted by the 546 nm *r*-CP source for 30 min, followed by *l*-CP light source for 30 and 60 min; (**e**) the value of $g_{CD}$ at 540 nm of the **11** conducted by the 546 nm *r*-CP and *l*-CP light sources for 60 min as a function of $n_D$ of CHCl$_3$-MeOH cosolvents; (**f**) the chiroptical stability of the **11** generated by the 546 nm *r*-CP light at 25 °C in the dark. The $g_{CD}$ value is plotted as a function of time. Modified from an original article [55] and the book chapter [212].

The CIE-CD spectra of the **11** are thermally stable and remain unchanged at 25 °C for at least seven days (Figure 11f). For comparison, non-colloidal **11** homogeneously dissolved in CHCl$_3$ solution does not provide any detectable CD signals before and after prolonged irradiation of the 546 nm *r*-CP light. The largely restricted rotational freedom associated with efficient confinement of the CP light source in the colloid as the optical resonator is essential in conducting CP light-driven CAPC experiments.

According to a theoretical study [235], chiroptical enhancement is possible when an ideal chiral sphere efficiently interacts with the surrounding chiral molecules. This means that the CIE-CPL and CIE-CD signals from the colloids are further enhanceable by tuning chiral fluidic medium. CP light is key in the migration and delocalization of photoexcited energy in optically active macro-colloids containing $\sim 10^8$ of chlorophyll upon excitation of unpolarized sunlight [171–174]. Note that chlorophylls contain three stereogenic centers at the peripheral positions of chlorophyll rings and two stereogenic centers in the alkyl chain tail.

### 4.5.3. Achiral Polymethacrylate Carrying Azobenzene Pendants Endowed with Excitation Wavelength Dependent Circularly Polarized Light Chirality

The concept of CAPC with light covers absolute asymmetric photosynthesis, photodestruction, and photoresolution. In the 1970s, Calvin et al. reported an anomaly of an excitation wavelength dependent CP light-driven photosynthesis-mode AAS recognized as switching product chirality of [8]-helicene in homogeneous toluene solution [52]. In 2014, Meinert et al. reported the wavelength dependent CP light-driven photodestruction-mode AAS revealing switching chirality when *rac*-alanine film is decomposed upon irradiation of two vacuum-UV light sources (184 and 200 nm) [53]. Our results of CAPC from the **14** and the **11** as their colloidal states prompted to address an apt question, as to whether the excitation wavelength dependent switching chirality conducted by CP light-driven photoresolution mode AAS is generalizable; whether restricted rotational modes are crucial for CP-light driven CAPC [56].

We chose the colloidal **21** bearing achiral azobenzene moieties as pendants to test the excitation wavelength dependent CP light-driven photoresolution mode AAS (Figure 12). The experiment of CAPC with light was designed to use three wavelengths (313 nm, 365 nm, 436 nm) for *r*- and *l*-CP light sources to efficiently excite π–π* and/or n–π* transitions of azobenzene moieties in restricted rotational states in the **21**. Contrarily, the azobenzene moieties are non-restricted, free rotational states in homogeneous solution. Similarly, restricted rotations along C–C single bonds in the main chain of **21** are possible in the colloids, while non-restricted rotations of those C–C single bonds occur in homogeneous solutions.

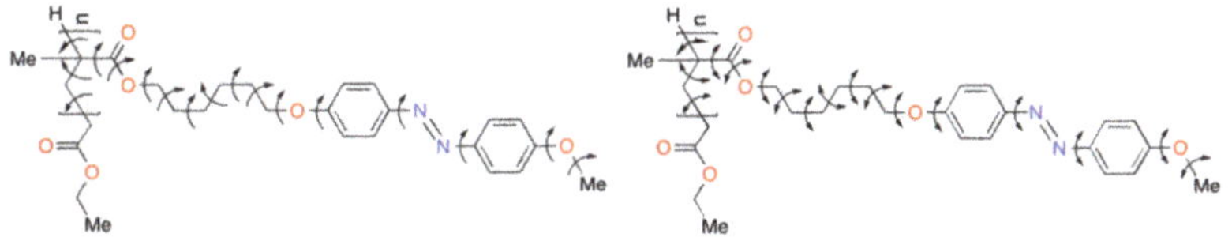

**Figure 12.** Chemical structures of the colloidal **21** to employ wavelength-dependent CP light driven CAPC experiments. By tuning $n_D$ value, a special mixture of 1,2-dichloroethane and methanol was used. Single-headed arrow means restricted rotation clockwise (CW) or counter-clockwise (CCW), while double-headed arrow is possible to freely rotate CW and CCW.

In Figure 13a, we show the values of CD ellipticity at 313 nm (in mdeg, left ordinate) and $g_{CD}$ at 313 nm (right ordinate) of the spherical shape colloidal **21** (125–300 nm in size) as a function of the $n_D$ of 1,2-dichloroethane (DCE, good solvent, $n_D = 1.444$) and methylcyclohexane (MCH, poor solvent, $n_D = 1.422$) cosolvents upon excitation of very weak *l*- and *r*-CP light sources at 365 nm for a short-period ($\approx 30$ μW·cm$^{-2}$, 60 s). Obviously, the values of CD ellipticity and $g_{CD}$ steeply are maximized at a very specific $n_D = 1.425$ (DCE/MCH = 0.5/2.5 (volume/volume) (v/v) regardless of the 365 nm *l*- and *r*-CP light sources.

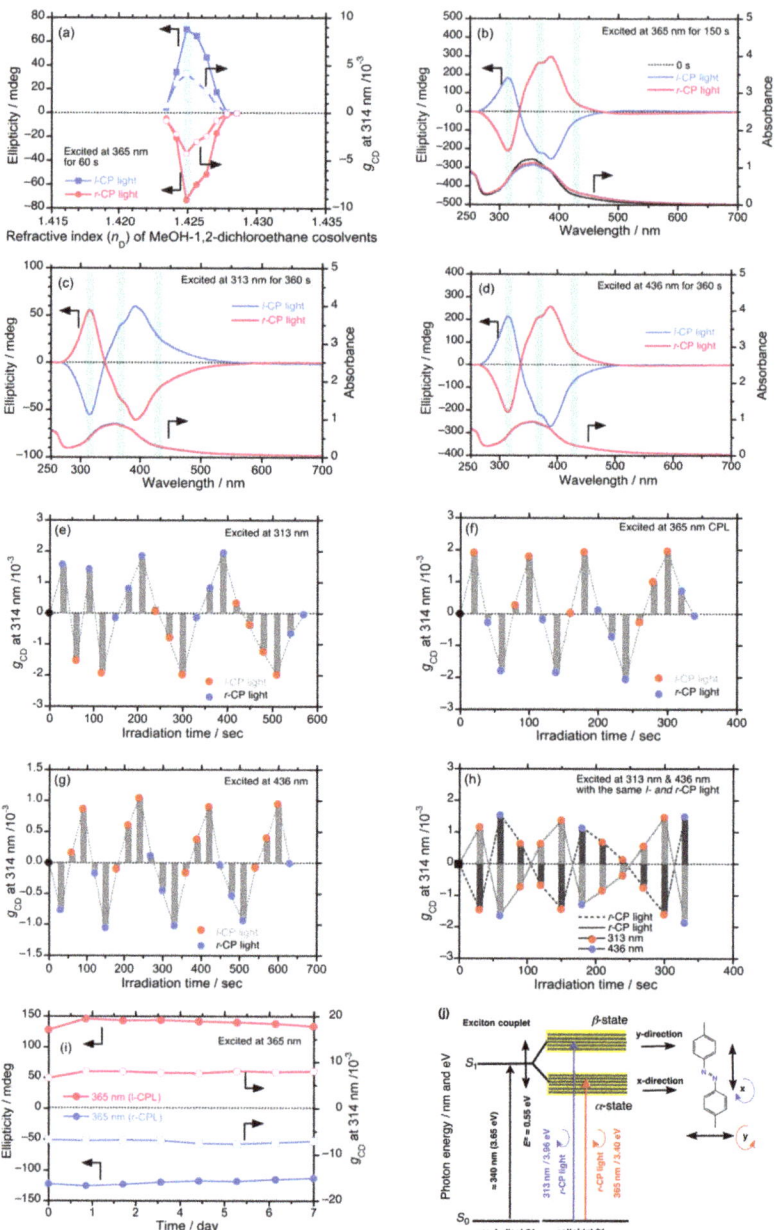

**Figure 13.** The CIE-CD and UV spectra of the colloidal **21** excited with: (**a**) *r*-CP light source at 546 nm and *l*-CP light source at 365 nm; and (**b**) *l*-CP light source at 546 nm and *l*-CP light source at 365 nm; (**c**) CIE-CPL and PL spectra of the **21** excited with *r*- and *l*-CP light source at 546 nm; (**d**) the $g_{CD}$ values as a function of $n_D$ of the cosolvents. The alteration in the $g_{CD}$ value at 314 nm of the **21** upon excitation with *r*- and *l*-CP light sources (**e**) at 313 nm; (**f**) at 365 nm; (**g**) at 436 nm; and (**h**) at 313 nm and 436 nm; (**i**) Thermal stability of the **14**. The $g_{CD}$ value as a function of solvent temperature; (**j**) Anti-Kasha's rule and Jablonski diagram of the colloidal **21**. Modified from the original article [56].

At the specific $n_D$ = 1.425 and volume fraction of DCE/MCH, we conducted *l*- and *r*-CP light driven CAPC experiments of the colloid **21** at 365 nm (150 s), 436 nm (360 s), and 313 nm (360 s). From Figure 13b, the 365 nm *r*-CP light source induces bisignate negative-couplet-like CD profile at 310 nm and 360–390 nm, conversely, the 365 nm *l*-CP light induces bisignate positive-couplet-like CD profile. Likewise, the 436 nm *r*-CP light source induces similar bisignate negative-couplet-like CD profile, as shown in Figure 13d.

Conversely, the 365 nm *l*-CP and the 436 nm *l*-CP light sources induce bisignate positive-couplet-like CD spectra (Figure 13b,d). On the other hand, it is evident from Figure 13c that the 313 nm *r*-CP and *l*-CP light sources induce bisignate positive and negative couplet-like CD profiles at 310 nm and 360–390 nm, respectively. For comparison, the 436 nm *r*-CP light-driven CAPC experiments to the colloids made of the starting monomer of **21** in MCH-DCE and **21** in homogeneous DCE solution did not induce any detectable CD signals [56].

These excitation-wavelength dependent CAPC experiments allow to verify a possibility of all chiroptical generation, erase, inversion, multiple switching, and a long-term memory characteristics of the **21** in the optofluidic medium with the $n_D$ = 1.425 using *l*- and *r*-CP light sources at 313 nm (Figure 13e), 365 nm (Figure 13f), 436 nm (Figure 13g), and a tandem of 313 nm and 436 nm (Figure 13h). All of the chiroptical modes are possible when the 313 nm, 365 nm, and 436 nm CP light sources are employed disregard of *l*- and *r*-CP light sources. The bisignate chiroptical sense of the **21** led by the 313 nm *l*-CP light source is absolutely opposite to the sense induced by the 365 nm and 436 nm *l*-CP light sources.

We re-confirmed that an alternative excitation of the 313 nm *l*-CP and 436 nm *l*-CP light sources to the **21** enables conducting all of the chiroptical modes. Likewise, the alternative excitation of the 313 nm *r*-CP and 436 nm *r*-CP light causes the opposite characteristics of the dual *l*-CP light sources. The resulting optically active **21** endowed with the 365 nm *r*-CP and *l*-CP light sources had long-term memory effects in the dark for at least seven days (Figure 13i). This uniqueness arises from the very restricted rotations of the azobenzene moieties and the C–C single bonds in the main chain of the **21** in the dark (Figure 12, left). The **16** in the ground state is a closed system that endures thermally activated energy at room temperature. Conversely, the **16** in the photoexcited state during CP-light irradiation works as the open-flow, soft-matter-based photonic resonator adaptable to the external CP-light energy by confining into the resonator with the help of the tuned RI fluidic medium.

From the spectra shifts at the three major bands of the **21**, which includes $v$(C–H) at ~ 3000 cm$^{-1}$, $v$(C = O) at ~1730 cm$^{-1}$, and $v$(C–O–C) at ~1250/~1150 cm$^{-1}$, a prolonged photoirradiation with *r*-CP at 365 nm for 10 min leads the **21** with *trans*-azobenzene pendants to *cis*-azobenzene **21** colloids. As a result, the *cis*-colloidal **21** prevents an efficient π–π stacking of azobenzene pendants, followed by production of an ill-organized, smaller size colloids due to a bent structure of *cis*-azobenzene. The *cis*-**21** colloids in the cosolvents were transparent by the naked eye. The small size colloids, less than ~100 nm, appear inconvenient to efficiently confine a traveling light at the colloid–liquid interface and prevent the multiple reflection mode of WGM.

In Figure 13j, we propose a hypothesis of the anti-Kasha rule that, by combining exciton couplet theory, Kasha's rule, Jablonski diagram, and x–y directions of azobenzene at the photoexcited state, two pathways as a spontaneous relaxation process from α-state (*y*-axis of azobenzene) with the lowest energy (~365 nm) and the β-state (*x*-axis of azobenzene) with higher energy (~313 nm) are possible. The *r*-CP light sources at 365 nm and 313 nm would make y- and x-axes of azobenzene twist CCW and CW, respectively. Likewise, *l*-CP light sources at 365 and 313 nm oppositely twist y- and x-axes of azobenzene CW and CCW, respectively [56]. The anti-Kasha rule is prominent to CP-light driven CAPC of colloidal **21**, **11**, and **14** in the RI-tuned optofluidic media. The anti-Kasha rule is applicable to characterize CIE-CD and CIE-CPL spectra in the controlled chirogenesis, as given in a Section 4.7.

## 4.6. Tempo-Spatial Chirogenesis
### 4.6.1. Changes in Colloidal Sizes of Diaryl Polysilane with Propagation Time

In this section, we highlight two topics: (a) multiple resonance effects in the $g_{CD}$ value of the colloidal **21** in the **10R**/CHCl$_3$/MeOH and **10S**/CHCl$_3$/MeOH tersolvents among the colloidal diaryl polysilanes **21–24** (Figure 14); and (b) tempo-spatial chirogenesis, including changes in colloidal sizes and CIE-CD signals at several volume fractions of **10R** and **10S** in the tersolvents with propagation time [207].

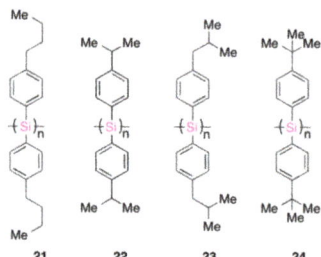

**Figure 14.** Chemical structures of CD-silent helical diaryl polysilanes revealing tempo-spatial CIE-CD effects with colloidal sizes in the limonene-containing optofluidic media.

According to the polymer consistent force-field (PCFF), diaryl polysilane **21** has four local minima from the potential energy surface due to phenyl–phenyl interaction (Figure 15a). Thus, **21** in a homogeneous solution reveals a CD-silent spectrum in the ground state (Figure 15b), and possibly a CPL-silent spectrum in the photoexcited state due to equal populations between two *P*-screw and two *M*-screw Si-Si backbones.

The $g_{CD}$ values of the colloidal **21** show two and three extrema at the specific volume fractions of **10R**/CHCl$_3$/MeOH and **10S**/CHCl$_3$/MeOH tersolvents (Figure 15e,f). This anomaly appears not unique for the **21** because a similar tendency in the $g_{CD}$–$n_D$ relations can be seen in the colloidal **14** (Figure 8d) and the colloidal **7** and **8** in the limonene-containing tersolvents (Figure 6e,f). The **21** at the optimized volume fraction of **10R** and **10S** in the tersolvents reveal clear bisignate CIE-CD spectra around 400 nm (Figure 15c) associated with mono-signate CIE-CPL spectra at 410 nm (Figure 15d). The (+)-sign CIE-CPL led by **10R** is identical to th (+)-sign of couplet-like CIE-CD at the first Cotton band of ~408 nm, conversely, the (−)-sign CIE-CPL with **10S** is the same of (−)-sign of CIE-CD with **10S** at 408 nm. The relaxation scenario from the $S_1$ to $S_0$ states of the **21** obeys the Kasha rule [216,217].

The reasons for two and/or three extrema in the $g_{CD}$ values at the specific volume fractions of **10R** and **10S** is ascribed to three rotamers of **10R** (and **10S**) [236,237]. Non-rigid chiral monocyclic terpene **10R** consisting of isopropenyl group and cyclohexene ring has one freely rotable C–C bond between the moieties. The rotable C–C bond generates three equatorial rotamers of **10R** (equatorial 1 (Eq1), equatorial 2 (Eq2), equatorial 3 (Eq3), Figure 16) with nearly equal populations with different ORD spectra with different signs in the range of 200–589 nm (Eq1; 0.39 with an intense (+)-ORD, Eq2; 0.31 with an intense (−)-ORD, Eq3; 0.30 with a weak (−)-ORD) along with a small fraction of energetically unstable axial rotamers [236,237]. We assume that the packing structure of **21** in the colloids is reorganizing in response to a preferential rotamer of **5R** and **5S**, whose fractions depend on a volume fraction of limonene in the tersolvents [207].

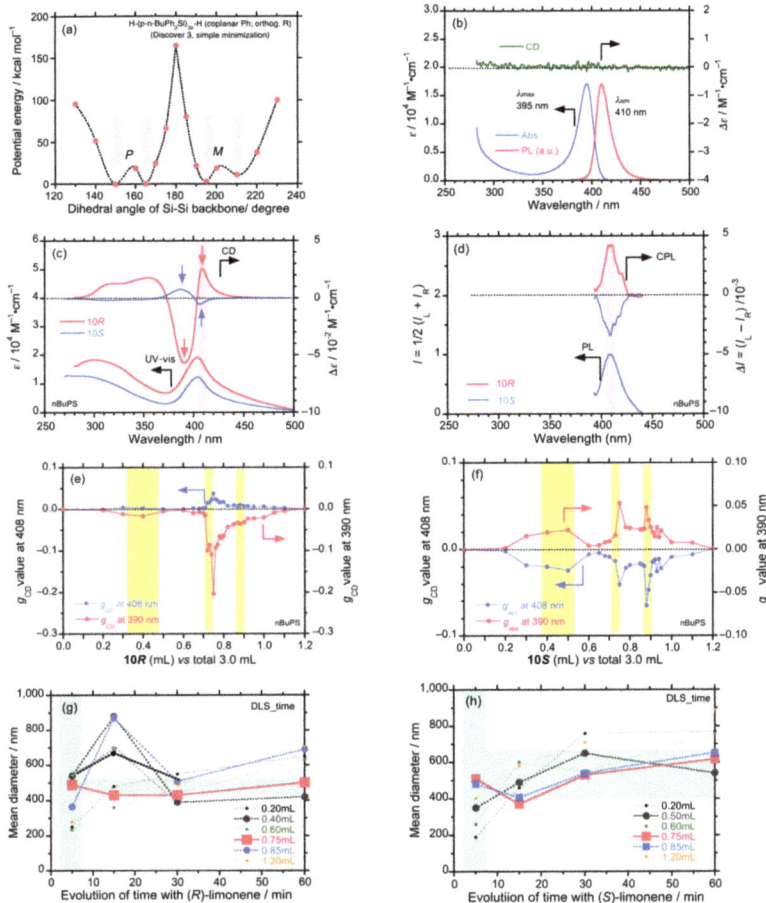

**Figure 15.** (a) Potential energy surface having four local minima of **21**; (b) CD, UV–visible, and PL spectra of **21** in a dilute CHCl$_3$ solution; (c) CIE-CD and UV-visible and (d) CIE-CPL and PL spectra of the colloidal **21** in the tersolvents of **10R** (or **10S**)/CHCl$_3$/MeOH = 0.75/0.30/1.95 (volume/volume/volume) (v/v/v). The values of $g_{CD}$ at 390 m and 408 nm of the **21** as a function of volume fraction of: (e) **10R** and (f) **10S** in the tersolvents of **10R** (or **10S**)/CHCl$_3$ (0.3 mL)/MeOH and total volume 3.0 mL. The colloidal sizes of **21** at several volume fraction of: (g) **10R** and (h) **10S** in the tersolvents as a function of propagation time. Modified from an original article [207].

**Figure 16.** Three equatorial rotamers (equatorial 1 (Eq1), equatorial 2 (Eq2), equatorial 3 (Eq3)) of **10R** at two local and one global minima optimized with Gaussian09 (DFT, B3LYP, 6–31G(d,p) basis set), suggested by two papers [236,237]. The calculated optical rotations at 365 nm with a relative population of Eq1, Eq2, Eq3 are +1501° (0.39), −428° (0.31), and −206° (0.30), respectively [237]. The other calculated populations of Eq1, Eq2, Eq3, and total of three axial rotamers are 0.32, 0.21, 0.43, and 0.04, respectively [236].

In Figure 15g,h, the tempo-spatial behaviors in the colloidal size of **21** at several volume fractions of **10R** and **10S** in their tersolvents are given by means of dynamic light scattering measurements. We can see the marked alteration in the colloidal sizes (200–900 nm) of **21** in the **10R** and **10S** tersolvents; the colloid sizes at two specific volume fractions (0.40 and 0.75 mL for **10R**; 0.75 and 0.85 mL for **10S**) appear unchanged with time, while those at other volume fractions seem to fluctuate and/or grow gradually with time. The resonant volume fractions in Figure 15e,f are possibly connected to the tempo-spatial behaviors of the colloidal sizes, followed by the maximizing $g_{CD}$ values.

4.6.2. Time-Dependent Evolution of CIE-CD and CIE-CPL in Co-Colloids of Achiral π-Conjugated Polymer and Helical Dialkyl Polysilanes

Based on the tempo-spatial behaviors of the colloidal **21** endowed with **10R** and **10S**, we designed co-colloids comprising achiral π-conjugated **20** (Figure 7) and rod-like helical dialkyl polysilanes, **5S** and **5R** (Figure 2). The optimized molar ratio as repeating units of **20** and **5R** (or **5S**) is found to be 1:1 (Figure 16h). The optimized $n_D$ value in the CHCl$_3$-MeOH cosolvents enabling the largest RI value of the co-colloid is found to be $n_D$ = 1.405 in a cosolvent of CHCl$_3$/MeOH = 1/2 (v/v) (Figure 17h). Note that the rigid rod-like helical **5R** and **5S** act as helical scaffolding to non-helical π-conjugated polymers.

The alterations of the CIE-CD/UV-visible-NIR spectra and the CIE-CPL/PL spectra of the 1:1 co-colloids of **20** and **5R** as a function of propagation time are displayed in Figure 17a,b. Likewise, the changes in the CIE-CD/UV-visible-NIR spectra and the CIE-CPL/PL spectra of the 1:1 co-colloids of **20** and **5S** with propagation time are given in Figure 17a,b. Obviously, the absolute magnitudes in CIE-CD and CIE-CPL increase and tend to level-off for a prolonged time of 24 h, regardless of **5R** and **5S**. The **20**/**5S** and **20**/**5R** co-colloids generate CD-activity and CPL-activity from optically inactive **20** with help from main chain helicity and/or side chain chirality of **5S** and **5R**. Time-dependent reorganization of **20**, with help from **5R**/**5S** in fluidic media occurs on the order of hours at room temperature.

To associate the $g_{CD}/g_{CPL}$ values with the colloidal size, we characterized the time-dependent co-colloidal sizes using a dynamic light scattering (DLS) method, as shown in Figure 17f,g. An approximately 650 nm size of the **20-5S** co-colloids progressively reach ~2000 nm within 24 h (Figure 17f). Likewise, an approximately 400 nm size of the **20-5R** co-colloids in the beginning attains ~1800 nm within 24 h (Figure 17e). The larger sizes of the co-colloids after prolonged propagation times are responsible for the increases in the larger $g_{CD}$ and $g_{CPL}$ values of the **20-5R** and **20-5S** co-colloids associated with a significant red-shift in CIE-CD and CIE-CPL spectra. It takes several hours to re-organize **20** with help from the Si–Si main chain helicity and/or side chain chirality of **5R** and **5S**.

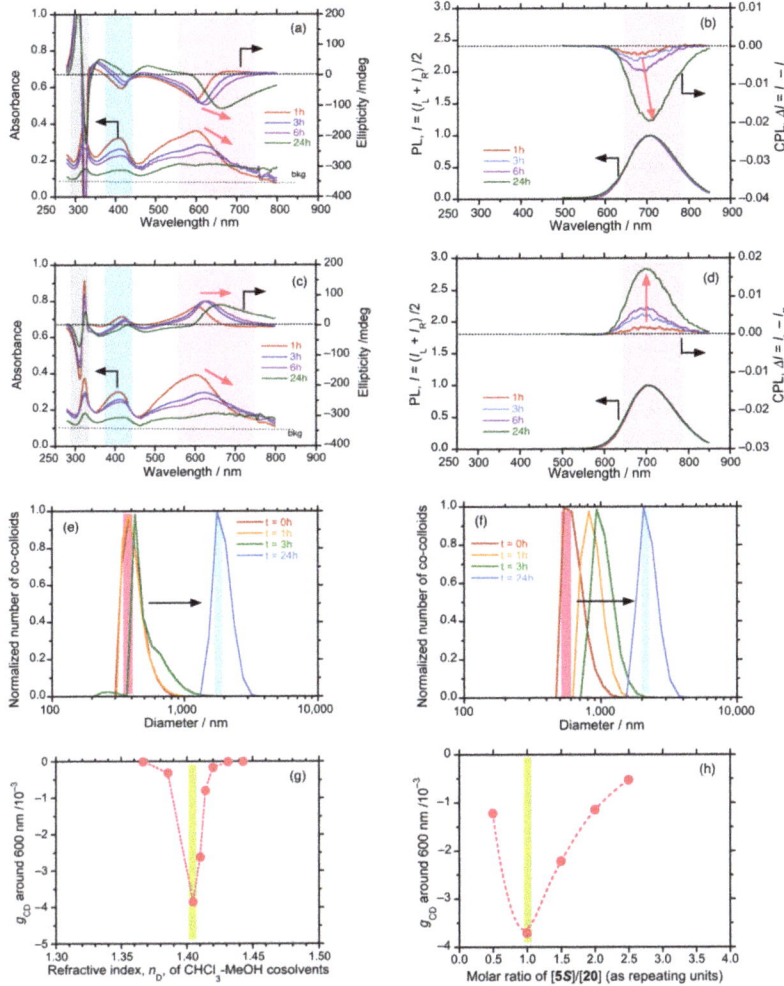

**Figure 17.** Alteration in the (**a**) CIE-CD and UV-visible-NIR spectra and (**b**) CIE-CPL and PL spectra excited at 400 nm for the **5S–20** co-colloids (1-to-1 molar ratio). Alteration in the (**c**) CIE-CD and UV-visible-NIR spectra and (**d**) CIE-CPL and PL spectra excited at 400 nm for the **5R–20** co-colloids (1-to-1 molar ratio). These co-colloids were produced in CHCl$_3$-MeOH (2/1 (v/v)). The hydrodynamic sizes of (**e**) **5S–20** and (**b**) **5R–20** co-colloids with the propagation time; (**g**) the $g_{CD}$ value around 600 nm of the **20–5S** co-colloid (1:1) as a function of $n_D$ in the CHCl$_3$–MeOH cosolvent; (**h**) the $g_{CD}$ value of the **20–5S** co-colloid as a function of the **5S**-to-**20** ratio generated in the CHCl$_3$–MeOH (2:1) cosolvent. Modified from an original article [210].

### 4.7. Unveiling Anti-Kasha's Rule from CIE-CPL, CIE-CPLE, and CIE-CD in Co-Colloids of Achiral π-Conjugated Polymer and Helical Polysilanes

In this section, we deal with the origin of bisignate CIE-CPL associated with bisignate CIE-CD spectra characteristic of the co-colloids comprising **11** with **5R** (1:1 mol ratio) and **11** with **5S** (1:1 mol ratio) by connecting to the corresponding CIE-CPLE spectra based on a hypothesis of anti-Kasha's rule to explain the origin of dual CPL emission behaviors. Note that **5R** and **5S** can work as *P*-screw and *M*-screw helical scaffoldings in the range of 280–325 nm to achiral **11** in the range of 350–650 nm. Prior to a series of chiroptical experiments, the $n_D$ value of the CHCl$_3$-MeOH cosolvents is optimized (Figure 18e); $n_D$ = 1.410 of the

CHCl$_3$/MeOH (2.1/0.9 (v/v)) cosolvent for **11/5R** = 1:1 in mole ratio and $n_D$ = 1.415 of the CHCl$_3$–MeOH cosolvent (2.2/0.8 (v/v)) cosolvent for **11/5S** = 1:1 in mole ratio, respectively.

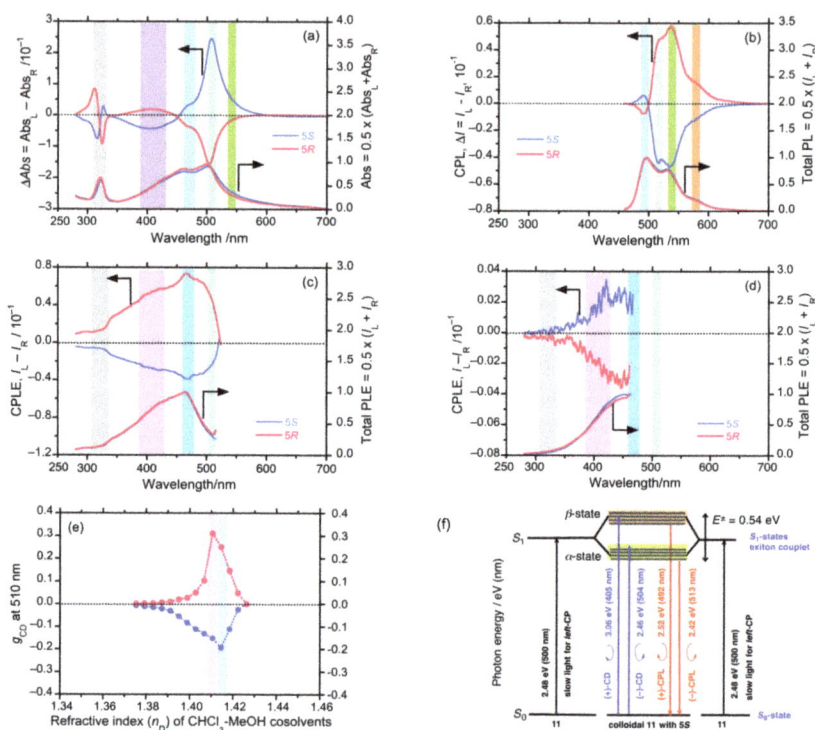

**Figure 18.** (a) The CIE-CD and UV-visible spectra and (b) the CIE-CPL and PL spectra excited at 420 nm of the co-colloids of **11** with **5S** and **5R** in a 1-to-1 molar ratio; (c) The CIE-CPLE and PLE spectra of the co-colloids of **11** with **5S** and **5R** monitored at (d) 575 nm and (e) 490 nm (**5R**) in the CHCl$_3$–MeOH cosolvent (2.2/0.8 (v/v)) and (d) 575 nm and (e) 495 nm (**5S**) in the CHCl$_3$–MeOH (2.1/0.9 (v/v)); (e) the $g_{CD}$ value at 500 nm of the co-colloidal **11**–**5S** (1:1) as a function of $n_D$ in the CHCl$_3$–MeOH cosolvent; (f) A hypothesis of anti-Kasha's rule and Jablonski diagram to explain bisignate CIE-CD, CIE-CPL, and CIE-CPLE spectra of the colloidal **11** endowed with **5S**. Modified from an original article [209].

Figure 18a,b depicts the CIE-CD/UV-visible and CIE-CPL/UV-PL spectra excited at 420 nm of the co-colloids consisting of **11/5R** (1:1) and **11/5S** (1:1), respectively. In Figure 18a, the weaker bisignate CIE-CD/UV spectra around 300 nm are attributed to Siσ–Siσ* transitions of **5R** and **5S**. In Figure 18a, it is obvious that the intense bisignate CIE-CD/UV-visible spectra at 400 and 500 nm are attributed to π–π* transitions of **11** led by **5R** and **11** led by **5S**, respectively. However, from Figure 18b, the intense bisignate, dual CIE-CPL bands associated with PL band located at 490 nm and 510–580 nm are assumed to originate from π–π* transitions of the colloidal **11** with **5R** and **5S**. One question is whether the bisignate CPL, not obvious from PL, arises from π–π* transitions at 400 nm and 500 nm from the colloidal **11**. To answer this query, we obtained the first CIE-CPLE/PLE spectra monitored at 575 nm and at 480–480 nm (Figure 15c,d). The (+)-sign CIE-CPLE/PLE spectra of the **11**–**5R** monitored at 575 nm are nearly identical to (+)-sign CD spectra of the π–π* transitions of the **11**–**5R** in the range of 300–500 nm and vice versa (Figure 15c). Conversely, the (−)-sign CIE-CPLE/PLE spectra of the **11**–**5R** monitored at 480–490 nm are nearly identical to (−)-sign CD spectra of the $\pi$–$\pi^*$ transitions of the **11–5R** in the range of 300–450 nm and vice versa (Figure 15d).

These spectral results allow us to propose the anti-Kasha's rule, similar to Section 4.5.3. Two pathways of spontaneous relaxation processes from the $\beta$-state in higher energy (~400 nm) and the $\alpha$-state with lower energy (~500 nm) are possible. The *r*-CP light in the range of 350 and 500 nm photoexcited the **11–5S** (and **11–5R**), followed by relaxation to the $\beta$- and $\alpha$-states that emit *r*-CPL and *l*-CPL, respectively. On the other hand, the *r*-CP light at 500 nm photoexcited the **11–5S** (and **11–5R**), followed by relaxation to the $\alpha$-states that emit *l*-CPL only. The photoexcited $\beta$-state with *r*-CP light has two relaxation pathways; from $\beta$-state with *r*-CP light and from $\alpha$-state with *l*-CP light.

The *r*- and *l*-CP light excitation processes, followed by spontaneous radiation processes, may not obey the conventional Kasha's rule, rather obey anti-Kasha's rule because dual CP luminescence characteristics are obvious. In recent years, anti-Kasha's rule became popular and is now a hot topic in the realm of photophysics and photochemistry when dual and multiple photoluminescence characteristics are observed [238–241]. Monitor-wavelength dependent CPLE spectroscopy associated with the corresponding CPL and CD spectroscopy is helpful to investigate the anti-Kasha's rule in solutions and in the solid film, and as solid powders.

## 5. Perspectives—Colloids Connecting to Light, Helix, Coacervate, Panspermia, Microoptics, Chiroptics, Radioisotopes, Nuclear Physics, Biology, Homochirality Question, and Cosmology

With assumptions based on the coacervate and Panspermia hypotheses, and WGM in µm-size colloidal polymers adaptable to external chemical and physical stimuli in a tuned optofluidic media, one remaining issue should be to address the effects of chiral physical sources and particles existing in the interstellar universe and on Earth, possibly causing the biomolecular handedness on Earth.

First, the author wishes to mention personal experience in my graduate school four decades ago. In 1977, Kunitake and Okahata found that didodecyldimethylammonium bromide in water, the simplest amphiphilic synthetic molecule (Figure 19, top left), spontaneously generates multilayer spherical colloids of ~100 nm in diameter from observation of TEM [242]. Under supervision of Kunitake and his three lab staffers (Takarabe, Okahata, and Shinkai), the author designed four new achiral, amphiphilic acrylic monomers (Figure 19, top right), polymerizable under influence of the $\beta^-$-decay process from $^{60}$Co→$^{60}$Ni nuclear fission reaction [243].

Among the four amphiphiles, $AcC_{10}C_{12}N^+2C_1Br$ in water, showing two phase-transition temperatures ($T_c$s) at $-16\,°C$ and $-9\,°C$, formed structureless colloids with no obvious multilayers, ranging from ~10 nm to ~100 nm in diameter (Figure 19a). Followed by polymerization of the colloids embedded to ice in the range of $-30\,°C$ and $-50\,°C$ with help from the $\beta^-$-decay nuclear reaction, the author serendipitously observed that the spiral fibers of amphiphilic polyacrylate (~10 nm in diameter) are formed in many places. Possibly, handedness of spiral fibers was left-handed (Figure 19b–d). Polymerization was confirmed by significant shifts in IR frequency in $v_{(C=O)}$ from 1715 cm$^{-1}$ (monomer) and 1725–1730 cm$^{-1}$ (polymer). Likewise, $AcC_{10}C_{18}N^+2C_1Br$ in water exhibited two phase-transition temperatures ($T_c$s) at 28 °C and 41 °C. Colloidal $AcC_{10}C_{18}N^+2C_1Br$ polymerized at 45 °C influenced by the $\beta^-$-decay reaction formed clearly loop-like structures (ca. 40–60 nm) made of spiral fiber (Figure 19e,f). The handed spiral loops from the polymerizable colloids by dosing handed electron, $\gamma$-rays, and anti-neutrino have very reminiscent circular DNAs and proteins/polypeptides influenced by high-energy cosmic rays under extraterrestrial and terrestrial conditions of primordial eons.

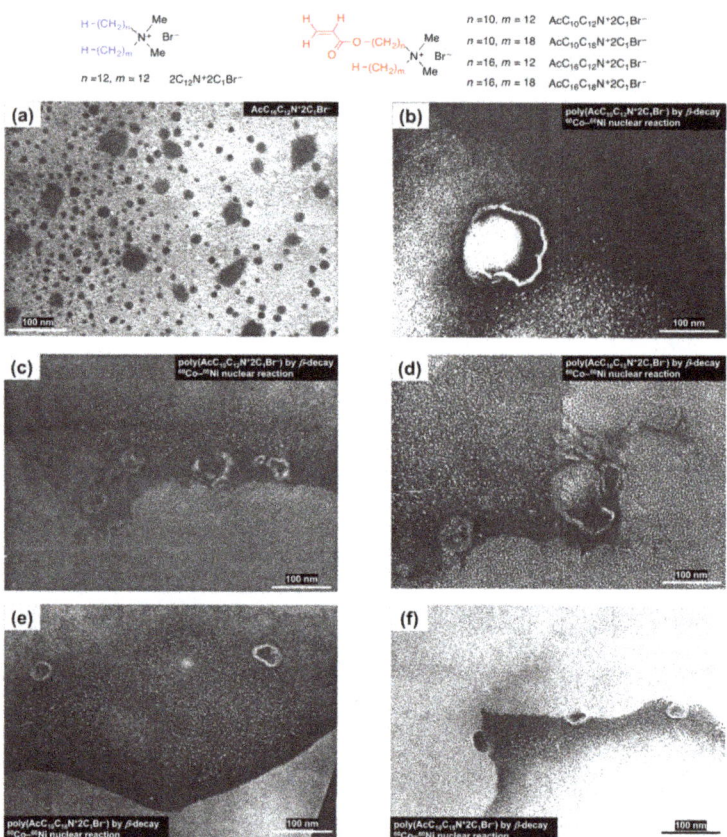

**Figure 19.** (Top, left) Chemical structures of the first amphiphilic didodecyldimethylammonium bromide to spontaneously form spherical shape colloids (~100 nm) with multilayer membrane [242]. (Top, right) Four amphiphilic acrylic monomers synthesized by the author (1978) (unpublished as and English paper, disclosed only as his Master thesis [243]. A TEM image (negative statue) of (**a**) colloidal $AcC_{10}C_{12}N^+2C1Br^-$ (20 mM) (negative statue) dispersed in pure water; (**b–d**) TEM images (positive statue) for the colloidal particles polymerized in the range of −30 °C and −50 °C endowed with $\beta$-decay of $^{60}Co \rightarrow ^{60}Ni$ nuclear fission reaction; dose rate: 20 R ($5.2 \times 10^{-3}$ C kg$^{-1}$) per s for 50 h, corresponding to $3.6 \times 10^6$ R ($9.3 \times 10^2$ C kg$^{-1}$); (**e,f**) TEM images (positive statue) for closed loop shape structures from colloidal $AcC_{10}C_{18}N^+2C1Br^-$ polymerized at 45 °C led by $\beta$-decay of $^{60}Co \rightarrow ^{60}Ni$ nuclear reaction; dose rate: 20 R ($5.2 \times 10^{-3}$ C kg$^{-1}$) per s for 3.5 h, corresponding to $2.6 \times 10^5$ R ($0.6 \times 10^2$ C kg$^{-1}$). These TEM images suggested spiral fibers dispatching from the colloidal polymers [243]. A Hitachi (Ibaraki, Japan) model H-500 TEM instrument was employed with an accelerating voltage of 100 kV at lab of Motoo Takayanagi (Kyushu University). All figures (**a–f**) were scanned from the Master thesis [243].

The author conjectured at that time that handed spiral from the colloids in the ultra-cold and hot conditions is a model of the cometary coacervates hypothesis in comets [123]. Three chiral physical sources are possible to connect to the spiral rings because the $\beta^-$-decay process from $^{60}Co$ radiates $l$-hand spinning $\beta$-electron, $r$-hand $\gamma$-ray (high-energy circularly polarized light), and $l$-hand anti-neutrino simultaneously from the source [69–76].

Amphiphilic colloidal particles in water may feel the forces led by these handed radioactive physical sources because atmospheric thundering and radioactive materials in Earth's crust and mantle might be ubiquitous in primordial eons, as well as the present days. Similarly, microdroplets in aqueous solution and aqueous aerosols in the atmosphere may be susceptible to these handed physical sources, leading to accelerated chemical reactions led by optofluidic WGM scenarios.

In recent years, several review papers on terrestrial and extraterrestrial life, chemical evolution, the origin of life, and the origin of chirality on Earth have been disclosed [11–21,244–247]. These papers comprehensively cover curious topics in the realms of astrochemistry, astrobiology, geoscience, space life science, synthetic biology, origin of life, central dogma, natural science, and materials science. The reviews should provide new insights in the realms of helical polymer chemistry [248], supramolecular chemistry [249], flow chemistry with light [250], natural product chemistry [251], and chiroptical polymer chemistry [252]. However, possible connections of these topics with non-equilibrium open-flow colloidal systems seem to be rare. An importance of (chir)optical characteristics of surrounding media, degree of smooth interfaces, micro-optics and optofluidics, photon confinement, resonance condition, WGM, and nature of light–matter interactions in the presence and absence of static electric field and static magnetic field are not well recognized. In my view, when one can setup open-flow, cell-wall-free μm-size colloidal polymers with a smooth surface dispersed in aqueous conditions with tuned RIs, any mirror-symmetrical physical forces and non-mirror-symmetrical inherently chiral particles, such as solar neutrinos and anti-neutrinos, are possible to induce and boost the left–right imbalance disregard of terrestrial and extraterrestrial conditions, even at cryogenic temperatures. The concept of the open-flow, non-equilibrium optofluidic colloidal and co-colloidal systems shed light at opening new science and engineering opportunities in chirogenesis and photochirogenesis, in addition to conventional stereochemistry, photochemistry, and photophysics in homogeneous solutions. The ideas are applicable to chirogenesis and photochirogenesis in μm-size water-based aerosols and microdroplets to address plausible scenarios for the homochirality questions on Earth.

## 6. Conclusions

Colloids and co-colloids with higher RI dispersed in optofluidic mediums with lower RI are capable of chiroptical resonators susceptible of external chiral chemical and CP light sources. Concerning μm-size colloids, made of π-/σ-conjugated polymers in optofluidic medium, with a tuned RI, several resonance effects cause efficient chirogenesis and photochirogenesis, revealed by gigantic, enhanced CD, ORD, CPL, and CPLE spectral datasets. The chirogenesis is susceptible to the nature of optofluidic medium, including chiral chemical substances and a ratio of π- and σ-conjugated polymers. Moreover, a fully controlled absolute photochirogenesis upon excitation of *left-* and *right-*CP light sources at several wavelengths is a possible disregard of photochromic and non-photochromic polymers at specific refractive indices of optofluidic media. The wavelength-dependent CP light source can fully drive all chiroptical generation, inversion, erase, switching, and short-/long-lived memories. New knowledge and deep understanding of CPL, CPLE, CD, and ORD spectroscopy for colloids and co-colloids can shed light on rationally designing elaborate CPL functioned π-/σ-conjugated polymers and metal-coordination polymers in the future. A tandem analysis of CPLE, CPL, and CD spectroscopy helps to discuss Kasha's rule and anti-Kasha's rule of radiation processes from the photoexcited optically active molecules, polymers, colloids, solids, crystals, and thin films. Our comprehensive studies, using artificial polymeric luminophores and chromophores, should provide new insight of the open-flow, non-equilibrium functional colloids, and co-colloids in the photoexcited and ground states in an RI-tuned optofluidic medium. Based on the results shown above, and my personal experiences, any chiral physical forces and particles could induce and boost the left–right imbalance, which is one possible answer to the homochirality question on Earth's disregard of terrestrial and extraterrestrial origin chirality. In particular, when open-flow, μm-size colloidal polymers with smooth surfaces, adaptable to external physical and chemical biases are surrounded by an optofluidic medium, such as an aqueous solution, the capabilities of chiral physical forces and chiral chemical sources are possible, to enhance by huge numbers of recycling times ($N$) in the optofluidic colloidal polymers, in line with the scheme of WGM. This means that a subtle left–right imbalance interior of the colloid, e.g., $10^{-6}\%$ ee, is boostable by $10^{-6} \times N$. This scenario is in contrast to a conventional stereochemistry, photochemistry, and photophysics in a homogeneous solution system, in which the light–matter interaction occurs in one time ($N = 1$) only by assuming the Beer–Lambert law in homogeneous solutions. When colloid-induced emission systems chiroptically fulfil the resonance condition, one can anticipate the significant enhancements in CPL and CD signals.

**Funding:** The author is grateful for the financial support from a Grant-in-Aid for Scientific Research (16655046, 21655041, 22350052, 23651092, 26620155, and 16H04155).

**Institutional Review Board Statement:** Not applicable.

**Informed Consent Statement:** Not applicable.

**Data Availability Statement:** Excepting Figures 16 and 19, no new data were created or analyzed in this study. Data sharing is not applicable to this article. Figures 16 and 19 are available on request from the corresponding author.

**Acknowledgments:** First of all, the author owes a debt of gratitude to Rui Tamura (Kyoto University) for the opportunity to contribute our CD/ORD/CPL/CPLE spectroscopic studies of the colloidal σ-/π-conjugated polymers at NTT and NAIST with my colleagues and students, over 20 years, to address the curious homochirality questions on Earth. The author expresses special thanks to his coworkers; Hiroshi Nakashima, Seiji Toyoda, Kyozaburo Takeda, Hiromi Takigawa-Tamoto, Masao Motonaga, Yoshihiro Kimura, Fumiko Ichiyanagi, Yoko Nakano, Yang Liu, Takashi Mori, Yoshifumi Kawagoe, Kana Yoshida, Ayako Nakao, Nozomu Suzuki, Keisuke Yoshida, Yuri Donguri, Yuka Kato, Makoto Taguchi, Sang Thi Duong, Mohamed Mehawed Abdellatif, Ai Yokokura, Nor Azura Abdul Rahim, Abd Jalil Jalilah, Laibing Wang, Shosei Yoshimoto, Sibo Guo, Yota Katsurada, Seiko Amazumi, Hailing Chen, Nanami Ogata, Kazuki Yamasaki, Asuka Okubo, Shun Okazaki, Hiroki Kamite, Puhup Puneet, Anubhav Saxena, Takumi Yamada, Nobuyuki Hara, Yuki Mimura, Wei Zhang, Yonggang Yang, Julian R. Koe, Ken Terao, the late Akio Teramoto, Yoshitane Imai, Kotohiro Nomura. The present paper, as my life work, was initiated by irreplaceable experiences, knowledge, serendipitous findings, and unanswered questions at Kyushu University (1976–1978) under supervisions of four scholars, Toyoki Kunitake, Kunihide Takarabe, Yoshio Okahata, and Seiji Shinkai. Finally, the author expresses special thanks to the three anonymous reviewers for their stimulating and critical comments.

**Conflicts of Interest:** The author declares no conflict of interest.

## References

1. Schrodinger, E. *What Is Life?: With Mind and Matter and Autobiographical Sketches*; Cambridge University Press: Cambridge, UK, 1944.
2. Mason, S.F. *Molecular Optical Activity and the Chiral Discriminations*; Cambridge University Press: Cambridge, UK, 1982.
3. Mason, S.F.; Tranter, G.E. The Electroweak Origin of Biomolecular Handedness. *Proc. R. Soc. Lond. A* **1985**, *397*, 45–65.
4. Bonner, W.A. Terrestrial and Extraterrestrial Sources of Molecular Homochirality. *Orig. Life Evol. Biosph.* **1992**, *21*, 407–420. [CrossRef]
5. Bailey, J.; Chrysostomou, A.; Hough, J.H.; Gledhill, T.M.; McCall, A.; Clark, S.; Ménard, F.; Tamura, M. Circular Polarization in Star-Formation Regions: Implications for Biomolecular Homochirality. *Science* **1998**, *281*, 672–674. [CrossRef] [PubMed]
6. Avalos, M.; Babiano, R.; Cintas, P.; Jiménez, J.L.; Palacios, J.C. From Parity to Chirality: Chemical Implications Revisited. *Tetrahedron Asymmetry* **2000**, *11*, 2845–2874. [CrossRef]
7. Nishino, H.; Kosaka, A.; Hembury, G.A.; Aoki, F.; Miyauchi, K.; Shitomi, H.; Onuki, H.; Inoue, Y. Absolute Asymmetric Photoreactions of Aliphatic Amino Acids by Circularly Polarized Synchrotron Radiation: Critically pH-Dependent Photobehavior. *J. Am. Chem. Soc.* **2002**, *124*, 11618–11627. [CrossRef] [PubMed]
8. Kawasaki, T.; Sato, M.; Ishiguro, S.; Saito, T.; Morishita, Y.; Sato, I.; Nishino, H.; Inoue, Y.; Soai, K. Enantioselective Synthesis of Near Enantiopure Compound by Asymmetric Autocatalysis Triggered by Asymmetric Photolysis with Circularly Polarized Light. *J. Am. Chem. Soc.* **2005**, *127*, 3274–3275. [CrossRef]
9. Guijarro, A.; Yus, M. *Origin of Chirality in the Molecules of Life: A Revision from Awareness to the Current Theories and Perspectives of this Unsolved Problem*; RSC Publishing: Cambridge, UK, 2008; ISBN 978-0-85404-156-5.

10. Plasson, R.; Kondepudi, D.K.; Bersini, H.; Commeyras, A.; Asakura, K. Emergence of Homochirality in Far-From-Equilibrium Systems: Mechanisms and Role in Prebiotic Chemistry. *Chirality* **2007**, *19*, 589–600. [CrossRef]
11. Macdermott, A.J. Chiroptical Signatures of Life and Fundamental Physics. *Chirality* **2012**, *24*, 764–769. [CrossRef]
12. Liu, M.; Zhang, L.; Wang, T. Supramolecular Chirality in Self-Assembled Systems. *Chem. Rev.* **2015**, *115*, 7304–7397. [CrossRef]
13. Davankov, V.A. Biological Homochirality on the Earth, or in the Universe? A Selective Review. *Symmetry* **2018**, *10*, 749. [CrossRef]
14. Famiano, M.; Boyd, R.; Kajino, T.; Onaka, T.; Mo, Y. Astrophysical Sites that Can Produce Enantiomeric Amino Acids. *Symmetry* **2019**, *11*, 23. [CrossRef]
15. Tamura, M.; Kandori, R.; Kusakabe, N.; Nakajima, Y.; Hashimoto, J.; Nagashima, C.; Nagata, T.; Nagayama, T.; Kimura, H.; Yamamoto, T.; et al. Near-infrared polarization images of the Orion nebula. *Astrophys. J. Lett.* **2006**, *649*, L29–L32. [CrossRef]
16. Takahashi, J.-I.; Kobayashi, K. Origin of Terrestrial Bioorganic Homochirality and Symmetry Breaking in the Universe. *Symmetry* **2019**, *11*, 919. [CrossRef]
17. Ribó, J.M.; Hochberg, D. Chemical Basis of Biological Homochirality during the Abiotic Evolution Stages on Earth. *Symmetry* **2019**, *11*, 814. [CrossRef]
18. Suzuki, N.; Itabashi, Y. Possible Roles of Amphiphilic Molecules in the Origin of Biological Homochirality. *Symmetry* **2019**, *11*, 966. [CrossRef]
19. Sang, Y.; Liu, M. Symmetry Breaking in Self-Assembled Nanoassemblies. *Symmetry* **2019**, *11*, 950. [CrossRef]
20. Soai, K.; Kawasaki, T.; Matsumoto, A. Role of Asymmetric Autocatalysis in the Elucidation of Origins of Homochirality of Organic Compounds. *Symmetry* **2019**, *11*, 694. [CrossRef]
21. Fujiki, M. Mirror Symmetry Breaking in Helical Polysilanes: Preference Between Left and Right of Chemical and Physical Origin. *Symmetry* **2010**, *2*, 1625–1652. [CrossRef]
22. Akabori, S.; Okawa, K.; Sato, M. Introduction of Side Chains into Polyglycine Dispersed on Solid Surface. I. *Bull. Chem. Soc. Jpn.* **1956**, *29*, 608–611. [CrossRef]
23. Rohlfing, D.L.; Oparin, A.I. (Eds.) *Molecular Evolution: Prebiological and Biological*; Plenum Press: New York, NY, USA, 1972.
24. Sparks, W.B.; Hough, J.H.; Kolokolova, L.; Germer, T.A.; Chen, F.; DasSarma, S.; DasSarma, P.; Robb, F.T.; Manset, N.; Reid, I.N.; et al. Circular Polarization in Scattered Light as a Possible Biomarker. *J. Quant. Spectrosc. Radiat. Transf.* **2009**, *110*, 1771–1779. [CrossRef]
25. Pizzarello, S. Molecular Asymmetry in Prebiotic Chemistry: An Account from Meteorites. *Life* **2016**, *6*, 18. [CrossRef] [PubMed]
26. Morrison, S.M.; Runyon, S.E.; Hazen, R.M. The Paleomineralogy of the Hadean Eon Revisited. *Life* **2018**, *8*, 64. [CrossRef] [PubMed]
27. Kirschvink, J.L.; Gaidos, E.J.; Bertani, L.E.; Beukes, N.J.; Gutzmer, J.; Maepa, L.N.; Steinberger, R.E. Paleoproterozoic Snowball Earth: Extreme Climatic and Geochemical Global Change and Its Biological Consequences. *Proc. Natl. Sci. Acad. USA* **2000**, *97*, 1400–1405. [CrossRef] [PubMed]
28. Rooney, A.D.; Macdonald, F.A.; Strauss, J.V.; Dudás, F.O.; Hallmann, C.; Selby, D. Re-Os Geochronology and Coupled Os-Sr Isotope Constraints on the Sturtian Snowball Earth. *Proc. Natl. Sci. Acad. USA* **2014**, *111*, 51–56. [CrossRef]
29. Ueno, Y.; Yamada, K.; Yoshida, N.; Maruyama, S.; Isozaki, Y. Evidence from Fluid Inclusions for Microbial Methanogenesis in the Early Archaean Era. *Nature* **2006**, *440*, 516–519. [CrossRef]
30. Nojzsis, S.J.; Arrhenius, G.; McKeegan, K.D.; Harrison, T.M.; Nutman, A.P.; Friend, C.R. Evidence for Life on Earth before 3800 Million Years Ago. *Nature* **1996**, *384*, 55–59.
31. Hoffman, P.F.; Schrag, D.P. The Snowball Earth Hypothesis: Testing the Limits of Global Change. *Terra Nova* **2002**, *14*, 129–155. [CrossRef]
32. Okamoto, Y.; Kakegawa, T.; Ishida, A.; Nagase, T.; Rosing, M.T. Evidence for Biogenic Graphite in Early Archaean Isua Metasedimentary Rocks. *Nat. Geosci.* **2014**, *7*, 25–28.
33. Tarduno, J.A.; Cottrell, R.D.; Davis, W.J.; Nimmo, F.; Bono, R.K. A Hadean to Paleoarchean Geodynamo Recorded by Single Zircon Crystals. *Science* **2015**, *349*, 521–524. [CrossRef]
34. Tang, F.; Taylor, R.J.; Einsle, J.F.; Borlina, C.S.; Fu, R.R.; Weiss, B.P.; Williams, H.M.; Williams, W.; Nagy, L.; Midgley, P.A.; et al. Secondary Magnetite in Ancient Zircon Precludes Analysis of a Hadean Geodynamo. *Proc. Natl Acad. Sci. USA* **2019**, *116*, 407–412. [CrossRef]
35. Witze, A. Greenland Rocks Suggest Earth's Magnetic Field Is Older Than We Thought. *Nature* **2019**, *576*, 347. [CrossRef] [PubMed]
36. Zawaski, M.J.; Kelly, N.M.; Orlandini, O.F.; Nichols, C.I.O.; Allwood, A.C.; Mojzsis, S.J. Reappraisal of Purported ca. 3.7 Ga Stromatolites from the Isua Supracrustal Belt (West Greenland) from Detailed Chemical and Structural Analysis. *Earth Planet. Sci. Lett.* **2020**, *545*, 116409. [CrossRef]
37. Epstein, S.; Krishnamurthy, R.V.; Cronin, J.R.; Pizzarello, S.; Yuen, G.U. Unusual Stable Isotope Ratios in Amino Acid and Carboxylic Acid Extracts from the Murchison Meteorite. *Nature* **1987**, *326*, 477–479. [CrossRef] [PubMed]
38. Engel, M.H.; Macko, S.A. Isotopic Evidence for Extraterrestrial Non-racemic Amino Acids in the Murchison Meteorite. *Nature* **1987**, *389*, 265–268. [CrossRef] [PubMed]
39. Frieden, B.R.; Plastino, A.; Plastino, A.R.; Soffer, B.H. Schrodinger Link Between Nonequilibrium Thermodynamics and Fisher Information. *Phys. Rev. E* **2002**, *66*, 046128. [CrossRef]
40. Ornes, S. Core Concept: How Nonequilibrium Thermodynamics Speaks to The Mystery of Life. *Proc. Natl. Acad. Sci. USA* **2017**, *114*, 423–424. [CrossRef]

41. Ribó, J.M.; Hochberg, D.; Crusats, J.; El-Hachemi, Z.; Moyano, A. Spontaneous Mirror Symmetry Breaking and Origin of Biological Homochirality. *J. R. Soc. Interface* **2017**, *14*, 20170699. [CrossRef]
42. Goldanskiĭ, V.I.; Kuz'min, V.V. Spontaneous Breaking of Mirror Symmetry in Nature and the Origin of Life. *Sov. Phys. Usp.* **1989**, *32*, 1–29. [CrossRef]
43. Avetisov, V.A.; Goldanskii, V.I.; Kuz'min, V.V. Handedness, Origin of Life and Evolution. *Phys. Today* **1991**, *44*, 33–41.
44. Avetisov, V.; Goldanskii, V. Mirror Symmetry Breaking at the Molecular Level. *Proc. Natl. Acad. Sci. USA* **1996**, *93*, 11435–11442. [CrossRef]
45. Frank, F.C. On Spontaneous Asymmetric Synthesis. *Biochim. Biophys. Acta* **1953**, *11*, 459–463. [CrossRef]
46. Kuhn, W.; Brown, E. Photochemische Erzeugung Optisch Aktiver Stoffe. *Naturwissenschaften* **1929**, *17*, 227–228. [CrossRef]
47. Rau, H. Asymmetric Photochemistry in Solution. *Chem. Rev.* **1983**, *83*, 535–547. [CrossRef]
48. Inoue, Y. Asymmetric Photochemical Reactions in Solution. *Chem. Rev.* **1992**, *92*, 741–770. [CrossRef]
49. Rau, H. Direct Asymmetric Photochemistry with Circularly Polarized light. In *Chiral Photochemistry: Molecular and Supramolecular Photochemistry*; Inoue, Y., Ramamurthy, V., Eds.; Marcel Dekker: New York, NY, USA, 2004; Chapter 1; pp. 1–44.
50. Meierhenrich, U.J.; Nahon, L.; Alcaraz, C.; Bredehöft, J.H.; Hoffmann, S.V.; Barbier, B.; Brack, A. Asymmetric Vacuum UV photolysis of the Amino Acid Leucine in the Solid State. *Angew. Chem. Int. Ed.* **2005**, *44*, 5630–5634. [CrossRef]
51. INOUE Photo-Chirogenesis. Available online: https://www.jst.go.jp/erato/en/research_area/completed/ihh_P.html (accessed on 22 January 2021).
52. Bernstein, W.J.; Calvin, M.; Buchardt, O. Absolute Asymmetric Synthesis. I. Mechanism of the Photochemical Synthesis of Nonracemic Helicenes with Circularly Polarized Light. Wavelength Dependence of the Optical Yield of Octahelicene. *J. Am. Chem. Soc.* **1972**, *94*, 494–498. [CrossRef]
53. Meinert, C.; Hoffmann, S.V.; Cassam-Chenaï, P.; Evans, A.C.; Giri, C.; Nahon, L.; Meierhenrich, U.J. Photonenergy-Controlled Symmetry Breaking with Circularly Polarized Light. *Angew. Chem. Int. Ed.* **2014**, *53*, 210–214. [CrossRef]
54. Fujiki, M.; Yoshida, K.; Suzuki, N.; Zhang, J.; Zhang, W.; Zhu, X. Mirror Symmetry Breaking and Restoration within µm-Sized Polymer Particles in Optofluidic Media by Pumping Circularly Polarised Light. *RSC Adv.* **2013**, *3*, 5213–5219. [CrossRef]
55. Fujiki, M.; Donguri, Y.; Zhao, Y.; Nakao, A.; Suzuki, N.; Yoshida, K.; Zhang, W. Photon Magic: Chiroptical Polarisation, Depolarisation, Inversion, Retention and Switching of Non-Photochromic Light-Emitting Polymers in Optofluidic Medium. *Polym. Chem.* **2015**, *6*, 1627–1638. [CrossRef]
56. Wang, L.; Yin, L.; Zhang, W.; Zhu, X.; Fujiki, M. Circularly Polarized Light with Sense and Wavelengths To Regulate Azobenzene Supramolecular Chirality in Optofluidic Medium. *J. Am. Chem. Soc.* **2017**, *139*, 13218–13226. [CrossRef]
57. Auzinsh, M.; Blushs, K.; Ferber, R.; Gahbauer, F.; Jarmola, A.; Tamanis, M. Electric Field Induced Symmetry Breaking of Angular Momentum Distribution in Atoms. *Phys. Rev. Lett.* **2006**, *97*, 043002. [CrossRef] [PubMed]
58. Bhattacharyya, K.; Surendran, A.; Chowdhury, C.; Datta, A. Steric and Electric Field Driven Distortions in Aromatic Molecules: Spontaneous and Non-Spontaneous Symmetry Breaking. *Phys. Chem. Chem. Phys.* **2016**, *18*, 31160–31167. [CrossRef] [PubMed]
59. Tielens, A.G.G.M. Interstellar Polycyclic Aromatic Hydrocarbon Molecules. *Annu. Rev. Astron. Astrophys.* **2008**, *46*, 289–337. [CrossRef]
60. Avalos, M.; Babiano, R.; Cintas, P.; Jiménez, J.L.; Palacios, J.C.; Barron, L.D. Absolute Asymmetric Synthesis under Physical Fields: Facts and Fictions. *Chem. Rev.* **1998**, *98*, 2391–2404. [CrossRef] [PubMed]
61. Zadel, G.; Eisenbrdun, C.; Wolff, G.-J.; Breitmaier, E. Enantioselective Reactions in a Static Magnetic Field-A False Alarm! *Angew. Chem. Int. Ed. Engl.* **1994**, *33*, 454–456. [CrossRef]
62. Kaupp, G.; Marquardt, T. Absolute Asymmetric Synthesis Solely under the Influence of a Static Homogeneous Magnetic Field? *Angew. Chem. Int. Ed.* **1994**, *33*, 1459–1461. [CrossRef]
63. Feringa, B.L.; Kellogg, R.M.; Hulst, R.; Zondervan, C.; Kruizinga, W.H. Attempts to Carry Out Enantioselective Reactions in a Static Magnetic Field. *Angew. Chem. Int. Ed.* **1994**, *33*, 1458–1459. [CrossRef]
64. Gölitz, P. Enantioselective Reactions in a Static Magnetic Field—False Alarm! *Angew. Chem. Int. Ed.* **1994**, *33*, 1457. [CrossRef]
65. Barron, L.D. Can A Magnetic Field Induce Absolute Asymmetric Synthesis? *Science* **1994**, *266*, 1491–1492. [CrossRef]
66. Naaman, R.; Wadeck, D.H. Chiral-Induced Spin Selectivity Effect. *J. Phys. Chem. Lett.* **2012**, *3*, 2178–2187. [CrossRef]
67. Banerjee-Ghosh, K.; Dor, O.B.; Tassinari, F.; Capua, E.; Yochelis, S.; Capua, A.; Yang, S.-H.; Parkin, S.S.P.; Sarkar, S.; Kronik, L.; et al. Separation of Enantiomers by Their Enantiospecific Interaction with Achiral Magnetic Substrates. *Science* **2018**, *360*, 1331–1334. [CrossRef] [PubMed]
68. Tassinari, F.; Steidel, J.; Paltiel, S.; Fontanesi, C.; Lahav, M.; Paltiel, Y.; Naaman, R. Enantioseparation by Crystallization Using Magnetic Substrates. *Chem. Sci.* **2019**, *10*, 5246–5250. [CrossRef] [PubMed]
69. Lee, T.D.; Yang, C.N. Question of Parity Conservation in Weak Interactions. *Phys. Rev.* **1956**, *104*, 254–258. [CrossRef]
70. Wu, C.S.; Ambler, E.; Hayword, R.W.; Hoppes, D.D.; Hudson, R.P. Experimental Test of Parity Conservation on Beta Decay. *Phys. Rev.* **1957**, *105*, 1413–1415. [CrossRef]
71. Ambler, E.; Hayward, R.W.; Hoppes, D.D.; Hudson, R.P.; Wu, C.S. Further Experiments on β Decay of Polarized Nuclei. *Phys. Rev.* **1957**, *106*, 1361–1363. [CrossRef]
72. Postma, H.; Huiskamp, W.J.; Miedema, A.R.; Steenland, M.J.; Tolhoek, H.A.; Gorter, C.J. Asymmetry of the Positron Emission by Polarized $^{58}$Co Nuclei. *Physica* **1957**, *23*, 259–260. [CrossRef]

73. Rodberg, L.S.; Weisskopf, V.F. Fall of Parity—Recent Discoveries Related to Symmetry of Laws of Nature. *Science* **1957**, *125*, 627–633. [CrossRef]
74. Goldhaber, M.; Grodzins, L.; Sunyar, A.W. Helicity of Neutrinos. *Phys. Rev.* **1958**, *109*, 1015–1017. [CrossRef]
75. Wróblewski, A.K. The Downfall of Parity—The Revolution That Happened Fifty Years Ago. *Acta. Phys. Pol. B* **2008**, *39*, 251–264.
76. Wu, C.S. The Discovery of the Parity Violation in Weak Interactions and Its Recent Developments. *Lect. Notes Phys.* **2008**, *746*, 43–49.
77. Stevenson, C.D.; Davis, J.P. Magnetars and Magnetic Separation of Chiral Radicals in Interstellar Space: Homochirality. *J. Phys. Chem. A* **2019**, *123*, 9587–9593. [CrossRef] [PubMed]
78. Kouveliotou, C.; Duncan, R.C.; Thompson, C. Magnetars. *Sci. Am.* **2003**, *288*, 34–41. [CrossRef] [PubMed]
79. Ohno, O.; Kaizu, Y.; Kobayashi, H. J-aggregate Formation of a Water-Soluble Porphyrin in Acidic Aqueous Media. *J. Chem. Phys.* **1993**, *99*, 4128–4139. [CrossRef]
80. Ribó, J.M.; Crusats, J.; Sagués, F.; Claret, J.; Rubires, R. Chiral Sign Induction by Vortices During the Formation of Mesophases in Stirred Solutions. *Science* **2001**, *292*, 2063–2066. [CrossRef]
81. Crusats, J.; El-Hachemi, Z.; Ribó, J.M. Hydrodynamic Effects on Chiral Induction. *Chem. Soc. Rev.* **2010**, *39*, 569–577. [CrossRef]
82. Okano, K.; Taguchi, M.; Fujiki, M.; Yamashita, T. Circularly Polarized Luminescence of Rhodamine B in a Supramolecular Chiral Medium Formed by a Vortex Flow. *Angew. Chem. Int. Ed.* **2011**, *50*, 12474–12477. [CrossRef]
83. Sun, J.; Li, Y.; Yan, F.; Liu, C.; Sang, Y.; Tian, F.; Feng, Q.; Duan, P.; Zhang, L.; Shi, X.; et al. Control over the Emerging Chirality in Supramolecular Gels and Solutions by Chiral Microvortices in Milliseconds. *Nat. Commun.* **2018**, *9*, 2599. [CrossRef]
84. Buckingham, A.D.; Stephens, P.J. Magnetic Optical Activity. *Annu. Rev. Phys. Chem.* **1966**, *17*, 399–432. [CrossRef]
85. Schatz, P.N.; McCaffery, A.J. Faraday effect. *Q. Rev. Chem. Soc.* **1969**, *23*, 552–584. [CrossRef]
86. Stephens, R.J. Magnetic Circular Dichroism. *Annu. Rev. Phys. Chem.* **1974**, *25*, 201–232. [CrossRef]
87. Kobayashi, N.; Muranaka, A.; Mack, J. *Circular Dichroism and Magnetic Circular Dichroism Spectroscopy for Organic Chemists*; RSC Publishing: Cambridge, UK, 2012.
88. Riehl, J.P.; Richardson, F.S. General Theory of Circularly Polarized Emission and Magnetic Circularly Polarized Emission from Molecular Systems. *J. Chem. Phys.* **1976**, *65*, 1011–1021. [CrossRef]
89. Wu, T.; Kapitán, J.; Andrushchenko, V.; Bouř, P. Identification of Lanthanide (III) Luminophores in Magnetic Circularly Polarized Luminescence Using Raman Optical Activity Instrumentation. *Anal. Chem.* **2017**, *89*, 5043–5049. [CrossRef] [PubMed]
90. Nelson, H.D.; Hinterding, S.O.M.; Fainblat, R.; Creutz, S.E.; Li, X.; Gamelin, D.R. A Selective Cation Exchange Strategy for the Synthesis of Colloidal $Yb^{3+}$-Doped Chalcogenide Nanocrystals with Strong Broadband Visible Absorption and Long-Lived Near-Infrared Emission. *J. Am. Chem. Soc.* **2017**, *139*, 6411–6421. [CrossRef] [PubMed]
91. Ivchenko, E.L. Magnetic Circular Polarization of Exciton Photoluminescence. *Phys. Solid State* **2018**, *60*, 1514–1526. [CrossRef]
92. Kaji, D.; Okada, H.; Hara, N.; Kondo, Y.; Suzuki, S.; Miyasaka, M.; Fujiki, M.; Imai, Y. Non-classically Controlled Sign in a 1.6 Tesla Magnetic Circularly Polarized Luminescence of Three Pyrenes in a Chloroform and a PMMA Film. *Chem. Lett.* **2020**, *49*, 674–676. [CrossRef]
93. Okada, H.; Hara, N.; Kaji, D.; Shizuma, M.; Fujiki, M.; Imai, Y. Excimer-Origin CPL vs. Monomer-Origin Magnetic CPL in Photo-Excited Chiral Binaphthyl-Ester-Pyrenes: Critical Role of Ester Direction. *Phys. Chem. Chem. Phys.* **2020**, *22*, 13862–13866. [CrossRef] [PubMed]
94. Yoshikawa, H.; Nakajima, G.; Mimura, Y.; Kimoto, T.; Kondo, Y.; Suzuki, S.; Fujiki, M.; Imai, Y. Mirror-Image Magnetic Circularly Polarized Luminescence (MCPL) from Optically Inactive $Eu^{III}$ and $Tb^{III}$ Tris(β-Diketonate). *Dalton Trans.* **2020**, *49*, 9588–9594. [CrossRef]
95. Raman, C.V.; Krishna, K.S. A New Type of Secondary Radiation. *Nature* **1928**, *121*, 501–502. [CrossRef]
96. Rikken, G.L.J.A.; Raupach, E. Enantioselective Magnetochiral Photochemistry. *Nature* **2000**, *405*, 932–934. [CrossRef]
97. Sharma, A. Enantiomeric Excess by Magnetic Circular Dichroism in Archaean Atmosphere. *Sci. Rep.* **2017**, *7*, 13295. [CrossRef]
98. Vester, F.; Ulbricht, T.L.V. Asymmetry: The Non-Conservation of Parity and Optical Activity. *Quart. Rev.* **1959**, *13*, 48–60.
99. Vester, F.; Ulbricht, T.L.V. Attempts to Induce Optically Activity with Polarized β-Radiation. *Tetrahedron* **1962**, *18*, 629–637.
100. Bonner, W.A. Experimental Evidence for β-Decay as a Source of Chirality by Enantiomer Analysis. *Orig. Life* **1984**, *14*, 383–390. [CrossRef] [PubMed]
101. Gidley, D.W.; Rich, A.; Van House, J.; Zitzewitz, P.W. β Decay and the Origins of Biological Chirality: Experimental Results. *Nature* **1982**, *297*, 639–643. [CrossRef]
102. Van House, J.; Rich, A.; Zitzewitz, P.W. Beta Decay and The Origin of Biological Chirality: New Experimental Results. *Orig. Life* **1984**, *14*, 413–420. [CrossRef]
103. Dreiling, J.; Gay, T. Chirally Sensitive Electron-Induced Molecular Breakup and the Vester-Ulbricht Hypothesis. *Phys. Rev. Lett.* **2014**, *113*, 118103. [CrossRef]
104. Electroweak Interaction. Available online: https://en.wikipedia:wiki/Electroweak_interaction (accessed on 20 December 2020).
105. Janoschek, R. Theories on the origin of biomolecular Homochirality. In *Chirality—From Weak Bosons to the α-Helix*; Janoschek, R., Ed.; Springer: Berlin, Germany, 1991; ISBN 978-3-642-76569-8.
106. Close, E.E. Parity Violation in Atoms? *Nature* **1976**, *264*, 505–506. [CrossRef]
107. Bucksbaum, P.H.; Commins, E.D.; Hunter, L.R. Observations of Parity Nonconservation in Atomic Thallium. *Phys. Rev. D* **1981**, *24*, 1134–1148. [CrossRef]

108. Emmons, T.P.; Reeves, J.M.; Fortson, E.N. Parity-Nonconserving Optical Rotation in Atomic Lead. *Phys. Rev. Lett.* **1983**, *51*, 2089–2091. [CrossRef]
109. Bouchiat, M.-A.; Pottier, L. Optical Experiments and Weak Interactions. *Science* **1986**, *234*, 1203–1210. [CrossRef]
110. World Scientific. *Parity Violation in Atoms and Polarized Electron Scattering*; Bernard, F., Bouchiat, M.-A., Eds.; World Scientific: Singapore, 1999; ISBN 9810237316.
111. Okun, L.B. Mirror Particles and Mirror Matter: 50 years of Speculation and Search. *Phys. Usp.* **2007**, *50*, 380–389. [CrossRef]
112. Tsigutkin, K.; Dounas-Frazer, D.; Family, A.; Stalnaker, J.E.; Yashchuk, V.V.; Budker, D. Observation of a Large Atomic Parity Violation Effect in Ytterbium. *Phys. Rev. Lett.* **2009**, *103*, 071601. [CrossRef]
113. Mason, A.; Tranter, G.E. Energy Inequivalence of Peptide Enantiomers from Parity Non-Conservation. *J. Chem. Soc. Chem. Commun.* **1983**, 117–119. [CrossRef]
114. Quack, M. Structure and Dynamics of Chiral Molecules. *Angew. Chem. Int. Ed.* **1989**, *28*, 571–586. [CrossRef]
115. Quack, M.; Stohner, J.; Willeke, M. High-Resolution Spectroscopic Studies and Theory of Parity Violation in Chiral Molecules. *Annu. Rev. Phys. Chem.* **2008**, *59*, 741–769. [CrossRef]
116. Quack, M. How Important is Parity Violation for Molecular and Biomolecular Chirality? *Angew. Chem. Int. Ed.* **2002**, *41*, 4618–4630. [CrossRef]
117. Schwerdtfeger, P.; Gierlich, J.; Bollwein, T. Large Parity-Violation Effects in Heavy-Metal-Containing Chiral Compounds. *Angew. Chem. Int. Ed.* **2003**, *42*, 1293–1296. [CrossRef]
118. Darquié, B.; Stoeffler, C.; Shelkovnikov, A.; Daussy, C.; Amy-Klein, A.; Chardonnet, C.; Zrig, S.; Guy, L.; Crassous, J.; Soulard, P.; et al. Progress Toward the First Observation of Parity Violation in Chiral Molecules by High-Resolution Laser Spectroscopy. *Chirality* **2010**, *22*, 870–884. [CrossRef]
119. Berger, R. Molecular Parity Violation in Electronically Excited States. *Phys. Chem. Chem. Phys.* **2003**, *5*, 12–17. [CrossRef]
120. Fujiki, M.; Koe, J.R.; Mori, T.; Kimura, Y. Questions of Mirror Symmetry at the Photoexcited and Ground States of Non-Rigid Luminophores Raised by Circularly Polarized Luminescence and Circular Dichroism Spectroscopy: Part 1. Oligofluorenes, Oligophenylenes, Binaphthyls and Fused Aromatics. *Molecules* **2018**, *23*, 2606. [CrossRef]
121. Fujiki, M.; Koe, J.R.; Amazumi., S. Questions of Mirror Symmetry at the Photoexcited and Ground States of Non-Rigid Luminophores Raised by Circularly Polarized Luminescence and Circular Dichroism Spectroscopy. Part 2: Perylenes, BODIPYs, Molecular Scintillators, Coumarins, Rhodamine B, and DCM. *Symmetry* **2019**, *11*, 363. [CrossRef]
122. Panspermia. Available online: https://en.wikipedia.org/wiki/Panspermia#cite_note-cometary_panspermia-3 (accessed on 2 November 2020).
123. Wickramasinghe, N.C. *Search for Our Cosmic Ancestry*; World Scientific: Singapore, 2014.
124. Kawaguchi, Y.; Yang, Y.; Kawashiri, N.; Shiraishi, K.; Takasu, M.; Narumi, I.; Satoh, K.; Hashimoto, H.; Nakagawa, K.; Tanigawa, Y.; et al. The Possible Interplanetary Transfer of Microbes: Assessing the Viability of *Deinococcus* spp. Under the ISS Environmental Conditions for Performing Exposure Experiments of Microbes in the Tanpopo Mission. *Orig. Life. Evol. Biosph.* **2013**, *43*, 411–428. [CrossRef] [PubMed]
125. Kawaguchi, Y.; Shibuya, M.; Kinoshita, I.; Yatabe, J.; Narumi, I.; Shibata, H.; Hayashi, R.; Fujiwara, D.; Murano, Y.; Hashimoto, H.; et al. DNA Damage and Survival Time Course of Deinococcal Cell Pellets During 3 Years of Exposure to Outer Space. *Front. Microbiol.* **2020**, *11*, 2050. [CrossRef] [PubMed]
126. Paganini, L.; Villanueva, G.L.; Roth, L.; Mandell, A.M.; Hurford, T.A.; Retherford, K.D.; Mumma, M.J. A Measurement of Water Vapour Amid a Largely Quiescent Environment on Europa. *Nat. Astron.* **2020**, *4*, 266–272. [CrossRef]
127. Goesmann, F.; Rosenbauer, H.; Bredehöft, J.H.; Cabane, M.; Ehrenfreund, P.; Gautier, T.; Giri, C.; Krüger, H.; Roy, L.L.; MacDermott, A.J.; et al. Organic Compounds on Comet 67P/Churyumov-Gerasimenko Revealed by COSAC Mass Spectrometry. *Science* **2015**, *349*, aab0689. [CrossRef]
128. Bibring, J.-P.; Langevin, Y.; Carter, J.; Eng, P.; Gondet, B.; Jorda, L.; Le Mouélic, S.; Mottola, S.; Pilorget, C.; Poulet, F.; et al. 67P/Churyumov-Gerasimenko Surface Properties as Derived from CIVA Panoramic Images. *Science* **2015**, *349*, aab0671. [CrossRef]
129. Iess, L.; Stevenson, D.J.; Parisi, M.; Hemingway, D.; Jacobson, R.A.; Lunine, J.I.; Nimmo, F.; Armstrong, J.W.; Asmar, S.W.; Ducci, M.; et al. The Gravity Field and Interior Structure of Enceladus. *Science* **2014**, *344*, 78–80. [CrossRef]
130. Lorenz, R.D.; Kirk, R.L.; Hayes, A.G.; Anderson, Y.Z.; Lunine, J.I.; Tokano, T.; Turtle, E.P.; Malaska, M.J.; Soderblom, J.M.; Lucas, A.; et al. A Radar Map of Titan Seas: Tidal Dissipation and Ocean Mixing through the Throat of Kraken. *Icarus* **2014**, *237*, 9–15. [CrossRef]
131. Pepe, F.; Lovis, C.; Ségransan, D.; Benz, W.; Bouchy, F.; Dumusque, X.; Mayor, M.; Queloz, D.; Santos, N.C.; Udry, S. The HARPS Search for Earth-like Planets in the Habitable Zone. *Astron. Astrophys.* **2011**, *534*, A58. [CrossRef]
132. Quintana, E.V.; Barclay, T.; Raymond, S.N.; Rowe, J.F.; Bolmont, E.; Caldwell, D.A.; Howell, S.B.; Kane, S.R.; Huber, D.; Crepp, J.R.; et al. An Earth-sized Planet in the Habitable Zone of a Cool Star. *Science* **2014**, *344*, 277–280. [CrossRef]
133. McGuire, B.A.; Carroll, P.B.; Loomis, R.A.; Finneran, I.A.; Jewell, P.R.; Remijan, A.J.; Blake, G.A. Discovery of the Interstellar Chiral Molecule Propylene Oxide ($CH_3CHCH_2O$). *Science* **2016**, *352*, 1449–1452. [CrossRef] [PubMed]
134. Furukawa, Y.; Chikaraishi, Y.; Ohkouchi, N.; Ogawa, N.O.; Gavin, D.P.; Dworkin, J.P.; Abe, C.; Nakamura, T. Extraterrestrial Ribose and Other Sugars in Primitive Meteorites. *Proc. Natl. Acad. Sci. USA* **2019**, *116*, 24440–24445. [CrossRef] [PubMed]

135. Hayabusa-2 Project. Available online: http://www.hayabusa2.jaxa.jp/en/topics/20201209_capsulephotos/ (accessed on 9 December 2020).
136. Miller, S.L. Production of Amino Acids under Possible Primitive Earth Conditions. *Science* **1953**, *117*, 528–529. [CrossRef]
137. Miller, S.L.; Urey, H.C. Organic Compound Synthesis on the Primitive Earth. *Science* **1959**, *130*, 245–251. [CrossRef] [PubMed]
138. Bada, J.L. New Insights into Prebiotic Chemistry from Stanley Miller's Spark Discharge Experiments. *Chem. Soc. Rev.* **2013**, *42*, 2186–2196. [CrossRef]
139. Enoto, T.; Wada, Y.; Furuta, Y.; Nakazawa, K.; Yuasa, T.; Okuda, K.; Makishima, K.; Sato, M.; Sato, Y.; Nakano, T.; et al. Photonuclear Reactions Triggered by Lightning Discharge. *Nature* **2017**, *551*, 481–484. [CrossRef]
140. Rozanski, K.; Fröhlich, K. Radioactivity and Earth Sciences: Understanding the Natural Environment. *IAEA Bull.* **1996**, *38*, 9–15.
141. Araki, T.; Enomoto, S.; Furuno, K.; Gando, Y.; Ichimura, K.; Ikeda, H.; Inoue, K.; Kishimoto, Y.; Koga, M.; Koseki, Y.; et al. Experimental Investigation of Geologically Produced Antineutrinos with KamLAND. *Nature* **2005**, *436*, 499–503. [CrossRef]
142. Rosianna, I.; Nugraha, E.D.; Syaeful, H.; Putra, S.; Hosoda, M.; Akata, N.; Tokonami, S. Natural Radioactivity of Laterite and Volcanic Rock Sample for Radioactive Mineral Exploration in Mamuju, Indonesia. *Geosciences* **2020**, *10*, 376. [CrossRef]
143. Córdova, A.; Engqvist, M.; Ibrahem, I.; Casa, J.; Sundén, H. Plausible Origins of Homochirality in the Amino Acid Catalyzed Neogenesis of Carbohydrates. *Chem. Commun.* **2005**, 2047–2049. [CrossRef]
144. Sato, I.; Urabe, H.; Ishiguro, S.; Shibata, T.; Soai, K. Amplification of Chirality from Extremely Low to Greater than 99.5 % *ee* by Asymmetric Autocatalysis. *Angew. Chem. Int. Ed.* **2003**, *42*, 315–317. [CrossRef] [PubMed]
145. Haldane, J.B.S. An Exact Test for Randomness of Mating. *New Biol.* **1954**, *16*, 12–27.
146. Lu, T.; Spruijt, E. Multiphase Complex Coacervate Droplets. *J. Am. Chem. Soc.* **2020**, *142*, 2905–2914. [CrossRef] [PubMed]
147. Favorskiy, I.; Vu, D.; Peytavit, E.; Arscott, S.; Paget, D.; Rowe, A.C.H. Circularly Polarized Luminescence Microscopy for the Imaging of Charge and Spin Diffusion in Semiconductors. *Rev. Sci. Instrum.* **2010**, *81*, 103902. [CrossRef] [PubMed]
148. Kaminsky, W.; Claborn, K.; Kahr, B. Polarimetric Imaging of Crystals. *Chem. Soc. Rev.* **2004**, *33*, 514–525. [CrossRef] [PubMed]
149. Narushima, T.; Okamoto, H. Circular Dichroism Microscopy Free from Commingling Linear Dichroism via Discretely Modulated Circular Polarization. *Sci. Rep.* **2016**, *6*, 35731. [CrossRef]
150. Le, K.Q.; Hashiyada, S.; Kondo, M.; Okamoto, H. Circularly Polarized Photoluminescence from Achiral Dye Molecules Induced by Plasmonic Two-Dimensional Chiral Nanostructures. *J. Phys. Chem. C* **2018**, *122*, 24924–24932. [CrossRef]
151. Claborn, K.; Puklin-Faucher, E.; Kurimoto, M.; Kaminsky, W.; Kahr, B. Circular Dichroism Imaging Microscopy: Application to Enantiomorphous Twinning in Biaxial Crystals of 1,8-Dihydroxyanthraquinone. *J. Am. Chem. Soc.* **2003**, *125*, 14825–14831. [CrossRef]
152. Narushima, T.; Okamoto, H. Circular Dichroism Nano-Imaging of Two-Dimensional Chiral Metal Nanostructures. *Phys. Chem. Chem. Phys.* **2013**, *15*, 13805–13809. [CrossRef]
153. Hauser, E.A. The History of Colloid Science: In Memory of Wolfgang Ostwald. *J. Chem. Educ.* **1955**, *32*, 2–9. [CrossRef]
154. Le Chatelier, H. Crystalloids Against Colloids in the Theory of Cements. *Trans. Faraday Soc.* **1919**, *14*, 8–11. [CrossRef]
155. Hermann Staudinger. Available online: https://en.wikipedia.org/wiki/Hermann_Staudinger (accessed on 11 December 2020).
156. Terayama, H. Method of colloid titration (a new titration between polymer ions). *J. Polym. Sci.* **1952**, *8*, 243–253. [CrossRef]
157. Kondepudi, D.K.; Kaufman, R.J.; Singh, N. Chiral Symmetry Breaking Crystallization. *Science* **1990**, *250*, 975–976. [CrossRef] [PubMed]
158. Martin, B.; Tharrington, A.; Wu, X.-I. Chiral Symmetry Breaking in Crystal Growth: Is Hydrodynamic Convection Relevant? *Phys. Rev. Lett.* **1996**, *77*, 2826–2829. [CrossRef] [PubMed]
159. Viedma, C. Chiral Symmetry Breaking During Crystallization: Complete Chiral Purity Induced by Nonlinear Autocatalysis and Recycling. *Phys. Rev. Lett.* **2005**, *94*, 065504. [CrossRef] [PubMed]
160. Viedma, C.; Ortiz, J.E.; Torres, T.D.; Izumi, T.; Blackmond, D.G. Evolution of Solid Phase Homochirality for a Proteinogenic Amino Acid. *J. Am. Chem. Soc.* **2008**, *130*, 15274–15275. [CrossRef]
161. Noorduin, W.L.; Vlieg, E.; Kellogg, R.M.; Kaptein, B. From Ostwald Ripening to Single Chirality. *Angew. Chem. Int. Ed.* **2009**, *48*, 9600–9606. [CrossRef]
162. Viedma, C. Selective Chiral Symmetry Breaking during Crystallization: Parity Violation or Cryptochiral Environment in Control? *Cryst. Growth Des.* **2007**, *7*, 553–556. [CrossRef]
163. Qian, S.X.; Snow, J.B.; Tzeng, H.M.; Chang, R.K. Lasing Droplets: Highlighting the Liquid-Air Interface by Laser Emission. *Science* **1986**, *231*, 486–488. [CrossRef]
164. Psaltis, D.; Quack, S.R.; Yang, C. Developing Optofluidic Technology through the Fusion of Microfluidics and Optics. *Nature* **2006**, *442*, 381–386. [CrossRef]
165. Fan, X.; White, I.M. Optofluidic Microsystems for Chemical and Biological Analysis. *Nat. Photonics* **2011**, *5*, 591–597. [CrossRef] [PubMed]
166. Schmidt, A.H.; Hawkins, A.R. The Photonic Integration of Non-Solid Media using Optofluidics. *Nat. Photon.* **2011**, *5*, 598–604. [CrossRef]
167. Fainman, Y.; Lee, L.P.; Psaltis, D. *Optofluidics*; Yang, C., Ed.; McGraw-Hill: New York, NY, USA, 2010.
168. Tang, S.K.Y.; Li, Z.; Abate, A.R.; Agresti, J.J.; Weitz, D.A.; Psaltis, D.; Whitesides, G.M. A Multi-Color Fast-Switching Microfluidic Droplet Dye Laser. *Lab Chip* **2009**, *9*, 2767–2771. [CrossRef] [PubMed]

169. Schäfer, J.; Mondia, J.P.; Sharma, R.; Lu, Z.H.; Susha, A.S.; Rogach, A.L.; Wang, L.J. Quantum Dot Microdrop Laser. *Nano Lett.* **2008**, *8*, 1709–1712. [CrossRef]
170. Domachuk, P.; Littler, I.C.M.; Cronin-Golomb, M.; Eggletona, B.J. Compact Resonant Integrated Microfluidic Refractometer. *Appl. Phys. Lett.* **2006**, *88*, 093513. [CrossRef]
171. Steinberg, I.Z. Circular Polarization of Luminescence: Biochemical and Biophysical Applications. *Annu. Rev. Biophys. Bioeng.* **1978**, *7*, 113–137. [CrossRef]
172. Barzda, V.; Musthrdy, L.; Garab, G. Size Dependency of Circular Dichroism in Macrocolloids of Photosynthetic Pigment-Protein Complexes. *Biochemistry* **1994**, *33*, 10837–10841. [CrossRef]
173. Barzda, V.; Istokovics, A.; Simidjiev, I.; Garab, G. Structural Flexibility of Chiral Macrocolloids of Light-Harvesting Chlorophyll a/b Pigment-Protein Complexes. Light-Induced Reversible Structural Changes Associated with Energy Dissipation. *Biochemistry* **1996**, *35*, 8981–8985. [CrossRef]
174. Gussakovsky, E.E.; Shahak, Y.; van Amerongen, H.; Barzda, V. Circularly Polarized Chlorophyll Luminescence Reflects the Macro-Organization of Grana in Pea Chloroplasts. *Photosynth. Res.* **2000**, *65*, 83–92. [CrossRef]
175. Dobson, C.M.; Ellison, G.B.; Tuck, A.F.; Vaida, V. Atmospheric aerosols as prebiotic chemical reactors. *Proc. Natl. Acad. Sci. USA* **2000**, *97*, 11864–11868. [CrossRef]
176. Nam, I.; Lee, J.K.; Nam, H.G.; Zare, R.N. Abiotic Production of Sugar Phosphates and Uridine Ribonucleoside in Aqueous Microdroplets. *Proc. Natl. Acad. Sci. USA* **2017**, *114*, 12396–12400. [CrossRef] [PubMed]
177. Nam, I.; Nam, H.G.; Zare, R.N. Abiotic Synthesis of Purine and Pyrimidine Ribonucleosides in Aqueous Microdroplets. *Proc. Natl. Acad. Sci. USA* **2018**, *115*, 36–40. [CrossRef] [PubMed]
178. Lee, J.K.; Banerjee, S.; Nam, H.G.; Zare, R.N. Acceleration of Reaction in Charged Microdroplets. *Q. Rev. Biophys.* **2015**, *48*, 437–444. [CrossRef]
179. Mortensen, D.N.; Williams, E.R. Microsecond and Nanosecond Polyproline II Helix Formation in Aqueous Nanodrops Measured by Mass Spectrometry. *Chem. Commun.* **2016**, *52*, 12218–12221. [CrossRef] [PubMed]
180. Guardingo, M.; Busqué, F.; Ruiz-Molina, D. Reactions in Ultra-Small Droplets by Tip-Assisted Chemistry. *Chem. Commun.* **2016**, *52*, 11617–11626. [CrossRef] [PubMed]
181. Lee, J.K.; Samanta, D.; Nam, H.G.; Zare, R.N. Micrometer-Sized Water Droplets Induce Spontaneous Reduction. *J. Am. Chem. Soc.* **2019**, *141*, 10585–10589. [CrossRef] [PubMed]
182. Zhou, Z.; Yan, X.; Lai, Y.H.; Zare, R.N. Fluorescence Polarization Anisotropy in Microdroplets. *J. Phys. Chem. Lett.* **2018**, *9*, 2928–2932. [CrossRef]
183. Kageyama, Y. Robust Dynamics of Synthetic Molecular Systems as a Consequence of Broken Symmetry. *Symmetry* **2020**, *12*, 1688. [CrossRef]
184. Eliel, E.L.; Wilen, S.H. Chiroptical properties. In *Stereochemistry of Organic Compounds*; John Wiley & Sons, Inc.: New York, NY, USA, 1994; pp. 991–1118.
185. Fujiki, M.; Wang, L.; Ogata, N.; Asanoma, F.; Okubo, A.; Okazaki, S.; Kamite, H.; Jalilah, A.J. Chirogenesis and Pfeiffer Effect in Optically Inactive Eu$^{III}$ and Tb$^{III}$ Tris (β-diketonate) Upon Intermolecular Chirality Transfer From Poly- and Monosaccharide Alkyl Esters and α-Pinene: Emerging Circularly Polarized Luminescence (CPL) and Circular Dichroism (CD). *Front. Chem.* **2020**, *8*, 685.
186. Albano, G.; Pescitelli, G.; Di Bari, L. Chiroptical Properties in Thin Films of π-Conjugated Systems. *Chem. Rev.* **2020**, *120*, 10145–10243. [CrossRef]
187. Mei, J.; Leung, N.L.; Kwok, R.T.; Lam, J.W.; Tang, B.Z. Aggregation-Induced Emission: Together We Shine, United We Soar! *Chem. Rev.* **2015**, *115*, 11718–11940. [CrossRef] [PubMed]
188. Song, F.; Zhao, Z.; Liu, Z.; Lam, J.W.; Tang, B.Z. Circularly Polarized Luminescence from AIEgens. *J. Mater. Chem. C* **2020**, *8*, 3284–3301. [CrossRef]
189. Luo, J.; Xie, Z.; Lam, J.W.Y.; Cheng, L.; Chen, H.; Qiu, C.; Kwok, H.S.; Zhan, X.; Liu, Y.; Zhu, D.; et al. Aggregation-Induced Emission of 1-Methyl-1,2,3,4,5-Pentaphenylsilole. *Chem. Commun.* **2001**, 1740–1741. [CrossRef] [PubMed]
190. Chen, J.; Law, C.C.W.; Lam, J.W.Y.; Dong, Y.; Lo, S.M.F.; Williams, I.D.; Zhu, D.; Tang, B.Z. Synthesis, Light Emission, Nanoaggregation, and Restricted Intramolecular Rotation of 1,1-Substituted 2,3,4,5-Tetraphenylsiloles. *Chem. Mater.* **2003**, *15*, 1535–1546. [CrossRef]
191. Liu, J.; Su, H.; Meng, L.; Zhao, Y.; Deng, C.; Ng, J.C.Y.; Lu, P.; Faisal, M.; Lam, J.W.Y.; Huang, X.; et al. What Makes Efficient Circularly Polarised Luminescence in the Condensed Phase: Aggregation-induced Circular Dichroism and Light emission. *Chem. Sci.* **2012**, *3*, 2737–2747. [CrossRef]
192. Li, H.; Zheng, X.; Su, H.; Lam, J.W.Y.; Wong, K.S.; Xue, S.; Huang, X.; Huang, X.; Li, B.S.; Tang, B.Z. Synthesis, Optical Properties, and Helical Self-Assembly of a Bivaline-Containing Tetraphenylethene. *Sci. Rep.* **2016**, *6*, 19277.
193. Ghosh, A.; Fischer, P. Chiral Molecules Split Light: Reflection and Refraction in a Chiral Liquid. *Phys. Rev. Lett.* **2006**, *97*, 173002. [CrossRef]
194. Silverman, M.P.; Badoz, J.; Briat, B. Chiral Reflection from a Naturally Optically Active Medium. *Opt. Lett.* **1992**, *17*, 886–888. [CrossRef]
195. Pedersen, J.; Mortensen, N.A. Enhanced Circular Dichroism via Slow Light in Dispersive Structured Media. *Appl. Phys. Lett.* **2007**, *91*, 213501. [CrossRef]

196. Mortensen, N.A.; Xiao, S. Slow-Light Enhancement of Beer-Lambert-Bouguer Absorption. *Appl. Phys. Lett.* **2007**, *90*, 141108. [CrossRef]
197. Uemura, N.; Toyoda, S.; Shimizu, W.; Yoshida, Y.; Mino, T.; Sakamoto, M. Absolute Asymmetric Synthesis Involving Chiral Symmetry Breaking in Diels-Alder Reaction. *Symmetry* **2020**, *12*, 910. [CrossRef]
198. Kawagoe, K.; Fujiki, M.; Nakano, Y. Limonene Magic: Noncovalent Molecular Chirality Transfer Leading to Ambidextrous Circularly Polarised Luminescent π-Conjugated Polymers. *New J. Chem.* **2010**, *34*, 637–647. [CrossRef]
199. Nakano, Y.; Fujiki, M. Circularly Polarized Light Enhancement by Helical Polysilane Aggregates Suspension in Organic Optofluids. *Macromolecules* **2011**, *48*, 7511–7919. [CrossRef]
200. Nakano, Y.; Ichiyanagi, F.; Naito, M.; Yang, Y.-G.; Fujiki, M. Chiroptical Generation and Inversion during the Mirror-Symmetry-Breaking Aggregation of Dialkylpolysilanes due to Limonene Chirality. *Chem. Commun.* **2012**, *48*, 6636–6638. [CrossRef] [PubMed]
201. Nakano, Y.; Liu, Y.; Fujiki, M. Ambidextrous Circular Dichroism and Circularly Polarised Luminescence from Poly (9,9-di-n-decylfluorene) by Terpene Chirality Transfer. *Polym. Chem.* **2010**, *1*, 460–469. [CrossRef]
202. Zhang, W.; Yoshida, K.; Fujiki, M.; Zhu, X. Unpolarized-Light-Driven Amplified Chiroptical Modulation Between Chiral Aggregation and Achiral Disaggregation of an Azobenzene-*alt*-Fluorene Copolymer in Limonene. *Macromolecules* **2011**, *44*, 5105–5111. [CrossRef]
203. Fujiki, M.; Jalilah, A.J.; Suzuki, N.; Taguchi, M.; Zhang, W.; Abdellatif, M.M.; Nomura, K. Chiral Optofluidics: Gigantic Circularly Polarized Light Enhancement of *all-trans*-Poly (9,9-di-n-octylfluorene-2,7-vinylene) during Mirror-Symmetry-Breaking Aggregation by Optically Tuning Fluidic Media. *RSC Adv.* **2012**, *2*, 6663–6671. [CrossRef]
204. Fujiki, M.; Kawagoe, Y.; Nakano, Y.; Nakao, A. Mirror-Symmetry-Breaking in Poly [(9,9-di-n-octylfluorenyl-2,7-diyl)-*alt*-biphenyl] (PF8P2) is Susceptible to Terpene Chirality, Achiral Solvents, and Mechanical Stirring. *Molecules* **2013**, *18*, 7035–7057. [CrossRef]
205. Wang, L.; Suzuki, N.; Liu, J.; Matsuda, T.; Rahim, N.A.A.; Zhang, W.; Fujiki, M.; Zhang, Z.; Zhou, N.; Zhu, X. Limonene Induced Chiroptical Generation and Inversion during Aggregation of Achiral Polyfluorene Analogs: Structure-Dependence and Mechanism. *Polym. Chem.* **2014**, *5*, 5920–5927. [CrossRef]
206. Liu, J.; Zhang, J.; Zhang, S.; Suzuki, N.; Fujiki, M.; Wang, L.; Li, L.; Zhang, W.; Zhou, N.; Zhu, X. Chiroptical Generation and Amplification of Hyperbranched π-Conjugated Polymers in Aggregation States Driven by Limonene Chirality. *Polym. Chem.* **2014**, *5*, 784–791. [CrossRef]
207. Fujiki, M.; Yoshida, K.; Suzuki, N.; Rahim, N.A.A.; Jalil, J.A. Tempo-Spatial Chirogenesis. Limonene-Induced Mirror Symmetry Breaking of Si-Si Bond Polymers during Aggregation in Chiral Fluidic Media. *J. Photochem. Photobiol. A* **2016**, *331*, 120–129. [CrossRef]
208. Rahim, N.A.A.; Fujiki, M. Aggregation-induced Scaffolding: Photoscissable Helical Polysilane Generates Circularly Polarized Luminescent Polyfluorene. *Polym. Chem.* **2016**, *7*, 4618–4629. [CrossRef]
209. Duong, S.T.; Fujiki, M. The Origin of Bisignate Circularly Polarized Luminescence (CPL) Spectra from Chiral Polymer Aggregates and Molecular Camphor: Anti-Kasha's Rule Revealed by CPL Excitation (CPLE) Spectra. *Polym. Chem.* **2017**, *8*, 4673–4679. [CrossRef]
210. Fujiki, M.; Yoshimoto, S. Time-evolved, Far-red, Circularly Polarised Luminescent Polymer Aggregates Endowed with Sacrificial Helical Si–Si Bond Polymers. *Mater. Chem. Front.* **2017**, *1*, 1773–1785. [CrossRef]
211. Chen, H.; Yin, L.; Liu, M.; Wang, L.; Fujiki, M.; Zhang, W.; Zhu, X. Aggregation-Induced Chiroptical Generation and Photoinduced Switching of Achiral Azobenzene-alt-Fluorene Copolymer Endowed with Left-and Right-Handed Helical Poly-silanes. *RSC Adv.* **2019**, *9*, 4849–4856. [CrossRef]
212. Fujiki, M. Aggregation-Induced Chirogenesis of Luminescent Polymers. In *Aggregation-Induced Emission: Materials and Applications*; Fujiki, M., Liu, B., Tang, B.Z., Eds.; American Chemical Society: Columbus, OH, USA, 2016; Volume 2, pp. 63–92.
213. Beth, R.A. Mechanical Detection and Measurement of the Angular Momentum of Light. *Phys. Rev.* **1936**, *50*, 115–125. [CrossRef]
214. Saha, M.N.; Bhargava, Y. The Spin of the Photon. *Nature* **1931**, *128*, 870. [CrossRef]
215. Simpson, N.B.; Dholakia, K.; Allen, L.; Padgett, M.J. Mechanical Equivalence of Spin and Orbital Angular Momentum of Light: An Optical Spanner. *Opt. Lett.* **1997**, *22*, 52–54. [CrossRef]
216. Turro, N.J. *Modern Molecular Photochemistry*; University Science Books: Sausalito, CA, USA, 1991; ISBN 0935702717.
217. Calvert, J.G.; Pitts, J.N. *Photochemistry*; John Wiley & Sons: Hoboken, NJ, USA, 1973; ISBN 0471130907.
218. Franck-Condon-Principle. Available online: https://en.wikipedia.org/wiki/Franck--Condon_principle (accessed on 23 December 2020).
219. Nakashima, H.; Fujiki, M.; Koe, J.R.; Motonaga, M. Solvent and Temperature Effects on the Chiral Aggregation of Poly-(alkylarylsilane) s Bearing Remote Chiral Groups. *J. Am. Chem. Soc.* **2001**, *123*, 1963–1969. [CrossRef]
220. Terao, K.; Mori, Y.; Dobashi, T.; Sato, T.; Teramoto, A.; Fujiki, M. Solvent and Temperature Effects on the Chiral Aggregation of Optically Active Poly (dialkylsilane)s Confined in Microcapsules. *Langmuir* **2004**, *20*, 306–308. [CrossRef]
221. Fujiki, M. Ideal Exciton Spectra in Single- and Double-Screw-Sense Helical Polysilanes. *J. Am. Chem. Soc.* **1994**, *116*, 6017–6018. [CrossRef]
222. Fujiki, M. Optically Active Polysilylenes: State-Of-The-Art Chiroptical Polymers. *Macromol. Rapid Commun.* **2001**, *22*, 539–563. [CrossRef]

223. Fujiki, M.; Koe, J.R.; Terao, K.; Sato, T.; Teramoto, A.; Watanabe, J. Optically Active Polysilanes. Ten Years of Progress and New Polymer Twist for Nanoscience and Nanotechnology. *Polym. J.* **2003**, *35*, 297–344. [CrossRef]
224. Eliel, E.L.; Wilen, S.H.; Mander, L.N. The CD magnitude is normalized using the dimensionless Kuhn's anisotropy factor in the ground state, defined as $g_{CD} = \Delta\varepsilon/\varepsilon = (Ab_{sL} - Abs_R)/[(Ab_{sL} + Abs_R)/2]$ at extremum. In *Stereochemistry of Organic Compounds*; John Wiley & Sons, Inc.: New York, NY, USA, 1994.
225. Eliel, E.L.; Wilen, S.H.; Mander, L.N. The CPL magnitude is normalized by Kuhn's anisotropy in the photoexcited state. $g_{CPL} = (IL - IR)/[(IL + IR)/2]$ at extremum, where $IL$ and $IR$ denote the emission intensities of the *left-* and *right-*CP light under excitation of unpolarized light, re-spectively. John Wiley & Sons, Inc.: New York, NY, USA, 1994.
226. Nakashima, H.; Koe, J.R.; Torimitsu, K.; Fujiki, M. Transfer and Amplification of Chiral Molecular Information to Poly-silylene Aggregates. *J. Am. Chem. Soc.* **2001**, *123*, 4847–4848. [CrossRef] [PubMed]
227. Righini, G.C.; Dumeige, Y.; Féron, P.; Ferrari, M.; Conti, G.N.; Ristic, D.; Soria, S. Whispering Gallery Mode Microresonators: Fundamentals and Applications. *Riv. Nuovo Cimento* **2011**, *34*, 435–488.
228. Symes, R.; Sayer, R.M.; Reid, J.P. Cavity Enhanced Droplet Spectroscopy: Principles, Perspectives and Prospects. *Phys. Chem. Chem. Phys.* **2004**, *6*, 474–487. [CrossRef]
229. Li, Z.; Psaltis, D. Optofluidic Dye Lasers. *Microfluid Nanofluid.* **2008**, *4*, 145–158. [CrossRef]
230. Sun, Y.; Shopova, S.I.; Wu, C.S.; Arnold, S.; Fan, X. Bioinspired Optofluidic FRET Lasers via DNA Scaffolds. *Proc. Nat. Acad. Sci. USA* **2010**, *107*, 16039–16042. [CrossRef]
231. He, C.; Yang, G.; Kuai, Y.; Shan, S.; Yang, L.; Hu, J.; Zhang, D.; Zhang, Q.; Zou, G. Dissymmetry Enhancement in Enantioselective Synthesis of Helical Polydiacetylene by Application of Superchiral Light. *Nat. Commun.* **2018**, *9*, 5117. [CrossRef]
232. Yang, G.; Zhang, S.; Hu, J.; Fujiki, M.; Zou, G. The Chirality Induction and Modulation of Polymers by Circularly Polarized Light. *Symmetry* **2019**, *11*, 474. [CrossRef]
233. Kim, J.Y.; Yeom, J.; Zhao, G.; Calcaterra, H.; Munn, J.; Zhang, P.; Kotov, N. Assembly of Gold Nanoparticles into Chiral Superstructures Driven by Circularly Polarized Light. *J. Am. Chem. Soc.* **2019**, *141*, 11739–11744. [CrossRef] [PubMed]
234. Asano, T.; Okada, T.; Shinkai, S.; Shigematsu, K.; Kusano, Y.; Manabe, O. Temperature and Pressure Dependences of Thermal Cis-to-Trans Isomerization of Azobenzenes which Evidence an Inversion Mechanism. *J. Am. Chem. Soc.* **1981**, *103*, 5161–5165. [CrossRef]
235. Klimov, V.V.; Zabkov, I.V.; Pavlov, A.; Guzatov, D.V. Eigen Oscillations of a Chiral Sphere and Their Influence on Radiation of Chiral Molecules. *Opt. Exp.* **2014**, *22*, 18564–18578. [CrossRef] [PubMed]
236. Ureña, F.P.; Moreno, J.R.A.; González, J.J.L. Conformational Study of (R)-(+)-Limonene in the Liquid Phase using Vibrational Spectroscopy (IR, Raman, and VCD) and DFT Calculations. *Tetrahedron Asymmetry* **2009**, *20*, 89–97. [CrossRef]
237. Reinscheid, F.; Reinscheid, U.M. Stereochemical Analysis of (+)-Limonene using Theoretical and Experimental NMR and Chiroptical Data. *J. Mol. Struct.* **2016**, *1106*, 141–153. [CrossRef]
238. Gierschner, J.; Behera, S.K.; Park, S.Y. Dual Emission: Classes, Mechanisms and Conditions. *Angew. Chem. Int. Ed.* **2020**. [CrossRef]
239. Röhrs, M.; Escudero, D. Multiple Anti-Kasha Emissions in Transition-Metal Complexes. *J. Phys. Chem. Lett.* **2019**, *10*, 5798–5804. [CrossRef]
240. Del Valle, J.C.; Catalán, J. Kasha's rule: A reappraisal. *Phys. Chem. Chem. Phys.* **2019**, *21*, 10061–10069. [CrossRef]
241. Demchenko, A.P.; Tomin, V.I.; Chou, P.T. Breaking the Kasha Rule for More Efficient Photochemistry. *Chem. Rev.* **2017**, *117*, 13353–13381. [CrossRef]
242. Kunitake, T.; Okahata, Y. A Totally Synthetic Bilayer Membrane. *J. Am. Chem. Soc.* **1977**, *99*, 3860–3861. [CrossRef]
243. Fujiki, M. Studies on Hydride Abstraction Reaction by Trityl Salts and Synthesis/Polymerization of Dialkylammonium Acrylate. Master's Thesis, Kyushu University, Fukuoka, Japan, 1978. (In Japanese).
244. Krishnamurthy, R.; Hud, N.V. Introduction: Chemical Evolution and the Origins of Life. *Chem. Rev.* **2020**, *120*, 4613–4615. [CrossRef] [PubMed]
245. Sandford, S.A.; Nuevo, M.; Bera, P.P.; Lee, T.J. Prebiotic Astrochemistry and the Formation of Molecules of Astrobiological Interest in Interstellar Clouds and Protostellar Disks. *Chem. Rev.* **2020**, *120*, 4616–4659. [CrossRef] [PubMed]
246. Glavin, D.P.; Burton, A.S.; Elsila, J.E.; Aponte, J.C.; Dworkin, J.P. The Search for Chiral Asymmetry as a Potential Biosignature in our Solar System. *Chem. Rev.* **2020**, *120*, 4660–4689. [CrossRef] [PubMed]
247. Ribó, J.M. Chirality: The Backbone of Chemistry as a Natural Science. *Symmetry* **2020**, *12*, 1982. [CrossRef]
248. Yashima, E.; Ousaka, N.; Taura, D.; Shimomura, K.; Ikai, T.; Katsuhiro Maeda, K. Supramolecular Helical Systems: Helical Assemblies of Small Molecules, Foldamers, and Polymers with Chiral Amplification and Their Functions. *Chem. Rev.* **2016**, *116*, 3752–13990. [CrossRef]
249. Borovkov, V.V.; Hembury, G.A.; Inoue, Y. Origin, Control, and Application of Supramolecular Chirogenesis in Bisporphyrin-Based Systems. *Acc. Chem. Res.* **2004**, *37*, 449–459. [CrossRef]
250. Plutschack, M.B.; Pieber, B.; Gilmore, K.; Seeberger, P.H. The Hitchhiker's Guide to Flow Chemistry II. *Chem. Rev.* **2017**, *117*, 11796–11893. [CrossRef]

251. Brill, Z.G.; Condakes, M.L.; Ting, C.P.; Maimone, T.J. Navigating the Chiral Pool in the Total Synthesis of Complex Terpene Natural Products. *Chem. Rev.* **2017**, *117*, 11753–11795. [CrossRef]
252. Puneet, P.; Fujiki, M.; Nandan, B. Circularly Polarized Luminescent Polymers: Emerging Materials for Photophysical Applications. In *Reactive and Functional Polymers*; Gutierrez, T., Ed.; Springer: Cham, Switzerland, 2020; Volume 3, pp. 117–139.

*Review*

# Generation of Circularly Polarized Luminescence by Symmetry Breaking

Yoshitane Imai

Department of Applied Chemistry, Faculty of Science and Engineering, Kindai University, 3-4-1 Kowakae, Higashi-Osaka, Osaka 577-8502, Japan; y-imai@apch.kindai.ac.jp

Received: 8 October 2020; Accepted: 23 October 2020; Published: 28 October 2020

**Abstract:** Circularly polarized luminescence (CPL) has attracted significant attention in the fields of chiral photonic science and optoelectronic materials science. In a CPL-emitting system, a chiral luminophore derived from chiral molecules is usually essential. In this review, three non-classical CPL (NC-CPL) systems that do not use enantiomerically pure molecules are reported: (i) supramolecular organic luminophores composed of achiral organic molecules that can emit CPL without the use of any chiral auxiliaries, (ii) achiral or racemic luminophores that can emit magnetic CPL (MCPL) by applying an external magnetic field of 1.6 T, and (iii) circular dichroism-silent organic luminophores that can emit CPL in the photoexcited state as a cryptochiral CPL system.

**Keywords:** chiral; circularly polarized luminescence (CPL); magnetic circularly polarized luminescence (MCPL); spontaneous resolution

## 1. Introduction

The potential application of luminescent techniques to various systems, such as organic and organometallic electroluminescence devices and optoelectronic devices, has attracted considerable attention [1–6]. Analogous to the chirality associated with molecules, there exists chirality of light, which is referred to as circularly polarized luminescence (CPL). Unlike circular dichroism (CD), which indicates the chirality of the ground state, CPL spectroscopy elucidates conformational and structural information pertaining to optically active molecules in the photoexcited state. Optically active luminescent materials may produce either clockwise or anti-clockwise CPL. Chiral luminophores demonstrating CPL have attracted research attention, particularly in the fields of chiral photonic science and optoelectronic materials science [7–18].

CPL generally requires chiral organic or organometallic luminophores. In organometallic luminophores, chiral organic ligands coordinating with optically active metal ions induce chirality in the luminescent complex. In such a CPL-emitting system, a chiral organic molecule is indispensable. In addition, in practical applications of CPL, both right- and left-handed CPL are used, and their selective emission requires chiral organic or organometallic luminophores with opposite chirality. Chiral organic luminophores that can be prepared from achiral or racemic molecules are preferred as efficient and industrially useful chiral luminophores [19,20].

In this review, three types of non-classical CPL (NC-CPL) systems, including symmetry breaking CPL (SB-CPL) systems, are reported. The first is a spontaneous-resolution CPL system using achiral or racemic molecules. The second is a magnetic circularly polarized luminescence (MCPL) system based on achiral or racemic molecules under an external magnetic field. Finally, a cryptochiral CPL system based on CD-silent molecules in the photoexcited state is discussed.

## 2. CD and CPL from Achiral and Racemic Molecules by Spontaneous Resolution

Chiral organic luminophores are typically prepared via several steps from chiral molecules. Unfortunately, few chiral starting molecules are readily available. In addition, commercial chiral molecules are generally more expensive than achiral or racemic molecules. To eliminate such concerns, it would be economically and industrially highly useful if a chiral organic luminophore could be produced via an achiral or racemic compound.

The origin and amplification of chirality, which has led to an overwhelming enantio-enrichment of organic molecules on Earth, has been a significant topic of interest in this field of science for many decades. One proposed hypothesis for the origin of chirality is the spontaneous resolution of one of the two possible enantiomers of chiral crystals from achiral or racemic molecules [21–30]. For example, 2-anthracenecarboxylic acid (**1**) is an achiral luminescent molecule, and racemic 1-phenylethylamine (*rac*-**2**) is a racemic molecule (an equimolar mixture of (*R*)- and (*S*)-1-phenylethylamine) (Figure 1). When a mixture of **1** and *rac*-**2** is crystallized from solution, the chiral supramolecular organic luminophores **I** or **I′** can be preferentially obtained by spontaneous resolution. Chiral luminophores **I** and **I′** are an enantiomeric pair and are composed of **1** and (*R*)-**2** for **I**, and **1** and (*S*)-**2** for **I′**. **I** or **I′** can be obtained selectively by using the corresponding seed crystals.

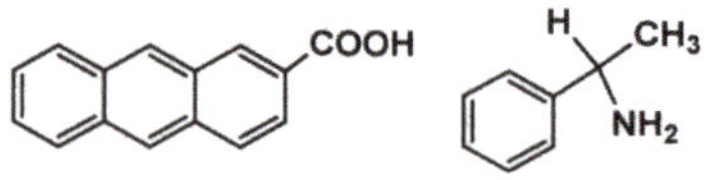

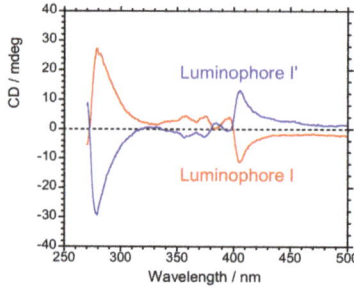

**Figure 1.** Achiral and racemic component molecules for chiral organic luminophores.

Unlike many organic luminophores, chiral luminophore **I** (or **I′**) is able to emit luminescence even in the solid state without aggregate quenching. The solid-state luminescences of **I** and **I′** are shifted to a shorter wavelength (by 34 nm) relative to that of their component luminescent molecule **1**. The solid-state luminescent maximum ($\lambda_{PL}$) of **I** (or **I′**) is 430 nm. The photoluminescence quantum yield ($\Phi_F$) increases from 4% for **1** to 20% for **I**. The solid-state CD spectra of **I** (indicated by the red line) and **I′** (indicated by the blue line) are mirror images (Figure 2). The CD signals derived from the fluorescent anthracene unit clearly appear between 330 and 450 nm. The circular anisotropy factor ($g_{CD}$) of the major CD Cotton band ($\lambda_{CD}$ = 404 nm) of **I** is approximately $-0.6 \times 10^{-3}$. This shows that the fluorescent anthracene units effectively exist in a chiral environment in the solid state.

**Figure 2.** Solid-state circular dichroism (CD) spectra of luminophores **I** (red line) and **I′** (blue line) in the solid state (KBr pellets).

As expected, X-ray analysis shows that luminophore **I** has a $P2_1$ chiral space group and a characteristic $2_1$-helical network column (Figure 3a). This characteristic column is mainly formed via ionic and hydrogen bonds composed of the carboxylic acid anions of **1** and the protonated amine cations of (*R*)-**2** (**1**:(*R*)-**2** = 1:1 component ratio). Luminophore **I** is constructed via the aggregation of these $2_1$-helical network columns by three types of edge-to-face interactions: anthracene–anthracene edge-to-face, benzene–anthracene edge-to-face, and anthracene–benzene edge-to-face interactions

(Figure 3b). This suggests that the formation of a $2_1$-helical network column is a key factor in the production of spontaneously resolved chiral organic luminophores [31].

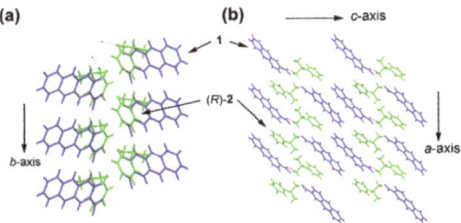

**Figure 3.** Crystal structures of luminophore **I**. (**a**) $2_1$-Helical columnar network structure along the *b*-axis. (**b**) Packing structure observed along the *b*-axis.

Chiral organic luminophore **I** (or **I′**) was successfully prepared by combining an achiral fluorescent molecule and a racemic molecule as a spontaneous-resolution system. Thus, by using only achiral molecules, chiral organic luminophores were formed.

When a mixture of the achiral luminescent molecule 2-anthracenecarboxylic acid (**1**) and the achiral molecule benzylamine (**3**) (Figure 4) is crystallized from solution at room temperature, spontaneous resolution results in the chiral supramolecular organic luminophore **II** or **II′**, which are composed of **1** and **3** with opposite chirality. In this case, a small amount of **III**, another polymorphic luminophore, is also obtained. Luminophores **II** or **II′** can also be selectively obtained by using the corresponding seed crystals.

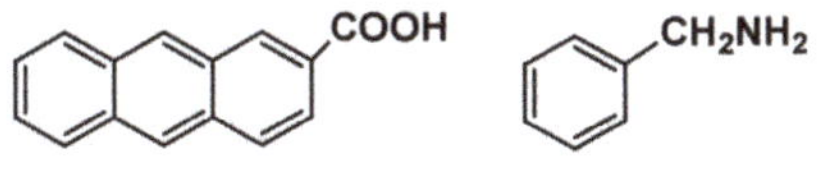

**Figure 4.** Achiral component molecules for circularly polarized luminescence (CPL).

As expected, chiral luminophore **II** also has an anion–cation and hydrogen bonded $2_1$-helical network column, similar to that of the spontaneously resolved chiral luminophore **I** (Figure 5a). The stoichiometry of components **1** and **3** in luminophore **II** is 1:1, and the space group is chiral ($P2_1$). Each network column is held in place through two types of edge-to-face interactions: anthracene–anthracene edge-to-face and benzene–anthracene edge-to-face interactions. The luminophore is created by the aggregation of this network column (Figure 5b).

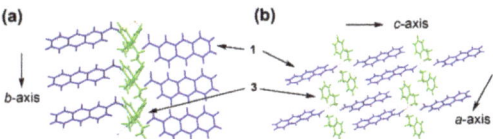

**Figure 5.** Crystal structures of luminophore **II**. (**a**) $2_1$-Helical columnar network along the *b*-axis. (**b**) Packing structure observed along the *b*-axis.

Chiral luminophores **II** and **II′** show CD in the solid state (Figure 6). Characteristic mirror-image CD spectra of **II** and **II′** are observed with a major peak around 416 nm (Figure 6). This peak originated from the fluorescent anthracene unit. The $g_{CD}$ value of the major CD Cotton band ($\lambda_{CD}$ = 416 nm) of **II** is approximately $-1.0 \times 10^{-3}$.

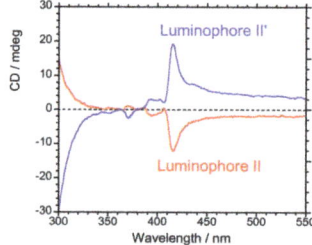

**Figure 6.** Solid-state CD spectra of luminophores **II** (red line) and **II'** (blue line) in the solid state (KBr pellets).

Luminophore **II** exhibits luminescence with a solid-state luminescent maximum ($\lambda_{PL}$) of 446 nm and a photoluminescence quantum yield ($\Phi_F$) of 16%, which is four times greater than that of the component luminophore **1** in the solid state. Interestingly, **II**, having a negative Cotton CD band, emits negative CPL with a circular anisotropy factor (Kuhn's dissymmetry ratio: $g_{CPL}$) of approximately $-1.1 \times 10^{-3}$, even though **II** is composed of achiral molecules (Figure 7). It can be concluded from the crystal structure and the theory of oscillator coupling that the CPL originates from the inter-columnar anthracene units between adjoining $2_1$-helical columns.

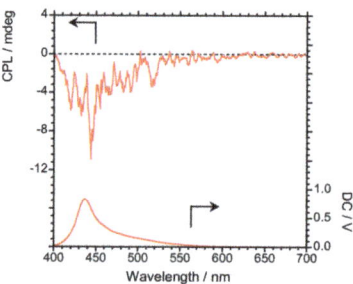

**Figure 7.** Solid-state CPL and the unpolarized photoluminescence (PL) spectra of luminophore **II** in the solid state (KBr pellet).

In this system, the production of a small amount of polymorphic luminophore **III** is an issue. Notably, this problem can be solved by changing the crystallization method. When solid-state luminescent molecule **1** is left to stand in the vapor of liquid molecule **3** at room temperature, only **III** is selectively obtained. In contrast, when solid **1** and liquid **3** are directly mixed and ground using an agate mortar, only chiral **II** or **II'** is produced.

Because many chiral molecules are not easily available and chiral molecules are more expensive than achiral or racemic molecules, chiral organic luminophores prepared from achiral or racemic molecules are preferred as industrial chiral luminophores. This study provides useful information for the creation of new spontaneously resolved CPL luminescent systems without using chiral factors [32].

## 3. CPL from Optically Inactive Organometallic and Organic Luminophores under a Magnetic Field

Organometallic luminophores have been a focus of CPL materials. Optically active lanthanide luminophores coordinated with chiral organic ligands in particular show an extremely narrow full-width at half-maximum CPL, with a high circular anisotropy factor from the visible to near-infrared regions [33–38]. In order to impart CPL characteristics to lanthanide luminophores, a chirality-inducible ligand causing antenna effects through ligand-to-metal charge transfer is inevitably needed. In recent years, an external static magnetic field has been used to act as a versatile chirality-inducible physical force in the ground and excited states, and to perturb chiral electronic structures toward several optically inactive and achiral luminophores [39–46]. Such an external magnetic influence at the molecular level causes the optically inactive luminophore to emit MCPL. In this section, the MCPL capabilities of optically inactive lanthanide luminophores Eu(III)(hfa)$_3$ and Tb(III)(hfa)$_3$ (Figure 8) are discussed.

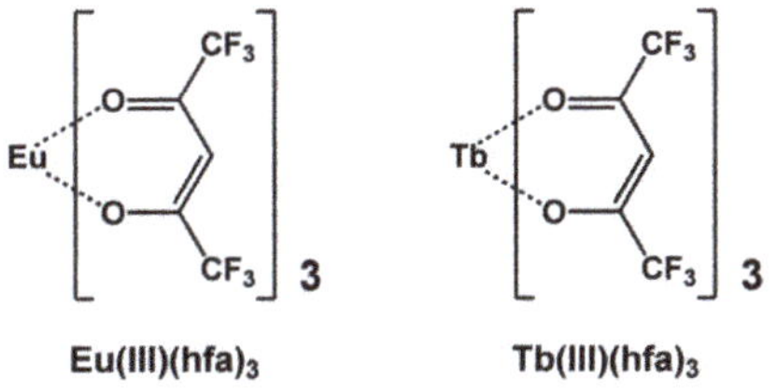

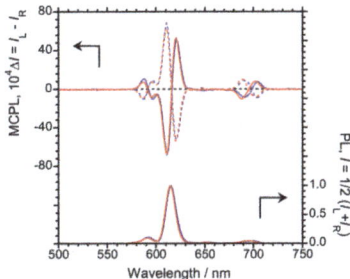

**Figure 8.** Optically inactive Eu(III)(hfa)$_3$ and Tb(III)(hfa)$_3$.

CPL was not observed for Eu(III)(hfa)$_3$ in the absence of an external magnetic field. In contrast, Eu(III)(hfa)$_3$ can emit MCPL and magnetic-field-induced unpolarized photoluminescence (hereafter denoted as PL, rather than MCPL) in CHCl$_3$ and acetone solutions under a 1.6 T magnetic field, as shown in Figure 9 (blue lines for CHCl$_3$ and red lines for acetone). The spectra under the N→S (N-up) magnetic field along the direction of the excitation light are shown via solid lines, and the spectra under the S→N (S-up) magnetic field along the direction of the excitation light are shown via dotted lines.

**Figure 9.** Magnetic circularly polarized luminescence (MCPL) (upper panel) and magnetic-field-induced unpolarized photoluminescence (PL) (lower panel) spectra of Eu(III)(hfa)$_3$ in N→S (solid lines) and S→N (dotted lines) configurations under a 1.6 T magnetic field in CHCl$_3$ (blue lines) and acetone (red lines) solutions (1.0 × 10$^{-3}$ M).

The MCPL spectra (upper panel in Figure 9) of Eu(III)(hfa)$_3$ in the two solutions are similar. The meaningful MCPL peaks ($\lambda_{MCPL}$) observed were ≈ 587, 596, 611, 621, 690 and 703 nm. They correspond to the 4f–4f transitions of Eu(III). The N→S magnetic field MCPL spectrum and S→N magnetic field MCPL spectrum are almost mirror images of each other. They show that the MCPL sign can be attributed to the direction of applied external magnetic field (N→S or S→N).

To discuss the quantitative MCPL efficiency, the $g_{MCPL}$ values are used. The $g_{MCPL}$ factor is defined as $g_{MCPL} = 2(I_L - I_R)/(I_L + I_R)$, and normalized by an external magnetic field (T$^{-1}$). In this equation, $I_L$ and $I_R$ show the left- and right-handed apparent MCPL intensities, respectively, when photoexcited by unpolarized light. The |$g_{MCPL}$| values of the main bands in CHCl$_3$ were 0.81 × 10$^{-2}$ T$^{-1}$ at 597 nm and 0.63 × 10$^{-2}$ T$^{-1}$ at 611 nm. The |$g_{MCPL}$| values of the two solutions are similar. These results indicate that the external static magnetic field can clearly induce CPL from optically inactive Eu(III)(hfa)$_3$, or possibly a mixture of the Δ- and Λ-forms of $D_3$-symmetric Eu(III)(hfa)$_3$, in the solution state. The MCPL of

Eu(III)(hfa)$_3$ was further investigated in three types of solid states—polymethylmethacrylate (PMMA) film, KBr pellet, and powder—under the same 1.6 T magnetic field as was used for the solution state measurements. The MCPL emitted from these solid-state luminophores was practically the same as that from the solution state.

Similarly, the 1.6 T magnetic field induces CPL in optically inactive Tb(III)(hfa)$_3$ luminophores, or possibly a mixture of the Δ- and Λ-forms of $D_3$-symmetric Tb(III) luminophores, in CHCl$_3$ and acetone solutions. The N→S and S→N magnetic fields can control the sign of the MCPL of the Tb(III) luminophores, as shown in Figure 10 (blue lines for CHCl$_3$ and red lines for acetone; solid lines for N→S and dotted lines for S→N configurations under 1.6 T). The $\lambda_{MCPL}$ maxima and the $|g_{MCPL}|$ values of Tb(III)(hfa)$_3$ in CHCl$_3$ are $0.53 \times 10^{-2}$ T$^{-1}$ at 484 nm, $0.46 \times 10^{-2}$ T$^{-1}$ at 493 nm, $0.16 \times 10^{-2}$ T$^{-1}$ at 538 nm, and $0.094 \times 10^{-2}$ T$^{-1}$ at 553 nm, which correspond to the characteristic 4f–4f transitions of Tb(III) in CHCl$_3$. The $|g_{MCPL}|$ values of the CHCl$_3$ and acetone solutions are similar.

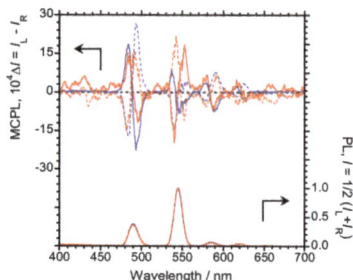

**Figure 10.** MCPL (upper panel) and PL (lower panel) spectra of Tb(III)(hfa)$_3$ in N→S (solid lines) and S→N (dotted lines) magnetic directions under a 1.6 T magnetic field in CHCl$_3$ (blue lines) and acetone (red lines) solutions ($1.0 \times 10^{-3}$ M).

Interestingly, the signs of the MCPL spectra of Tb(III)(hfa)$_3$ between the CHCl$_3$ and acetone solutions are partially inverted in the N→S and S→N magnetic fields. In CHCl$_3$, the N→S magnetic field spectrum clearly shows positive(+)-/negative(−)-sign MCPL signals at 538/553 nm, derived from the $^5D_4 \to {}^7F_5$ transitions. In contrast, the signs of the MCPL signals derived from the same $^5D_4 \to {}^7F_5$ transitions are reversed in acetone, and are clearly negative(−)-/positive(+)-sign MCPL. This sign inversion may be caused by the different coordination environment around Tb(III) of lone pair electrons on the C=O oxygen of acetone. It appears from these results that the direction of rotation of the MCPL from Tb(III)(hfa)$_3$ can be controlled by both the magnetic field direction and the nature of the solvent molecule. Similar to the chiroptical characteristics of Eu(III)(hfa)$_3$, MCPL from Tb(III)(hfa)$_3$ can also be emitted in solid states, such as a PMMA film, KBr pellet and powder, under a 1.6 T magnetic field [47].

Optically active π-conjugated organic luminophores demonstrating CPL are attracting attention in the fields of chiral photonic and optoelectronic materials science. Most organic CPL luminophores are prepared from chiral organic starting materials. Here, we describe CPL emitted from an achiral organic luminophore under an external magnetic field. Three typical π-conjugated luminophores—pyrene (**Py**), 1-pyrenol (**1-PyOH**) and 2-pyrenol (**2-PyOH**)—which do not possess any chiral sources were investigated (Figure 11).

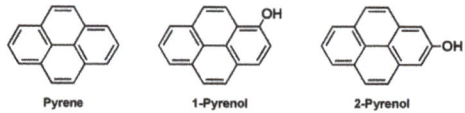

**Figure 11.** Achiral luminophores **Py**, **1-PyOH**, and **2-PyOH**.

Surprisingly, these achiral pyrene luminophores clearly emit mirror-image MCPLs under a 1.6 T magnetic field in CHCl$_3$ solution when Faraday-type N→S or S→N geometry is employed, similarly to lanthanoid luminophores, as shown in Figure 12 (solid lines for the N→S magnetic field and dotted lines for the S→N magnetic field). In addition, the signs of the MCPL can be controlled by the position (1- or 2-position) of the OH group on the peripheral pyrene ring under the same N→S or S→N geometry.

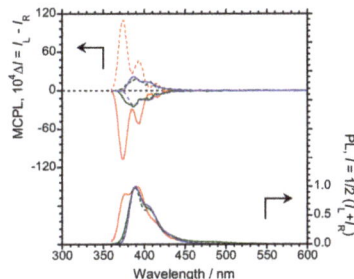

**Figure 12.** MCPL (upper panel) and PL (lower panel) spectra of **Py** (red lines), **1-PyOH** (green lines), and **2-PyOH** (blue lines) in N→S (solid lines) and S→N (dotted lines) magnetic directions under a 1.6 T magnetic field in CHCl$_3$ solution (1.0 × 10$^{-4}$ M).

In **Py**, luminescence peaks ($\lambda_{MCPL}$) can be observed at 374, 393 and 416 nm. These correspond to the 0–0′, 0–1′ and 0–2′ vibronic PL bands of the pyrenyl monomer species. The luminescence peaks of **1-PyOH** and **2-PyOH** are similar to those of **Py**. The |$g_{MCPL}$| value of **Py** in CHCl$_3$ is 0.82 × 10$^{-3}$ T$^{-1}$ at 374 nm. Pyrene has monomer and excimer luminescence. In the three pyrenes, the pyrenyl excimer-origin MCPL cannot be detected around 450–500 nm even at the high concentration of 1.0 × 10$^{-2}$ M. This indicates that pyrene excimers in the excited state apparently do not contribute to MCPL.

By using this method, various magnetic circularly polarized luminophores can be prepared from optically inactive and achiral molecules. In the future, various achiral luminophores sandwiched by two permanent magnetic fields should be able to emit MCPL from the visible to near-infrared wavelength regions, and the sign of the MCPL could be modulated by the alternating current magnetic field [48].

## 4. CPL from Cryptochiral Organic Luminophores

In 1977, Mislow reported on cryptochirality [49,50], which means hidden molecular chirality. This novel and unique chiroptical phenomenon contributes to molecular cryptography, steganography and watermarks. The most reported chiral CPL luminophores are CD-active/CPL-active luminophores. To overcome this limitation, [(4R,5R)-2,2-dimethyl-1,3-dioxolane-4,5-diyl]bis-methanolyl-bis-1-pyrene [(R,R)-4] and its enantiomer [(S,S)-4] were designed (Figure 13). These luminophores can be prepared from the corresponding enantiomerically pure 2,3-O-isopropylidene-threitol and 1-pyreneacetic acid in a single step.

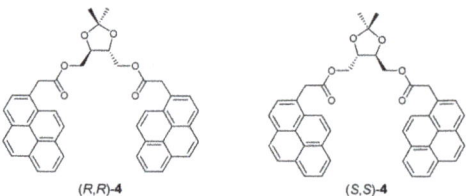

**Figure 13.** Cryptochiral dioxolane-bipyrene luminophores (*R,R*)-**4** and (*S,S*)-**4**.

A characteristic of these luminophores is that the chiral 2,2-dimethyl-1,3-dioxolane unit acts as a chiral inducer and conductor to induce chirality in the two distant fluorescent pyrene moieties. The two pyrenes are far from the central chirality of the 2,2-dimethyl-1,3-dioxolane unit and act as a hidden chiral excimer derived from photoexcitation, which is the heart of cryptochirality. Under the influence of this conductor, **4** can induce a preferential chiral twist of the fluorescent pyrenes upon an external bias on demand.

As shown in the lower panel of Figure 14, the UV–vis absorption spectrum of (*R,R*)-**4** (red line) (or (*S,S*)-**4** (blue line)) shows two major well-resolved vibronic π–π* transitions of isolated pyrene in the 250–290 and 300–370 nm regions in CHCl$_3$ solution. However, (*R,R*)-**4** (or (*S,S*)-**4**) does not provide any meaningful CD signals (Figure 14, upper panel). In contrast, (*R,R*)-**4** (red line) (or (*S,S*)-**4** (blue line)) can clearly emit CPL at 460 nm with an 18% $\Phi_F$ value, originating from the excimer pyrene in CHCl$_3$ solution (Figure 15, upper panel). Luminophores (*R,R*)-**4** and (*S,S*)-**4** provide nearly mirror-image CPL spectra. Their |$g_{CPL}$| value is ~8.9 × 10$^{-4}$ at approximately 460 nm in CHCl$_3$.

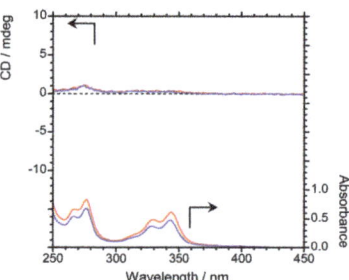

**Figure 14.** CD (upper panel) and UV–vis absorption (lower panel) spectra of (*R,R*)-**4** (red lines) and (*S,S*)-**4** (blue lines) in CHCl$_3$ solution (1.0 × 10$^{-4}$ M).

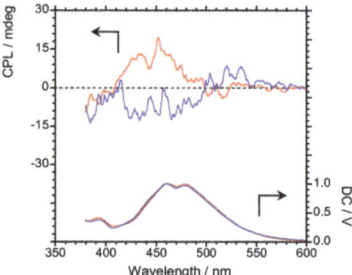

**Figure 15.** CPL (upper panel) and PL (lower panel) spectra of (*R,R*)-**4** (red lines) and (*S,S*)-**4** (blue lines) in CHCl$_3$ solution (1.0 × 10$^{-4}$ M).

The most probable structures of (R,R)-**4** in both the ground and photoexcited states were calculated. From these calculations, a model of this mechanism was developed and is shown in Figure 16. The structure obtained for the ground state does not adopt face-to-face pyrenyl π-π stacking conformers or slip pyrenyl π−π dimer conformers, but adopts the T-shaped conformers of pyrenes. These unique pyrene conformations may be the reason for the nonexistent or weak CD signals in the ground state. In contrast, (R,R)-**4** adopts chiral pyrenyl π−π-stacked conformers in the photoexcited state. Thus, the CPL magnitudes may be enhanced by the chiral π−π pyrenyl stacks, providing chiroptically detectable signals [51].

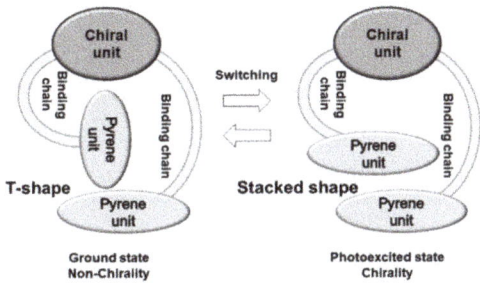

**Figure 16.** Mechanism of cryptochiral CPL system.

Based on this mechanism, instead of using a central chiral 2,2-dimethyl-1,3-dioxolane unit that acts as a controller for the expression of chirality of two pyrenes, a cryptochiral CD-silent/CPL-active luminophore was designed using an axially chiral binaphthyl unit. Chiral luminophores (R)-**5** and (S)-**5** are composed of the corresponding chiral 5,5′,6,6′,7,7′,8,8′-octahydro-1,1′-bi-2-naphthyl as the chiral unit, and two pyrenes as fluorescent units connected through flexible ester tethers (Figure 17).

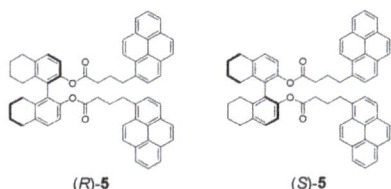

**Figure 17.** Cryptochiral binaphthyl-bipyrene luminophores (R)-**5** and (S)-**5**.

To study the ground state chiroptical properties of **5**, the CD and UV–vis absorption spectra of (R)-**5** and (S)-**5** were measured, in the same manner as **4** (Figure 18). In the UV–vis absorption spectra, three main π-π* vibronic transitions ($^1L_a$ transitions) derived from two pyrene moieties were observed between 315 and 360 nm. On the other hand, the intensities of the CD signals of **5**, corresponding to these UV bands, were noticeably weak.

The PL and CPL spectra of (R)-**5** and (S)-**5** in CHCl$_3$ are shown in Figure 19. From luminophore **5**, strong excimer PL is emitted at 468 nm ($\lambda_{PL}$) from the pyrenes. In addition, luminophores (R)-**5** and (S)-**5** emit clear CPL, and each CPL sign is positive and negative, respectively. The absolute $g_{CPL}$ value for (R)-**5** is $+2.5 \times 10^{-3}$ at 454 nm. This suggests that because of the moderately controlled flexible framework, chiral luminophore **5** functions as a cryptochiral CPL system whose CD properties cannot be detected.

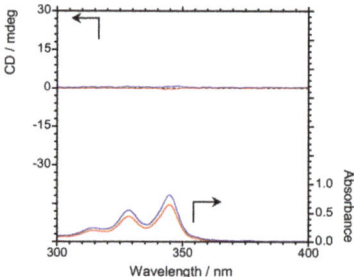

**Figure 18.** CD (upper panel) and UV–vis absorption (lower panel) spectra of (R)-**5** (red lines) and (S)-**5** (blue lines) in CHCl$_3$ solution (1.0 × 10$^{-5}$ M).

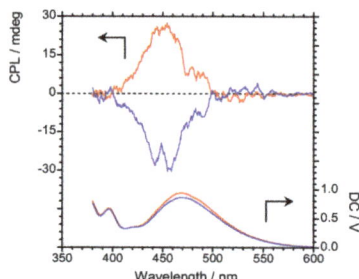

**Figure 19.** CPL (upper panel) and PL (lower panel) spectra of (R)-**5** (red lines) and (S)-**5** (blue lines) in CHCl$_3$ solution (1.0 × 10$^{-5}$ M).

This cryptochiral CPL system is similar to the cryptochiral dioxolane–bipyrene luminescent system **4**. The system simultaneously controls two chiral points (binaphthyl and pyrene units) in the photoexcited state through suitable choices of rotamer, linkers and fluorophores. In luminophore **5**, the two fluorescent pyrene units are in the conformation of an almost achiral T-shape in the ground state. Moreover, the $\theta$ value of the axially chiral octahydrobinaphthyl unit is approximately 80°–90°. The unique arrangements and conformations of the two pyrene and one binaphthyl units make the CD spectrum almost undetectable [52]. In the photoexcited state, however, the conformation of two pyrene units changes to a chiral π-stacked spatial arrangement. These configuration transfers are responsible for the cryptochiral properties, meaning the silent CD and active CPL derived from the excimer pyrene moiety [53].

## 5. Conclusions

In most CPL-emitting systems, a chiral luminophore derived from chiral molecules is essential. In addition, the selective emission of right- and left-handed CPL requires chiral organic or organometallic luminophores with opposite chirality. However, non-classical CPL (NC-CPL) systems that use no enantiomerically pure molecules have been reported. Such systems include: (i) a spontaneous-resolution CPL system, in which supramolecular organic luminophores prepared from achiral organic molecules can emit CPL without the use of any chiral auxiliaries; (ii) achiral or racemic luminophores that can emit CPL, viz. MCPL, by applying a 1.6 T external magnetic field in both the solution and solid states; and (iii) a cryptochiral CPL system, in which CD-silent organic luminophores can emit CPL in the photoexcited state. These systems indicate that CPL can be emitted from achiral or racemic molecules by suitable symmetry breaking.

**Funding:** This review was supported by a Grant-in-Aid for Scientific Research (nos. 18K05094, 19H02712, 19H04600, and 20H04678) from the MEXT/Japan Society for the Promotion of Science and the Research Foundation. This review was also supported by JST, CREST (JPMJCR2001), Japan.

**Acknowledgments:** Y.I. thanks Michiya Fujiki (Nara Institute of Science and Technology) and Reiko Kuroda (Chubu University) for helpful discussions regarding the origin of circularly polarized luminescence phenomena.

**Conflicts of Interest:** The authors declares no conflict of interest.

## References

1. Yang, X.; Zhou, G.; Wong, W.Y. Functionalization of phosphorescent emitters and their host materials by main-group elements for phosphorescent organic light-emitting devices. *Chem. Soc. Rev.* **2015**, *44*, 8484–8575. [PubMed]
2. Veldhuis, S.A.; Boix, P.P.; Yantara, N.; Li, M.; Sum, T.C.; Mathews, N.; Mhaisalkar, S.G. Perovskite Materials for Light-Emitting Diodes and Lasers. *Adv. Mater.* **2016**, *28*, 6804–6834. [PubMed]
3. Im, Y.; Kim, M.; Cho, Y.J.; Seo, J.; Yook, K.S.; Lee, J.Y. Molecular Design Strategy of Organic Thermally Activated Delayed Fluorescence Emitters. *Chem. Mater.* **2017**, *29*, 1946–1963. [CrossRef]
4. Huang, T.; Jiang, W.; Duan, L. Recent progress in solution processable TADF materials for organic light-emitting diodes. *J. Mater. Chem. C* **2018**, *6*, 5577–5596.
5. Ma, X.; Wang, J.; Tian, H. Assembling-Induced Emission: An Efficient Approach for Amorphous Metal-Free Organic Emitting Materials with Room-Temperature Phosphorescence. *Acc. Chem. Res.* **2019**, *52*, 738–748.
6. Zhang, D.-W.; Li, M.; Chen, C.-F. Recent advances in circularly polarized electroluminescence based on organic light-emitting diodes. *Chem. Soc. Rev.* **2020**, *49*, 1331–1343.
7. Field, J.E.; Muller, G.; Riehl, J.P.; Venkataraman, D. Circularly Polarized Luminescence from Bridged Triarylamine Helicenes. *J. Am. Chem. Soc.* **2003**, *125*, 11808–11809.
8. Maeda, H.; Bando, Y. Recent progress in research on stimuliresponsive circularly polarized luminescence based on π-conjugated molecules. *Pure Appl. Chem.* **2013**, *85*, 1967–1978. [CrossRef]
9. Sanchez-Carnerero, E.M.; Agarrabeitia, A.R.; Moreno, F.; Maroto, B.L.; Muller, G.; Ortiz, M.J.; Moya, S. Circularly Polarized Luminescence from Simple Organic Molecules. *Chem. Eur. J.* **2015**, *21*, 13488–13500. [PubMed]
10. Kumar, J.; Nakashima, T.; Kawai, T. Circularly Polarized Luminescence in Chiral Molecules and Supramolecular Assemblies. *J. Phys. Chem. Lett.* **2015**, *6*, 3445–3452. [PubMed]
11. Longhi, G.; Castiglioni, E.; Kosyoubu, J.; Mazzeo, G.; Sergio, A. Circularly Polarized Luminescence: A Review of Experimental and Theoretical Aspects. *Chirality* **2016**, *28*, 696–707. [CrossRef]
12. Sun, Z.; Suenaga, T.; Sarkar, P.; Sato, S.; Kotani, M.; Isobe, H. Stereoisomerism, crystal structures, and dynamics of belt-shaped cyclonaphthylenes. *Proc. Natl. Acad. Sci. USA* **2016**, *113*, 8109–8114. [CrossRef]
13. Tanaka, H.; Inoue, Y.; Mori, T. Circularly Polarized Luminescence and Circular Dichroisms in Small Organic Molecules: Correlation between Excitation and Emission Dissymmetry Factors. *ChemPhotoChem* **2018**, *2*, 386–402. [CrossRef]
14. Pop, F.; Zigon, N.; Avarvari, N. Main-Group-Based Electro- and Photoactive Chiral Materials. *Chem. Rev.* **2019**, *119*, 8435–8478. [CrossRef] [PubMed]
15. Ma, J.-L.; Peng, Q.; Zhao, C.-H. Circularly Polarized Luminescence Switching in Small Organic Molecules. *Chem. Eur. J.* **2019**, *25*, 15441–15454. [CrossRef] [PubMed]
16. Ohishi, Y.; Inouye, M. Circularly polarized luminescence from pyrene excimers. *Tetrahedron Lett.* **2019**, *60*, 151232. [CrossRef]
17. Gao, J.; Zhang, W.Y.; Wu, Z.G.; Zheng, Y.X.; Fu, D.W. Enantiomorphic Perovskite Ferroelectrics with Circularly Polarized Luminescence. *J. Am. Chem. Soc.* **2020**, *142*, 4756–4761. [CrossRef] [PubMed]
18. He, C.; Feng, Z.; Shan, S.; Wang, M.; Chen, X.; Zou, G. Highly enantioselective photo-polymerization enhanced by chiral nanoparticles and in situ photopatterning of chirality. *Nat. Commun.* **2020**, *11*, 1188. [CrossRef]
19. Jin, Q.; Chen, S.; Sang, Y.; Guo, H.; Dong, S.; Han, J.; Chen, W.; Yang, X.; Li, F.; Duan, P. Circularly polarized luminescence of achiral open-shell π-radicals. *Chem. Commun.* **2019**, *55*, 6583–6586. [CrossRef]

20. Zhao, J.; Zhang, T.; Dong, X.-Y.; Sun, M.-E.; Zhang, C.; Li, X.; Zhao, Y.S.; Zang, S.-Q. Circularly Polarized Luminescence from Achiral Single Crystals of Hybrid Manganese Halides. *J. Am. Chem. Soc.* **2019**, *141*, 15755–15760. [CrossRef]
21. Mason, S. Biomolecular homochirality. *Chem. Soc. Rev.* **1988**, *17*, 347–359. [CrossRef]
22. Girard, C.; Kagan, H.B. Nonlinear Effects in Asymmetric Synthesis and Stereoselective Reactions: Ten Years of Investigation. *Angew. Chem. Int. Ed.* **1998**, *37*, 2923–2959. [CrossRef]
23. Feringa, B.L.; Delden, R.A. Absolute Asymmetric Synthesis: The Origin, Control, and Amplification of Chirality. *Angew. Chem. Int. Ed.* **1999**, *38*, 3418–3438. [CrossRef]
24. Green, M.M.; Park, J.-W.; Sato, T.; Teramoto, A.; Lifson, S.; Selinger, R.L.B.; Selinger, J.V. The Macromolecular Route to Chiral Amplification. *Angew. Chem. Int. Ed.* **1999**, *38*, 3139–3154. [CrossRef]
25. Eschenmoser, A. Chemical Etiology of Nucleic Acid Structure. *Science* **1999**, *284*, 2118–2124. [CrossRef]
26. Soai, K.; Osanai, S.; Kadowaki, K.; Yonekubo, S.; Shibata, T.; Sato, I. *d*- and *l*-Quartz-Promoted Highly Enantioselective Synthesis of a Chiral Organic Compound. *J. Am. Chem. Soc.* **1999**, *121*, 11235–11236. [CrossRef]
27. Sato, I.; Kadowaki, K.; Soai, K. Asymmetric synthesis of an organic compound with high enantiomeric excess induced by inorganic ionic sodium chlorate. *Angew. Chem. Int. Ed.* **2000**, *39*, 1510–1512. [CrossRef]
28. Kondepudi, D.K.; Asakura, K. Chiral Autocatalysis, Spontaneous Symmetry Breaking, and Stochastic Behavior. *Acc. Chem. Res.* **2001**, *34*, 946–954. [CrossRef]
29. Zepik, H.; Shavit, E.; Tang, M.; Jensen, T.R.; Kjaer, K.; Bolbach, G.; Leiserowitz, L.; Weissbuch, I.; Lahav, M. Chiral amplification of oligopeptides in two-dimensional crystalline self-assemblies on water. *Science* **2002**, *295*, 1266–1269. [CrossRef]
30. Sato, I.; Kadowaki, K.; Ohgo, Y.; Soai, K. Highly enantioselective asymmetric autocatalysis induced by chiral ionic crystals of sodium chlorate and sodium bromate. *J. Mol. Catal. A Chem.* **2004**, *216*, 209–214. [CrossRef]
31. Imai, Y.; Kamon, K.; Murata, K.; Harada, T.; Nakano, Y.; Sato, T.; Fujiki, M.; Kuroda, R.; Matsubara, Y. Preparation of a spontaneous resolution chiral fluorescent system using 2-anthracenecarboxylic acid. *Org. Biomol. Chem.* **2008**, *6*, 3471–3475. [CrossRef] [PubMed]
32. Imai, Y.; Murata, K.; Asano, N.; Nakano, Y.; Kawaguchi, K.; Harada, T.; Sato, T.; Fujiki, M.; Kuroda, R.; Matsubara, Y. Selective Formation and Optical Property of a $2_1$-Helical Columnar Fluorophore Composed of Achiral 2-Anthracenecarboxylic Acid and Benzylamine. *Cryst. Growth Des.* **2008**, *8*, 3376–3379. [CrossRef]
33. Rexwinkel, R.B.; Meskers, S.C.J.; Riel, J.P.; Dekkers, H.P.J.M. Analysis of enantioselective quenching of tris(2,6-pyridinedicarboxylate)terbate(3-) luminescence by resolved tris(1,10-phenanthroline)ruthenium(2+) in methanol and in water. *J. Phys. Chem.* **1992**, *96*, 1112–1120. [CrossRef]
34. Petoud, S.; Muller, G.; Moore, E.G.; Xu, J.; Sokolnicki, J.; Riehl, J.P.; Le, U.N.; Cohen, S.M.; Raymond, K.N. Brilliant Sm, Eu, Tb, and Dy Chiral Lanthanide Complexes with Strong Circularly Polarized Luminescence. *J. Am. Chem. Soc.* **2007**, *129*, 77–83. [CrossRef] [PubMed]
35. Lunkley, J.L.; Shirotani, D.; Yamanari, K.; Kaizaki, S.; Muller, G. Extraordinary Circularly Polarized Luminescence Activity Exhibited by Cesium Tetrakis(3-heptafluoro-butylryl-(+)-camphorato) Eu(III) Complexes in EtOH and $CHCl_3$ Solutions. *J. Am. Chem. Soc.* **2008**, *130*, 13814–13815. [CrossRef]
36. Walton, J.W.; Carr, R.; Evans, N.H.; Funk, A.M.; Kenwright, A.M.; Parker, D.; Yufit, D.S.; Botta, M.; Pinto, S.D.; Wong, K.-L. Isostructural Series of Nine-Coordinate Chiral Lanthanide Complexes Based on Triazacyclononane. *Inorg. Chem.* **2012**, *51*, 8042–8056. [CrossRef]
37. Zinna, F.; Bari, L.D. Lanthanide Circularly Polarized Luminescence: Bases and Applications. *Chirality* **2015**, *27*, 1–13. [CrossRef]
38. Zinna, F.; Giovanella, U.; Bari, L.D. Highly Circularly Polarized Electroluminescence from a Chiral Europium Complex. *Adv. Mater.* **2015**, *27*, 1791–1795. [CrossRef]
39. Richardson, F.; Brittain, H.G. A structural study of tris(β-diketonate)europium(III) complexes in solution using magnetic circularly polarized luminescence spectroscopy. *J. Am. Chem. Soc.* **1981**, *103*, 18–24. [CrossRef]
40. Foster, D.R.; Richardson, F.S. Magnetic circularly polarized luminescence of 9-coordinate europium(III) complexes in aqueous solution. *Inorg. Chem.* **1983**, *22*, 3996–4002. [CrossRef]
41. Foster, D.R.; Richardson, F.S.; Vallarino, L.M.; Shilladt, D. Magnetic circularly polarized luminescence spectra of Eu(β-diketonate)$_3$X$_2$ complexes in nonaqueous solution. *Inorg. Chem.* **1983**, *22*, 4002–4009. [CrossRef]

42. Glover-Fischer, D.P.; Metcalf, D.H.; Hopkins, T.A.; Pugh, V.J.; Chisdes, S.J.; Kankare, J.; Richardson, F.S. Excited-State Enantiomer Interconversion Kinetics Probed by Time-Resolved Chiroptical Luminescence Spectroscopy. The Solvent and Temperature Dependence of $\Lambda$-Eu(dpa)$_3^{3-}$ ⇌ $\Delta$-Eu(dpa)$_3^{3-}$ Enantiomer Interconversion Rates in Solution. *Inorg. Chem.* **1998**, *37*, 3026–3033. [CrossRef]
43. Okutani, K.; Nozaki, K.; Iwamura, M. Specific Chiral Sensing of Amino Acids Using Induced Circularly Polarized Luminescence of Bis(diimine)dicarboxylic Acid Europium(III) Complexes. *Inorg. Chem.* **2014**, *53*, 5527–5537. [CrossRef]
44. Nelson, H.D.; Hinterding, S.O.M.; Fainblat, R.; Creutz, S.E.; Li, X.; Gamelin, D.R. Mid-Gap States and Normal vs Inverted Bonding in Luminescent Cu$^+$- and Ag$^+$-Doped CdSe Nanocrystals. *J. Am. Chem. Soc.* **2017**, *139*, 6411–6421. [CrossRef]
45. Jalilah, A.J.; Asanoma, F.; Fujiki, M. Unveiling controlled breaking of the mirror symmetry of Eu(fod)$_3$ with $\alpha$-/$\beta$-pinene and BINAP by circularly polarised luminescence (CPL), CPL excitation, and $^{19}$F-/$^{31}$P{$^1$H}-NMR spectra and Mulliken charges. *Inorg. Chem. Front.* **2018**, *5*, 2718–2733. [CrossRef]
46. Okada, H.; Hara, N.; Kaji, D.; Shizuma, M.; Fujuiki, M.; Imai, Y. Excimer-origin CPL vs monomer-origin magnetic CPL in photo-excited chiral binaphthyl-ester-pyrenes: Critical role of ester direction. *Phys. Chem. Chem. Phys.* **2020**, *22*, 13862–13866. [CrossRef] [PubMed]
47. Yoshikawa, H.; Nakajima, G.; Mimura, Y.; Kimoto, T.; Kondo, S.; Suzuki, A.; Fujiki, M.; Imai, Y. Mirror-image magnetic circularly polarized luminescence (MCPL) from optically inactive Eu$^{III}$ and Tb$^{III}$ tris($\beta$-diketonate). *Dalton Trans.* **2020**, *49*, 9588–9594. [CrossRef]
48. Kaji, D.; Okada, H.; Hara, N.; Kondo, Y.; Suzuki, S.; Miyasaka, M.; Fujiki, M.; Imai, Y. Non-classically Controlled Sign in a 1.6 Tesla Magnetic Circularly Polarized Luminescence of Three Pyrenes in a Chloroform and a PMMA Film. *Chem. Lett.* **2020**, *49*, 674–676. [CrossRef]
49. Mislow, K.; Bickart, P. An epistemological note on chirality. *Isr. J. Chem.* **1977**, *15*, 1–6. [CrossRef]
50. Mislow, K. Absolute asymmetric synthesis: A commentary. *Collect. Czech. Chem. Commun.* **2003**, *68*, 849–864.
51. Amako, T.; Nakabayashi, K.; Suzuki, N.; Guo, S.; Rahim, N.A.A.; Harada, T.; Fujiki, M.; Imai, Y. Pyrene magic: Chiroptical enciphering and deciphering 1,3-dioxolane bearing two wirepullings to drive two remote pyrenes. *Chem. Commun.* **2015**, *51*, 8237–8240. [CrossRef] [PubMed]
52. Kimoto, T.; Tajima, N.; Fujiki, M.; Imai, Y. Control of Circularly Polarized Luminescence by Using Open- and Closed-Type Binaphthyl Derivatives with the Same Axial Chirality. *Chem. Asian J.* **2012**, *7*, 2836–2841. [CrossRef] [PubMed]
53. Hara, N.; Yanai, M.; Kaji, D.; Shizuma, M.; Tajima, N.; Fujiki, M.; Imai, Y. A Pivotal Biaryl Rotamer Bearing Two Floppy Pyrenes that Exhibits Cryptochiral Characteristics in the Ground State. *ChemistrySelect* **2018**, *3*, 9970–9973. [CrossRef]

**Publisher's Note:** MDPI stays neutral with regard to jurisdictional claims in published maps and institutional affiliations.

© 2020 by the author. Licensee MDPI, Basel, Switzerland. This article is an open access article distributed under the terms and conditions of the Creative Commons Attribution (CC BY) license (http://creativecommons.org/licenses/by/4.0/).

*Review*

# Photochemical Methods for the Real-Time Observation of Phase Transition Processes upon Crystallization

**Fuyuki Ito**

Department of Chemistry, Institute of Education, Shinshu University, Nagano 380-8544, Japan; fito@shinshu-u.ac.jp; Tel.: +81-26-238-4114

Received: 25 September 2020; Accepted: 16 October 2020; Published: 19 October 2020

**Abstract:** We have used the fluorescence detection of phase transformation dynamics of organic compounds by photochemical methods to observe a real-time symmetry breaking process. The organic fluorescent molecules vary the fluorescence spectra depending on molecular aggregated states, implying fluorescence spectroscopy can be applied to probe the evolution of the molecular-assembling process. As an example, the amorphous-to-crystal phase transformation and crystallization with symmetry breaking at droplet during the solvent evaporation of mechanofluorochromic molecules are represented in this review.

**Keywords:** amorphous-to-crystal phase transformation; detection of real-time symmetry breaking; mechanofluorochromism; fluorescence spectroscopy; liquid-like cluster; evaporative crystallization; quartz crystal microbalance; two-step nucleation model

---

## 1. Introduction

The phenomena of phase transformation with structural changes are important as both basic findings and applications for materials science. Studies have been focused on thermodynamics for the macroscopic phase or phase transition dynamics based on computer simulations. Based on the molecular science of chemistry, various phase transformation dynamics are needed to understand the variety depending on the molecular individuality and to develop smart materials. We have been utilizing fluorescence spectroscopy to clarify phase transformation dynamics of organic fluorescent molecules, which relies on the molecular aggregated state. Therefore, fluorescence spectroscopy can be applied to probe the process of molecular assembly. As an example, Yu et al. [1] demonstrated the fluorescence visualization of an amorphous-to-crystalline transformation in situ microscopic observation of the crystallization of molecules in microparticles through fluorescence color changes. Heterogeneous crystallization of amorphous microparticles was clearly observed by this method. This study can provide a picture based on real-time detection of the crystallization kinetics that occur spontaneously by external stimuli, such as mechanochromic behavior and solid–solid transitions.

Based on this research report on the phase transition phenomenon evaluation by fluorescence detection, we utilized mechanofluorochromic molecules to evaluate the transition state dynamics during the amorphous-to-crystal phase transition process. In addition, by utilizing this knowledge, we have started research on the solvent evaporative crystallization process of organic fluorescent molecules, which will be reviewed.

## 2. Thermodynamic Evaluation of Amorphous-to-Crystal Phase Transformation Process by Fluorescence Spectral Changes

A number of molecules have been reported that exhibit emission color changes due to the mechanical stimulation of organic solids, namely mechanofluorochromism effects, over the past decade. [2,3] Mechanochromic phenomena are generally based on the electronic state modulation caused by the change of the intermolecular interaction due to the change of the intermolecular distance by the mechanical stimulation of the solid. In particular, the fluorescence color changes are ascribed to the amorphous-to-crystal phase transformation. The dibenzoylmethane boron difluoride complex (BF$_2$DBM) exhibits a fluorescence spectral change by smearing in the solid state, depending on its concentration of the doped in the polymer films [4–7]. Fraser et al. first reported the reversible mechanofluorochromic behavior of the 4-*tert*-butyl-4′-methoxydibenzoylmethane (trade name: avobenzone) boron difluoride complex solid based on the amorphous-to-crystal phase transformation [8]. The emission color significantly shifts to the longer wavelength (red) region upon smearing the samples. The samples then spontaneously return back to the original fluorescence with the elapse of time under ambient temperature. In this section, we review the quantitative evaluation of the thermodynamic parameters for a thermally backward reaction after smearing to probe the fluorescence change.

The molecular structures of BF$_2$DBM derivatives are shown in Figure 1. These compounds were synthesized according to a previous report [8]. The sample used for mechanochromic studies was prepared by dropping a $2.0 \times 10^{-3}$ mol·dm$^{-3}$ dichloromethane solution onto a paraffin-coated weighing paper with a pipette, and then the solvent-evaporated BF$_2$DBM derivatives on the weighing papers were rubbed with a spatula to apply a mechanical perturbation. The fluorescence spectra and their spectral changes were monitored on a Shimadzu RF-5300PC fluorescence spectrophotometer. The temperature controller was home-made and was combined with a rubber heater (Hakko Co. Ltd., Nagano, Japan) and a digital temperature controller (Omron E5CN-QT).

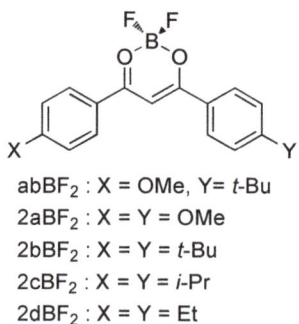

**abBF$_2$** : X = OMe, Y= *t*-Bu
**2aBF$_2$** : X = Y = OMe
**2bBF$_2$** : X = Y = *t*-Bu
**2cBF$_2$** : X = Y = *i*-Pr
**2dBF$_2$** : X = Y = Et

**Figure 1.** Molecular structures of BF$_2$DBM derivatives.

Figure 2 shows fluorescence spectra of powder **abBF$_2$** on weighing paper at 303 K, the spectra of which were normalized at the maximum value. The fluorescence of **abBF$_2$** showed a blue emission and the peak was located at 460 nm. The fluorescence spectrum originated from a dendric solid as previously reported [8]. A new fluorescence band built up around 500 nm with a shoulder at 550 nm after smearing with a spatula, suggesting the generation of the amorphous phase of **abBF$_2$** [8]. The intensity around 550 nm was increased with increasing smearing time and applied force [9]. With the elapse of time, the intensity over 530 nm was decreased. After 1030 min, the emissive color appeared green under the UV lamp and the fluorescence peaks were around 460 nm and 500 nm. These observations indicated that the yellow fluorescent amorphous state was changed to green emission at room temperature, demonstrating that **abBF$_2$** has a mechanofluorochromic property due to the thermally backward reaction.

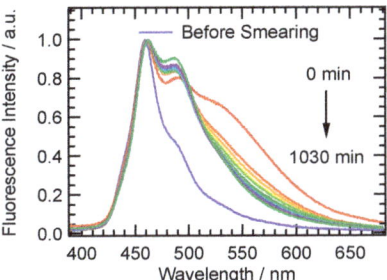

**Figure 2.** Fluorescence spectra of **abBF₂** excited with 370 nm at 303 K before and time evolution after the smearing.

In order to quantify the thermally backward reaction kinetics after smearing, we measured the intensity change of **abBF₂** fluorescence as a function of time. Figure 3 shows a kinetic trace of fluorescence intensity at 550 nm excited with 370 nm by temperature. The fluorescence intensity steeply decreased with elapsed time, obeying first-order kinetics with a double-exponential decay function. The rate constants were determined by least-squares fitting, assuming exponential decay of two components (faster ($k_F$) and slower ($k_S$)) based on first-order kinetics of the thermally backward reaction. Both rate constants increased with increasing temperature.

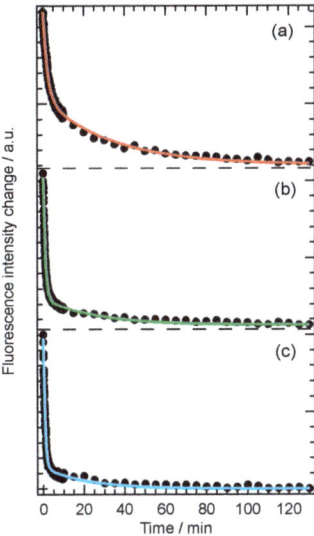

**Figure 3.** Changes in fluorescence intensity of **abBF₂** as a function of time after smearing at (**a**) 296 K, (**b**) 303 K, and (**c**) 313 K monitored at 550 nm. The best-fitting curves based on a double-exponential decay function are indicated by solid lines.

From the temperature dependence of the rate constant, the activation parameter of the amorphous–crystal phase transition of the BF₂DBM derivatives can be determined [9,10]. First, the activation parameters of the reaction can be estimated by Arrhenius plots, which are based on the rate law of the reaction: $k = A\exp(-E_a/RT)$ can be calculated and the activation parameters estimated, where $A$, $E_a$, $R$, and $T$ are the pre-exponential factor, activation energy, gas constant, and temperature, respectively. Figure 4a is the Arrhenius plot of rate constants of **abBF₂**, with the line determined by least-squares fitting. The activation energies of enthalpy ($\Delta H^{\ddagger}$) and entropy ($\Delta S^{\ddagger}$)

for the thermally backward reaction of **abBF$_2$** were also evaluated by the Eyring plot as shown in Figure 4b, the equation of which is $k = (k_B T/h)\exp(\Delta S^\ddagger/R)\exp(-\Delta H^\ddagger/RT)$, where $k_B$ is the Boltzmann constant. The calculated results are listed in Table 1. The thermodynamic parameters relate to the thermally backward reaction of **abBF$_2$** originating from the amorphous–crystal phase transformation.

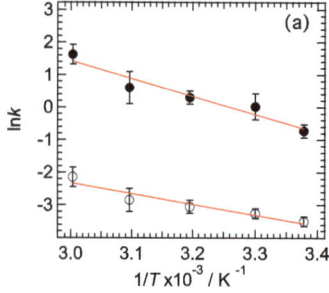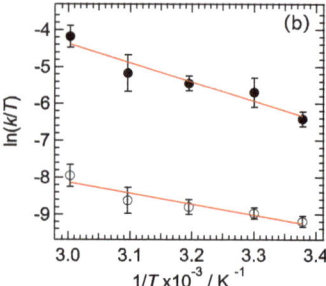

**Figure 4.** Arrhenius (**a**) and Eyring (**b**) plots for thermally backward reaction for **abBF$_2$** as a function of temperature ($T$). The rate constants ($k$) values for faster ($k_F$) and slower ($k_S$) components are indicated by closed and open circles.

**Table 1.** Thermodynamic parameters of BF$_2$DBM derivatives at the transition state. $A$, $E_a$, $\Delta H^\ddagger$, and $\Delta S^\ddagger$ are the pre-exponential factor, activation energy, activation energies of enthalpy, and activation energies of entropy, respectively.

|  | Component | $E_a$/kJ·mol$^{-1}$ | $A$/s$^{-1}$ | $\Delta H^\ddagger$/kJ·mol$^{-1}$ | $\Delta S^\ddagger$/J·K$^{-1}$·mol$^{-1}$ |
|---|---|---|---|---|---|
| **abBF$_2$** | Faster | 45.8 | $1.05 \times 10^6$ | 43.2 | $-104$ |
|  | Slower | 27.2 | $3.00 \times 10^1$ | 24.6 | $-191$ |
| **2aBF$_2$** | Faster | 44.1 | $2.36 \times 10^5$ | 41.6 | $-116$ |
|  | Slower | 33.6 | $2.82 \times 10^2$ | 31.1 | $-172$ |
| **2bBF$_2$** | Faster | 23.1 | $2.65 \times 10^1$ | 20.5 | $-192$ |
|  | Slower | 33.6 | $2.70 \times 10^{-1}$ | 14.5 | $-230$ |
| **2cBF$_2$** | Faster | 25.8 | $3.44 \times 10^1$ | 23.2 | $-190$ |
|  | Slower | 23.5 | $8.62 \times 10^{-1}$ | 21.9 | $-221$ |
| **2dBF$_2$** | Faster | 21.6 | $2.06 \times 10^1$ | 19.0 | $-194$ |
|  | Slower | 23.5 | $7.37 \times 10^0$ | 20.9 | $-203$ |

Next, thermodynamic parameters for the substituent effects were investigated. Based on the $E_a$ values for $k_F$, the backward reaction for BF$_2$DBM derivatives could be classified into two categories: category **a** (**abBF$_2$** and **2aBF$_2$**), and category **b** (**2bBF$_2$**, **2cBF$_2$**, and **2dBF$_2$**). The $k_F$ values for category **a** compounds are larger than those of category **b**. These results suggest that the methoxy group, which is a common substituent in category **a**, influences the thermally backward reaction. The pre-exponential factor ($A$ values) of $k_F$ is much larger than that of $k_S$, suggesting that the reaction frequency of $k_S$ is small. Therefore, only the $k_F$ values will be discussed here. The estimated activation energies of entropy $\Delta S^\ddagger$ values are negative, therefore suggesting that the order of the transition state (activated) complex is higher than that of the amorphous state just after mechanical perturbation. The thermodynamic parameters are derived from the phase transition from amorphous to crystalline with symmetry breaking by the thermally backward reaction, which is described below [8]. Transition state formation is influenced by any substituent and controls activation energies of enthalpy ($\Delta H^\ddagger$) and $\Delta S^\ddagger$. The $\Delta H^\ddagger$ values for $k_F$ ($\Delta H_F^\ddagger$) and activation energy ($E_a$) are similar to category **a** species. These values are then found to be twice those estimated from category **b**. All $\Delta H^\ddagger$ values are greater than the energy from van der Waals interactions (generally 1 kJ·mol$^{-1}$), and comparable to hydrogen bond interactions (about 17–63 kJ·mol$^{-1}$). It is indicated that hydrogen-bonding is dominant for intermolecular interactions to form transition states (activated complex) [11]. These estimates indicate that the existence of a methoxy group in category **a** correlates with the $\Delta H_F^{\ddagger}$ values, which is likely to correspond to the excess energy produced by the cleavage of the C(arene)–H···O(methoxy) bond [8,12]. The $S^{\ddagger}$ values for $k_F$ ($\Delta S^{\ddagger}_F$) of categories **a** and **b** are about −110 J·K$^{-1}$·mol$^{-1}$ and −190 J·K$^{-1}$·mol$^{-1}$, respectively, suggesting that the activation complexes of category **b** are a higher barrier than in category **a**. The Gibbs energy barriers ($\Delta G^{\ddagger}$) of BF$_2$DBM at 303 K were within the range of 74.8–80.8 kJ·mol$^{-1}$. Although similar for all BF$_2$DBM series reported here in this study, $\Delta H^{\ddagger}$ and $\Delta S^{\ddagger}$ around room temperature depended strongly on the substituents. Furthermore, there is no significant effect on of $\Delta H^{\ddagger}$ on $\Delta G^{\ddagger}$ in the room temperature range. According to these findings, the important driving force of the activation complex formation is not only $\Delta H^{\ddagger}$ but also $\Delta S^{\ddagger}$, therefore we concluded that the substituents exert entropic control in the solid-phase reaction because of the excess energy from the breakage of the C(arene)–H···O(methoxy) bond in category **a** [8,12]. Therefore, we propose that the substituent-dependent change in $\Delta S^{\ddagger}$ is also common to the mechanofluorochromic behavior of BF$_2$DBM derivatives based on the amorphous–crystalline phase transition and it is one of the important parameters in molecular design.

Next, we acquired differential scanning calorimetry (DSC) curves to clarify the thermodynamic parameters for the crystallization process. The melting points ($T_m$) estimated from endothermic peaks of **2aBF$_2$**, **2bBF$_2$**, **2cBF$_2$**, and **2dBF$_2$** are 508 K, 545 K, 486 K, and 477 K, respectively. Exothermic peaks correspond to the crystallization temperature ($T_c$) as listed in Table 2 We estimated the enthalpy ($\Delta H_c$) and entropy ($\Delta S_c$) of crystallization by using $\Delta H_c = T\Delta S_c$ from the peak area of the DSC curves. The $\Delta H_c$ value is related to intermolecular interactions. In order to evaluate the intermolecular interaction, we compared the existence of short contact regions smaller than the sum of van der Waals radii of neighboring molecules from the results of X-ray structure analysis of BF$_2$DBM crystals [8,13–15]. Difluoride interacts with two phenyl rings of other molecules. Short contact with the boron difluoride coordinate does not only occur with the two phenyl rings, but also with the methoxy group in the case of **abBF$_2$** and **2aBF$_2$**. Notably, $\Delta S_c$ is strongly affected by the intermolecular interaction with the methoxy group; we concluded this interaction priority promotes crystal reformation and enables it to contribute to rotational motion around the C–O group in addition to the C(arene)–H···F interaction [16]. Sket et al. reported that a BF2DBM derivative with a methoxy group has two crystal polymorphisms because the rotational energy of the C–O bond of the methoxy group is a relatively low energy barrier [17]. The degree of freedom of molecular motion of the BF$_2$DBM derivative is enhanced by the methoxy group, suggesting that the entropy values become large. The intermolecular interaction, depending on the substituents, is consistent with thermodynamic parameters. External stimuli may promote the entropic term and the activation of rotational motion around the C–O bonds will enable control of the crystal formation process.

**Table 2.** Thermodynamic parameters concerning the crystallization. $T_c$, $\Delta H_c$, and $\Delta S_c$ are crystallization temperature, the enthalpy of crystallization, and entropy of crystallization, respectively.

|  | $T_c$/K | $\Delta H_c$/kJ·mol$^{-1}$ | $\Delta S_c$/J·K$^{-1}$·mol$^{-1}$ |
|---|---|---|---|
| **abBF$_2$** | 445 | −33.8 | −76.0 |
| **2aBF$_2$** | 495 | −33.5 | −67.7 |
| **2bBF$_2$** | 519 | −25.8 | −49.8 |
| **2cBF$_2$** | 437 | −13.6 | −31.1 |
| **2dBF$_2$** | 461 | −11.6 | −25.2 |

Crystallization Gibbs energy ($\Delta G_c$) of the BF$_2$DBM derivative was then estimated by $\Delta G_c = \Delta H_c - T\Delta S_c$. The $\Delta G_c$ values at 303 K are −10.8, −13.0, −10.7, −4.2, and −4.0 kJ·mol$^{-1}$ for **abBF$_2$**, **2aBF$_2$**, **2bBF$_2$**, **2cBF$_2$**, and **2dBF$_2$**, respectively, and it can be placed in the following order: **2cBF$_2$** ≈ **2dBF$_2$** > **2bBF$_2$** ≈ **abBF$_2$** ≈ **2aBF$_2$**. This order is different from the order of $\Delta H_c$ and $\Delta S_c$. To discuss this order, the stacking properties of BF2DBM derivatives were compared [14,15]. On the basis of X-ray crystallography, the overlap between adjacent molecules with plane–plane orientation in the crystal can be classified into two groups: the overlap between benzene (B) and dihydrodioxaborinine (D) rings

(B-on-D overlap) or two benzene rings (B-on-B overlap), depending on the substituents [15]. As shown in the inset of Figure 5, **abBF₂**, **2aBF₂**, and **2bBF₂** are packed in the B-on-D type, while those in **2cBF₂** and **2dBF₂** are packed in the B-on-B type in the crystal. These findings suggest that $\Delta G_c$ depends on the molecular packing form in the crystal and is regulated by a balance between the $\Delta H_c$ and $\Delta S_c$ values. Higher degrees of overlap are proposed to result in stronger intermolecular interactions (π–π interaction), which enables us to interpret the fluorescence properties of BF₂DBM derivatives in the solid state [5,14,15].

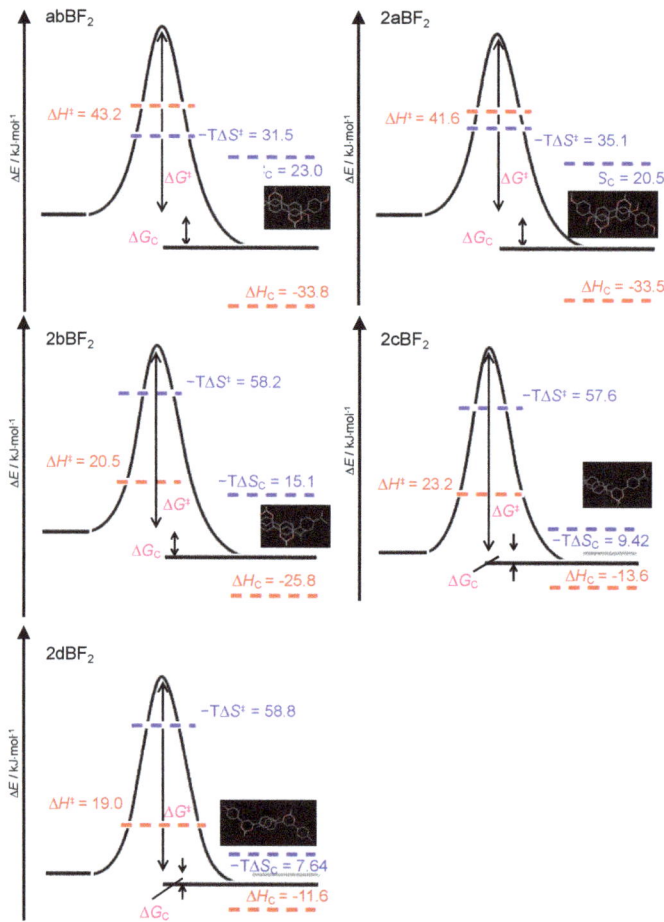

**Figure 5.** The reaction coordinate diagrams and thermodynamic parameters at 303 K of thermally backward reaction coordinates of BF₂DBM derivatives and the molecular packing obtained by X-ray crystallography. $\Delta E$, $\Delta H^\ddagger$, $\Delta S^\ddagger$, $\Delta G^\ddagger$, $\Delta H_c$, $\Delta S_c$, $\Delta G_c$ and T are the energy of the system, activation energies of enthalpy, activation energies of entropy, Gibbs energy of the transition state, the enthalpy of crystallization, entropy of crystallization, and crystallization Gibbs energy, respectively

Based on these findings, Figure 5 presents a scheme of the reaction coordinates of the amorphous-to-crystal transformation of BF₂DBM derivatives. These values of thermodynamic parameters are estimated at 303 K. Although the values of $\Delta G^\ddagger$ were similar for all BF₂DBM derivatives at 303 K, $\Delta H^\ddagger$ and $\Delta S^\ddagger$ differ depending on the substituents. We show that the substituents of the BF2DBM derivatives not only change the energy barrier of the system, but also affect the rate of the thermally backward reaction by compensating between enthalpy and entropy terms. The formation of the transition state is governed by the entropy term associated with C(arene)–H···O(methoxy).

In summary, the thermodynamic parameters for the thermally backward reaction in the amorphous-to-crystal phase transformation of BF$_2$DBM derivatives were strongly dependent on the substituents which affect not only the mode of molecular packing or stacking in the crystals, but also the thermodynamic parameters in the transition states. Thermodynamic studies based on the fluorescence changes will be significant to design organic molecules. We also think that the amorphous-to-crystal transformation correlates with the crystal growth process from the melt states.

## 3. Fluorescence Visualization of the Solvent Evaporative Crystallization Process via the Mutual State

Based on the findings from the amorphous-crystal phase transition observed by fluorescence change, we have probed the crystallization process from solution, particularly evaporative crystallization, which will be summarized in this section. We have utilized **2bBF$_2$** (Figure 4a) solution for the fluorescence observation during solvent evaporation [13]. The detection of the amorphous-like state before crystallization based on the fluorescence color change means that visualization of the two-step nucleation model can be achieved. Yu et al. recently reported that monitored amorphous-to-crystalline phase transition processes were observed in real-time based on fluorescence color changes [1].

The fluorescence images of **2bBF$_2$** in dilute solution, crystalline and amorphous states is purple, blue and green-orange, respectively, which are shown in Figure 6b–d. The fluorescence spectra previously reported [13] showed sharp peaks at 413 nm, 430 nm, and 460 nm, which are attributed to the vibrational structure of the monomer fluorescence. The emission of the crystal was observed at 445 nm and 470 nm, and in the amorphous state at around 550 nm. The two phases (crystal and amorphous) can be characterized by X-ray diffraction measurements, meaning that molecular forms and the aggregation states can be distinguished by the fluorescence color [13].

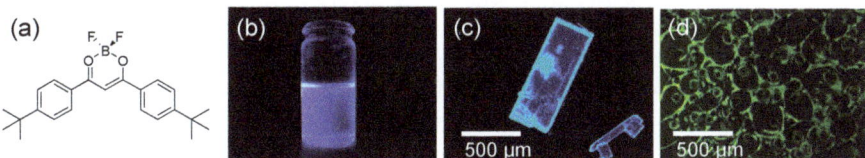

**Figure 6.** (a) Molecular structure of **2bBF$_2$**. Fluorescence images of **2bBF$_2$** in (b) 1,2-dichloromethane, (c) crystalline state, and (d) amorphous state under 365 nm UV irradiation.

Next, we attempted to measure the fluorescence color and spectral changes during evaporative crystallization from solution. The observation of the molecular assembly state by fluorescence change can be used to identify liquid clusters proposed in the two-step nucleation mechanism. Figure 7a shows captured images from a video taken under UV irradiation during the solvent evaporation from $3.1 \times 10^{-2}$ mol·dm$^{-3}$ **2bBF$_2$** in a 1,2-dichloroethane droplet. This video was uploaded in the supporting information of a previous report [13]. The fluorescence color of the droplet is purple just after dropping. The emission color changes to orange from the edge of the droplet, and at about 32 s, the entire droplet turns orange. A doughnut-shaped droplet with orange emission formed with a decreasing purple emission region. When evaporation of the solvent was completed, the entire region turned to blue emission, and some of the region showed orange emission. This orange emission was not observed even in a highly concentrated solution of **2bBF$_2$**; it is exhibited only in a supersaturated droplet during the evaporation. We measured fluorescence spectral changes as a function of time as shown in Figure 7b. The fluorescence spectrum acquired just after dropping corresponds to the monomer emission. The peak around 550 nm monotonically increased with decreasing monomer peaks. Finally, the peaks at 445 nm and 470 nm appeared due to the crystalline state. The series of fluorescence spectral changes correspond to the fluorescence color change observed in the images taken under UV irradiation. Based on the fluorescence properties of **2bBF$_2$**, we can explain the molecular assembly by evaporative crystallization from solution. The crystal of **2bBF$_2$** generated from solution by way of the amorphous-like state was found.

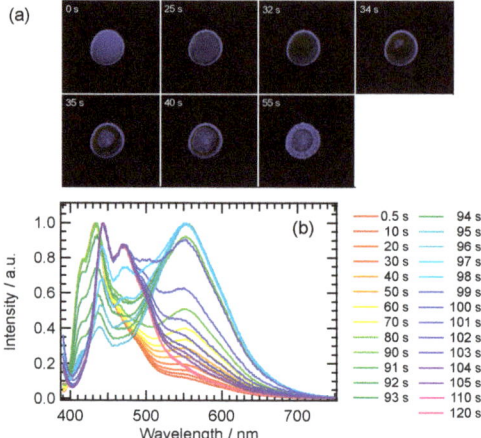

**Figure 7.** (**a**) Fluorescence images of **2bBF$_2$** in 1,2-dichloroethane droplet (diameter is approximately 5 mm) during evaporation under 365 nm UV irradiation. (**b**) Fluorescence spectral changes of **2bBF$_2$** during solvent evaporation.

In order to analyze the fluorescence change, we simulated the fluorescence spectra by non-linear least-squares fitting based on six Gaussians [13]. Figure 8a shows the relative abundances by the spectral fitting as a function of time. The initial monomer fraction (0.9) monotonically decreased during the solvent evaporation, whereas the amorphous fraction was increased up to 95 s. The abundance was about 0.6 at 95 s, which then decreased dramatically. After 95 s, the fraction of the crystal suddenly increased with the decreasing amorphous state. It is indicated that the crystal can be formed from the isolated monomer state via the amorphous-like state, showing a hierarchical change such as a consecutive reaction. We have proposed the scheme of evaporative crystallization of molecular-assembling process, shown in Figure 8b.

According to the observation of fluorescence changes during the solvent evaporative crystallization of **2bBF$_2$**, it was confirmed that the fluorescence color changed from purple to blue via orange, corresponding to the formation of crystals from monomers through amorphous-like states in the crystallization process. These findings suggested that the amorphous-like state is transiently formed prior to the crystal formation. Observation of the amorphous-like aggregates in the mutual step suggests the existence of a mutual state prior to the crystallization existing in the supersaturated region during the solvent evaporation. In the present case, the orange emission from the amorphous-like species implies the presence of liquid-like clusters with highly-dense aggregates, which was proposed in the two-step nucleation model for the crystallization. The existence of the liquid-like cluster before nucleation is a key factor for the two-step nucleation model, which has been established based on NMR spectroscopy [18,19], electron microscopy [20], induction time of crystal formation [21], and non-photochemical laser induced crystallization [22]. The relative abundance changes over time of the molecular form of **2bBF$_2$** with solvent evaporation clearly reveal that the formation of the amorphous-like states works as a precursor to nucleation. We have verified that the fluorescence visualization during solvent evaporative crystallization agrees with the previously known two-step model for crystal formation [23,24]. In conclusion, we have achieved fluorescence visualization of the existence of the nucleation precursor (highly dense liquid-like cluster state) proposed by the two-step nucleation model [13].

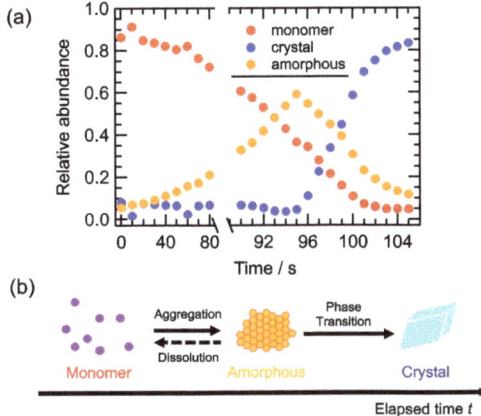

**Figure 8.** (a) Temporal changes in the relative abundance of monomers, amorphous-like, and crystalline states, which was calculated from fluorescence spectral analysis assuming six Gaussians. (b) Scheme of molecular assembly process based on fluorescence spectral changes.

## 4. Optical and Viscoelastic Properties of the Mutual State during the Phase Transformation

Many reports support the two-step nucleation model, in which the intermediate phase plays an important role in the crystal formation process [25,26]. Tsarfati et al. mentioned that the crystallization pathway involves three main steps. The three steps are i: initial densification from the solvent-rich precursor, ii: early ordering, and iii: concurrent evolution of order and morphology [27]. This finding implies that the liquid-like cluster state contains the solvent in the solution. However, it is not clear whether the orange emission of BF$_2$DBM is in a state containing a solvent as a highly dense aggregate or in an amorphous state as a solid aggregate. Therefore, in order to clarify the state of orange emission as an intermediate phase before nucleation, we focused on the real-time change of optical and viscoelastic properties of the droplet during evaporation crystallization [28].

The fluorescence images and polarized optical image were simultaneously observed during the solvent evaporation with (upper side) and without (lower side) UV irradiation as shown in Figure 9. The droplet was put in the two polarizers arranged by the cross-Nicol condition. The upper images of the droplet correspond to the fluorescence color, which is purple just after the dropping. No transparency in the image without UV irradiation indicates that the crystalline region in the droplet of **2bBF$_2$** solution is absent. The fluorescence color changed from purple to orange over time. After 60 s, birefringence was observed in the polarized optical image without UV irradiation. The texture in the image indicates the formation of a crystalline state. Compared with both images with and without UV irradiation, particularly at 85 s, no birefringence was observed in the orange emission region, suggesting the optically isotropic phase, which can be identical to the liquid-like cluster state, as an intermediate proposed in the two-step nucleation model.

We adapted the quartz crystal microbalance (QCM) in order to evaluate the mass and the viscoelastic changes based on the changes of the resonance frequency of quartz ($\Delta f$) and resistance ($\Delta R$) from the adsorption onto the quartz substrate in real-time [29]. QCM was also adapted to the monitoring of the deposited film thickness in the vacuum deposition process. We attempted to use it to investigate the dynamic viscoelastic property changes for the evaporative crystallization of the **2bBF$_2$** droplet. Figure 10a shows the results of the QCM measurement ($\Delta f$ and $\Delta R$ changes) after the dropping of the **2bBF$_2$** solution on the Au electrode as functions of time. As an overall tendency, the $\Delta f$ and $\Delta R$ values changed in two steps during the evaporative crystallization. It is possible to identify the three main stages concerning the fluorescence changes of **2bBF$_2$**: purple to blue via orange emission.

We estimated the mass change ($\Delta m$) of the **2bBF$_2$** droplet during the solvent evaporation on the Au electrode by the Sauerbrey equation [30]. $\Delta m$ evolution during solvent evaporation is shown in Figure 10b. Just after dropping, $\Delta m$ was 2 μg until 80 s, a value comparable with that for the 1,2-dichroloethane solvent only. $\Delta m$ temporarily decreased to 7.3 μg and then reached 15 μg at 95 s from 80 to 86 s. The increase in $\Delta m$ is ascribed to the adsorption and precipitation of **2bBF$_2$** onto the Au electrode.

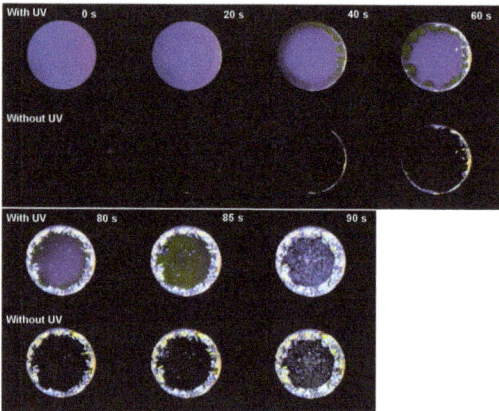

**Figure 9.** Images of the **2bBF$_2$** in 1,2-dichroloethane droplet under the cross-Nicol condition with and without UV irradiation during the solvent evaporation. From [31]; reprinted with permission from the Chemical Society of Japan.

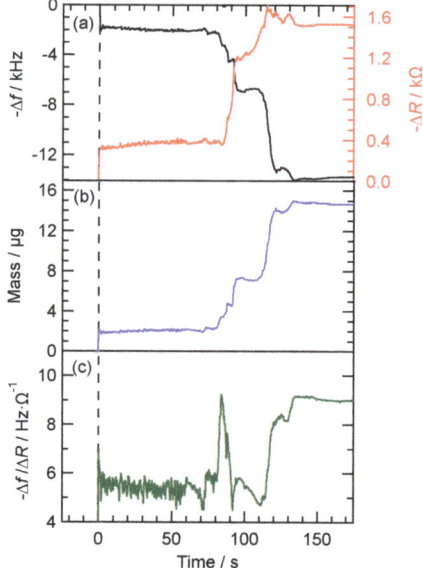

**Figure 10.** Time courses of (**a**) the resonance frequency of quartz ($\Delta f$) and resistance ($\Delta R$) monitored by quartz crystal microbalance (QCM), (**b**) the mass change calculated by the Sauerbrey equation, and (**c**) $-\Delta f/\Delta R$ during the solvent evaporation of **2bBF$_2$**. The background color indicates the corresponding fluorescence changes during the solvent evaporation of the droplet. From [31]; reprinted with permission from the Chemical Society of Japan.

The QCM results can be used to evaluate not only the $\Delta m$, but also the viscoelastic properties of the adsorbed materials. Kanazawa et al. [31] and Muramatsu et al. [32] reported that the viscosity coefficient can be estimated based on the $\Delta f$ and $\Delta R$. However, it is complicated to evaluate both the viscosity and elasticity separately in the viscoelastic medium, because $\Delta f$ depends on both the $\Delta m$ and viscosity. The ratio $-\Delta f/\Delta R$ was proposed by Kubono et al. to semi-quantitatively separate the viscosity and elasticity [33]. The $-\Delta f/\Delta R$ value of only solvent is about 5.5. Figure 10c shows the time evolution of the $-\Delta f/\Delta R$ value after the dropping of **2bBF$_2$** solution. From just after dropping until 70 s, the $-\Delta f/\Delta R$ value was maintained at 5.5. From 70 to 86 s, $-\Delta f/\Delta R$ temporarily increased to 9.2 at 84 s, which most probably originated from the artifact due to the adhesion of the aggregates. It then recovered to 5.5 until 112 s, suggesting that the amorphous-like state as an intermediate has a similar viscosity to the solution. Finally, the $-\Delta f/\Delta R$ value increased to 9.2, which is ascribed to the transformation to elastic crystals. The condensed monomer molecules in the solution form an amorphous-like state with an optically isotropic and viscous fluid. Then it adsorbs onto the substrate as a liquid-like cluster. The $\Delta m$ continually increases with time, which can be explained by the dissolved monomer molecules being further adsorbed onto the amorphous-like aggregates with increasing solution concentration. The polarized optical images and QCM measurements reveal that the intermediate liquid-like cluster state has high viscosity through solvent evaporation, but nevertheless the emission color is similar to that of the amorphous solid.

## 5. Summary

Our group has examined the fluorescence detection of phase transformation dynamics with symmetry breaking of organic compounds by photochemical methods. The phase transformation process was confirmed based on the fluorescence color change in real-time. In particular, the intermediate state (the liquid-like cluster) was visualized during the evaporative crystallization for the first time. Since this method can be carried out with a fluorescence microscope or a general optical system, it is advantageous that the experimental apparatus is relatively inexpensive. To clarify the inhomogeneity and propagation of the phase transformation process in real-time symmetry breaking, we will utilize a time- and space-resolved phase transformation detection system based on the hyperspectral camera imaging of the fluorescence spectra.

**Funding:** This research was funded by the Nanotechnology Platform Program and by JSPS KAKENHI grant numbers JP21750021, JP15H01081, JP17H05253, JP19H02686.

**Acknowledgments:** The authors thank Takehiro Sagawa, Jun-ichi Fujimori, Yukino Suzuki, Mai Saigusa, and Naoki Kanayama (Shinshu University); Atsushi Sakai and Hiroshi Ikeda (Osaka Prefecture University); Yudai Ogata and Keiji Tanaka (Kyushu University); Mitsuo Hara and Takahiro Seki (Nagoya University); and Ryohei Yasukuni and Marc Lamy de la Chapelle (Paris 13 University) for their support with the experiments and kind discussion. We also thank Kennosuke Itoh (Shinshu University) for the NMR measurements.

**Conflicts of Interest:** The authors declare no conflict of interest.

## References

1. Ye, X.; Liu, Y.; Lv, Y.; Liu, G.; Zheng, X.; Han, Q.; Jackson, K.A.; Tao, X. In Situ Microscopic Observation of the Crystallization Process of Molecular Microparticles by Fluorescence Switching. *Angew. Chem. Int. Ed.* **2015**, *54*, 7976–7980. [CrossRef] [PubMed]
2. Sagara, Y.; Kato, T. Mechanically induced luminescence changes in molecular assemblies. *Nat. Chem.* **2009**, *1*, 605–610. [CrossRef] [PubMed]
3. Chi, Z.; Zhang, X.; Xu, B.; Zhou, X.; Ma, C.; Zhang, Y.; Liu, S.; Xu, J. Recent advances in organic mechanofluorochromic materials. *Chem. Soc. Rev.* **2012**, *41*, 3878–3896. [CrossRef]
4. Мирочник, А.; Fedorenko, E.; Kaidalova, T.; Merkulov, E.; Kuryavyi, V.; Galkin, K.N.; Karasev, V. Reversible luminescence thermochromism and phase transition in crystals of thiophenylacetylacetonatoboron difluoride. *J. Lumin.* **2008**, *128*, 1799–1802. [CrossRef]

5. Sakai, A.; Tanaka, M.; Ohta, E.; Yoshimoto, Y.; Mizuno, K.; Ikeda, H. White light emission from a single component system: Remarkable concentration effects on the fluorescence of 1,3-diaroylmethanatoboron difluoride. *Tetrahedron Lett.* **2012**, *53*, 4138–4141. [CrossRef]
6. Zhang, G.; Singer, J.P.; Kooi, S.E.; Evans, R.E.; Thomas, E.L.; Fraser, C.L. Reversible solid-state mechanochromic fluorescence from a boron lipid dye. *J. Mater. Chem.* **2011**, *21*, 8295–8299. [CrossRef]
7. Liu, T.; Chien, A.D.; Lu, J.; Zhang, G.; Fraser, C.L. Arene effects on difluoroboron β-diketonate mechanochromic luminescence. *J. Mater. Chem.* **2011**, *21*, 8401–8408. [CrossRef]
8. Zhang, G.; Lu, J.; Sabat, M.; Fraser, C.L. Polymorphism and Reversible Mechanochromic Luminescence for Solid-State Difluoroboron Avobenzone. *J. Am. Chem. Soc.* **2010**, *132*, 2160–2162. [CrossRef] [PubMed]
9. Sagawa, T.; Ito, F.; Sakai, A.; Ogata, Y.; Tanaka, K.; Ikeda, H. Substituent-dependent backward reaction in mechanofluorochromism of dibenzoylmethanatoboron difluoride derivatives. *Photochem. Photobiol. Sci.* **2016**, *15*, 420–430. [CrossRef]
10. Ito, F.; Sagawa, T. Quantitative evaluation of thermodynamic parameters for thermal back-reaction after mechanically induced fluorescence change. *RSC Adv.* **2013**, *3*, 19785. [CrossRef]
11. Steiner, T. The hydrogen bond in the solid state. *Angew. Chem. Int. Ed.* **2002**, *41*, 48–76. [CrossRef]
12. Sun, X.; Zhang, X.; Li, X.; Liu, S.; Zhang, G. A mechanistic investigation of mechanochromic luminescent organoboron materials. *J. Mater. Chem.* **2012**, *22*, 17332. [CrossRef]
13. Ito, F.; Suzuki, Y.; Fujimori, J.-I.; Sagawa, T.; Hara, M.; Seki, T.; Yasukuni, R.; De La Chapelle, M.L. Direct Visualization of the Two-step Nucleation Model by Fluorescence Color Changes during Evaporative Crystallization from Solution. *Sci. Rep.* **2016**, *6*, 22918. [CrossRef]
14. Tanaka, M.; Ohta, E.; Sakai, A.; Yoshimoto, Y.; Mizuno, K.; Ikeda, H. Remarkable difference in fluorescence lifetimes of the crystalline states of dibenzoylmethanatoboron difluoride and its diisopropyl derivative. *Tetrahedron Lett.* **2013**, *54*, 4380–4384. [CrossRef]
15. Sakai, A.; Ohta, E.; Yoshimoto, Y.; Tanaka, M.; Matsui, Y.; Mizuno, K.; Ikeda, H. New Fluorescence Domain "Excited Multimer" Formed upon Photoexcitation of Continuously Stacked Diaroylmethanatoboron Difluoride Molecules with Fused π-Orbitals in Crystals. *Chem. Eur. J.* **2015**, *21*, 18128–18137. [CrossRef] [PubMed]
16. Zhang, X.; Yan, C.-J.; Pan, G.-B.; Zhang, R.-Q.; Wan, L.-J. Effect of C–H···F and O–H···O Hydrogen Bonding in Forming Self-Assembled Monolayers of BF2-Substituted β-Dicarbonyl Derivatives on HOPG: STM Investigation. *J. Phys. Chem. C* **2007**, *111*, 13851–13854. [CrossRef]
17. Galer, P.; Korošec, R.C.; Vidmar, M.; Šket, B. Crystal Structures and Emission Properties of the BF2 Complex 1-Phenyl-3-(3,5-dimethoxyphenyl)-propane-1,3-dione: Multiple Chromisms, Aggregation or Crystallization-Induced Emission, and the Self-Assembly Effect. *J. Am. Chem. Soc.* **2014**, *136*, 7383–7394. [CrossRef] [PubMed]
18. Hughes, C.E.; Harris, K.D. A Technique for In Situ Monitoring of Crystallization from Solution by Solid-State13C CPMAS NMR Spectroscopy. *J. Phys. Chem. A* **2008**, *112*, 6808–6810. [CrossRef]
19. Hughes, C.E.; Williams, P.A.; Keast, V.L.; Charalampopoulos, V.G.; Edwards-Gau, G.R.; Harris, K.D. New In Situ solid-state NMR techniques for probing the evolution of crystallization processes: Pre-nucleation, nucleation and growth. *Faraday Discuss.* **2015**, *179*, 115–140. [CrossRef] [PubMed]
20. Harano, K.; Homma, T.; Niimi, Y.; Koshino, M.; Suenaga, K.; Leibler, L.; Nakamura, E. Heterogeneous nucleation of organic crystals mediated by single-molecule templates. *Nat. Mater.* **2012**, *11*, 877–881. [CrossRef]
21. Knezic, D.; Zaccaro, J.; Myerson, A.S. Nucleation Induction Time in Levitated Droplets. *J. Phys. Chem. B* **2004**, *108*, 10672–10677. [CrossRef]
22. Garetz, B.A.; Matić, J.; Myerson, A.S. Polarization Switching of Crystal Structure in the Nonphotochemical Light-Induced Nucleation of Supersaturated Aqueous Glycine Solutions. *Phys. Rev. Lett.* **2002**, *89*, 175501. [CrossRef] [PubMed]
23. Erdemir, D.; Lee, A.Y.; Myerson, A.S. Nucleation of Crystals from Solution: Classical and Two-Step Models. *Acc. Chem. Res.* **2009**, *42*, 621–629. [CrossRef] [PubMed]
24. Vekilov, P.G. Nucleation. *Cryst. Growth Des.* **2010**, *10*, 5007–5019. [CrossRef]
25. Guo, C.; Wang, J.; Li, J.; Wang, Z.; Tang, S. Kinetic Pathways and Mechanisms of Two-Step Nucleation in Crystallization. *J. Phys. Chem. Lett.* **2016**, *7*, 5008–5014. [CrossRef]

26. Zhang, T.H.; Liu, X.Y. How Does a Transient Amorphous Precursor Template Crystallization. *J. Am. Chem. Soc.* **2007**, *129*, 13520–13526. [CrossRef]
27. Tsarfati, Y.; Rosenne, S.; Weissman, H.; Shimon, L.J.W.; Gur, D.; Palmer, B.A.; Rybtchinski, B. Crystallization of Organic Molecules: Nonclassical Mechanism Revealed by Direct Imaging. *ACS Central Sci.* **2018**, *4*, 1031–1036. [CrossRef]
28. Ito, F.; Saigusa, M.; Kanayama, N. Evaporative Crystallization of Dibenzoylmethanato Boron Difluoride Probed by Time-resolved Quartz Crystal Microbalance Responses with Fluorescence Changes. *Chem. Lett.* **2019**, in press. [CrossRef]
29. Ariga, K.; Endo, K.; Aoyama, Y.; Okahata, Y. QCM analyses on adsorption of gaseous guests to cast films of porphyrin-resorcinol derivatives. *Colloids Surfaces A* **2000**, *169*, 177–186. [CrossRef]
30. Sauerbrey, G. Verwendung von Schwingquarzen zur Wägung dünner Schichten und zur Mikrowägung. *Eur. Phys. J. A* **1959**, *155*, 206–222. [CrossRef]
31. Kanazawa, K.K.; Gordon, J.G. Frequency of a quartz microbalance in contact with liquid. *Anal. Chem.* **1985**, *57*, 1770–1771. [CrossRef]
32. Muramatsu, H.; Tamiya, E.; Karube, I. Computation of equivalent circuit parameters of quartz crystals in contact with liquids and study of liquid properties. *Anal. Chem.* **1988**, *60*, 2142–2146. [CrossRef]
33. Kubono, A.; Akiyama, R. Viscoelastic Analysis in the Formation of Organic Thin Films. *Mol. Cryst. Liq. Cryst.* **2006**, *445*, 213–222. [CrossRef]

**Publisher's Note:** MDPI stays neutral with regard to jurisdictional claims in published maps and institutional affiliations.

© 2020 by the author. Licensee MDPI, Basel, Switzerland. This article is an open access article distributed under the terms and conditions of the Creative Commons Attribution (CC BY) license (http://creativecommons.org/licenses/by/4.0/).

*Review*

# Robust Dynamics of Synthetic Molecular Systems as a Consequence of Broken Symmetry

Yoshiyuki Kageyama

Faculty of Science, Hokkaido University, Hokkaido 060-0810, Japan; y.kageyama@sci.hokudai.ac.jp;
Tel.: +81-11-706-3532

Received: 8 September 2020; Accepted: 10 October 2020; Published: 14 October 2020

**Abstract:** The construction of molecular robot-like objects that imitate living things is an important challenge for current chemists. Such molecular devices are expected to perform their duties robustly to carry out mechanical motion, process information, and make independent decisions. Dissipative self-organization plays an essential role in meeting these purposes. To produce a micro-robot that can perform the above tasks autonomously as a single entity, a function generator is required. Although many elegant review articles featuring chemical devices that mimic biological mechanical functions have been published recently, the dissipative structure, which is the minimum requirement for mimicking these functions, has not been sufficiently discussed. This article aims to show clearly that dissipative self-organization is a phenomenon involving autonomy, robustness, mechanical functions, and energy transformation. Moreover, it reports the results of recent experiments with an autonomous light-driven molecular device that achieves all of these features. In addition, a chemical model of cell-amplification is also discussed to focus on the generation of hierarchical movement by dissipative self-organization. By reviewing this research, it may be perceived that mainstream approaches to synthetic chemistry have not always been appropriate. In summary, the author proposes that the integration of catalytic functions is a key issue for the creation of autonomous microarchitecture.

**Keywords:** dissipative structure; energy conversion; mechanical work; self-oscillation; collective dynamics; autonomous motion; self-replication; autocatalysis; molecular motor; molecular robot

## 1. Introduction

It is one of the dreams of chemists to create life or its imitative system using a synthetic chemical method [1,2]. Life is thought to be a collection of nanomachines, and their cooperative behavior is capable of performing work continually in a stationary environment. The process termed 'dissipative self-organization' is one of the key processes involved in this framework [3]. On the other hand, typical inanimate objects in which modules are not self-organized can only perform work passively under transient environmental conditions. Here, work is broadly defined to denote the transfer of energy to the surroundings in a form other than thermal motion; the mechanical work this involves is defined in Section 2. The function of continuously developing in a stationary environment is defined as an autonomous function; autonomy is the realization of continuity by the internal factors of the system.

For example, fluorescent molecules emit light continuously in a photostationary state; catalyst molecules continuously convert reactive substrates into reaction products. The continuous function of these molecules (broadly defined as 'work') is derived from the fact that the molecules consist of self-organized nuclei and electrons. Focusing on such electronic characteristics at the molecular scale, a substantial number of chemical studies have been conducted. Because molecular structures effectively represent the self-organized forms of nuclei and electrons, chemists can predict and consider various functions by examining molecular structural formulas and calculating the density of their electron distributions, while studying phenomena from the perspective of these electronic properties.

This paper describes the autonomous function of molecular assembly generated by molecular self-organization. This concept realizes continuous work at a larger level than the molecular level. It is a concept of nanotechnology that is expected to be applied to material transfer devices and molecular robots and computers and to drive innovation in the field of energy research. Dissipative self-organization remains an attractive keyword for current chemistry. On the other hand, the present author is uncertain whether or not our community genuinely values the concept of dissipative self-organization. One reason for this doubt lies in the fact that the conceptual vocabulary and approaches of nonequilibrium thermodynamics are challenging for chemists who perform their experiments in a flask. Another reason is that clearly realized studies of dissipative self-organization in synthetic chemistry are limited in number. Besides, capturing a temporally developing phenomenon in a printed research paper is not easy. In addition, the terms of self-organization and self-assembly are frequently synonymous; in Japanese, for instance, they are the same word. In this paper, the basic concept of dissipative self-organization is defined in order to underline its importance to current chemical research, and recent developments in synthetic studies are reviewed. It should be noted that the origin of symmetry-breaking is not discussed in this paper.

## 2. Dissipative Self-Organization is a Concept for Autonomous Mechanical Work

Atkins' *Physical Chemistry* describes mechanical work as 'the transfer of energy that makes use of organized motion in the surroundings' [4] and distinguishes it from the transfer of energy to disordered thermal motion. In other words, to create molecular devices capable of performing useful work, it is necessary to provide the devices with proportions that will create organized movements in the surroundings (Figure 1a). Hence, a single synthetic molecular motor is too small to achieve beneficial mechanical work independently and, therefore, the motor molecule must either be positioned in a heterogeneous field or the molecules must be assembled [5]. This text considers the latter approach in detail. If a large number of molecules are merely assembled to increase the device's size, the movement of molecules is time-averaged and the molecules show no macroscopic changes under a steady condition (Figure 1b). This state is referred to as a chemical or near-chemical equilibrium, and under it, the molecules cannot create organized movements in their surroundings. In other words, to create molecular devices capable of performing beneficial work, it is necessary to introduce 'a mechanism in which the devices continually demonstrate macroscopic changes even if the environment is in a steady state' (Figure 1c). This mechanism is referred to as dissipative self-organization. In short, it is a mechanism for generating macroscopic motions via movement at the molecular level. The generated motion also has the potential to create motion in a larger hierarchy. Furthermore, due to the ability of the self-continuous action to an external object, devices that exhibit dissipative self-organization can act as an oscillation generator to actuate the repetitive motion of equilibrial materials. The creation of such molecular technology will lead to the invention of autonomous molecular robots, autonomous molecular pumps, and autonomous molecular information processors.

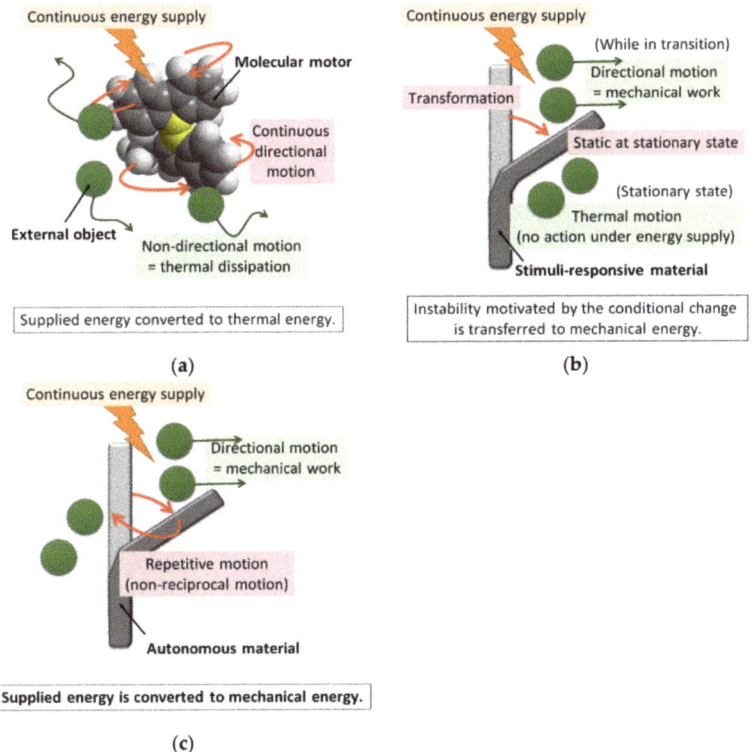

**Figure 1.** How to compel molecular materials to work. (**a**) A small molecular motor cannot carry out mechanical work for several reasons, one of which is that the transferred energy is converted to thermal energy immediately. (**b**) A stimuli-responsive material can perform work only while relaxing the instability motivated by a shift in an external condition. (**c**) Autonomous transformation of a molecular system can perform work via energy transformation. The green circles denote objects subjected to force in the surroundings.

## 3. Dissipative Self-Organization for Robustness and Energy Conversion

In chemistry, transient aspects of the system are generally described by reaction rate equations. Here, we hypothetically consider chemical reactions in a homogeneous solution, where (a) the symmetry of reactions is broken, and (b) the kinetic constants ($k$s) are consistent. The flaws of the hypothesis will be discussed below and in [6]. An autocatalytic reaction, in which the A → B reaction is accelerated by B, is introduced as a symmetry-broken reaction [7]. The reaction with the co-existence of a non-autocatalytic process, the kinetics equation of which is described in Appendix A, is shown in Figure 2a [6]. After introducing the logistic curve, general textbooks of nonlinear chemistry describe Lotka–Volterra-type dynamics: the open reaction of A → B → C → D with two autocatalytic steps demonstrates oscillatory behavior in composition (Figure 2b) [3,8]. On the other hand, we consider a triangular type reaction of A → B → C → A, aiming to distinguish the system's boundary.

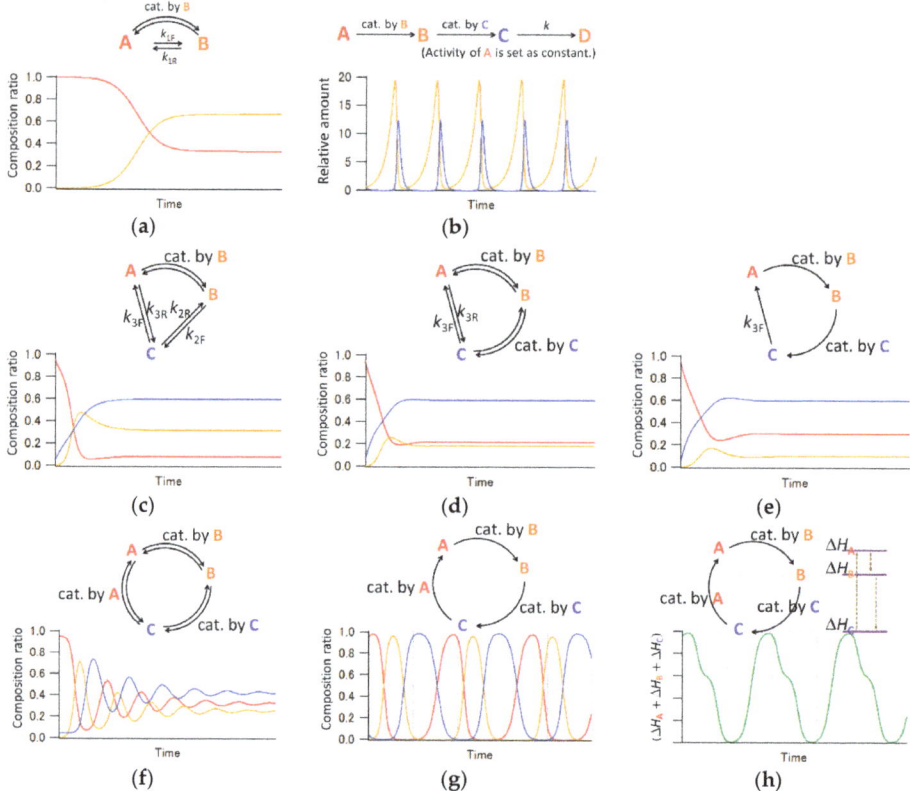

**Figure 2.** Numerical analyses for composition curves of reactions involving autocatalysis. (**a**) Logistic-type composition curve. (**b**) Lotka–Volterra-type composition oscillation. (**c**) Composition curve for a triangle reaction including one autocatalytic process. (**d**,**e**) Composition curves for triangle reactions, including two autocatalytic processes with and without considering reverse processes, respectively. (**f**,**g**) Oscillatory composition curves in triangle reactions, including three autocatalytic processes with and without considering reverse processes, respectively. (**h**) The variation in total heat of formations of A, B, and C while the change is shown in (**g**). Because the exteriorization of energy from the system occurs when the sum of the heat of formations decreases, an external energy supply is required to maintain the oscillation. Similarly, all calculations assumed consistent kinetic constants, meaning that an adequate energy supply and relevant emissions exist, although the fact is not indicated in the schemes. The red, orange, and blue lines denote compounds A, B, and C, respectively. The equations for the numerical analysis are described in Appendix A.

To simplify the discussion, we present here the simulation results while ignoring co-existing non-autocatalytic sub-routes within the autocatalytic steps [6]. In triangular-type reactions with one or two autocatalytic steps, it is easy to achieve equilibrium, as shown in Figure 2c,d. Even if we ignore the reverse processes, the reaction reaches equilibrium, as shown in Figure 2e. On the other hand, in reactions with three autocatalytic processes, periodic oscillations in the composition ratio are autonomously demonstrated, as shown in Figure 2f (taking into account reverse processes of autocatalytic paths) and Figure 2g (ignoring them). As described here, the kinetic equations indicate that multi-molecular oscillations may be generated by the contribution of plural nonlinear processes, and that the feature is emasculated by the coexistence of reverse processes. Note that autocatalysis is employed only because it is a well-known and easy-to-calculate nonlinear chemical process—other symmetry-broken phenomena, especially those with large nonlinearity and irreversibility, may contribute to realizing the oscillatory system.

However, the enthalpy of the system varies with oscillation (Figure 2h). The actual system dissipates thermal energy outside the system when the enthalpy decreases, and the periodic behavior is relaxed and terminated. This fact suggests that a supply of convertible energy from outside the system is required to develop the oscillation. In summary, a system with autonomous multi-molecular oscillation can be realized as shown in Scheme 1. The reaction scheme is somewhat similar to a catalytic reaction and a fluorescence scheme; the difference is that the change is in the distribution of components or in the molecular-level structure, and that the nonlinear phenomena are involved in the scheme and act inter- or intra-molecularly.

This temporal pattern in a multi-molecular system under continuous energy supply and periodic energy dissipation is referred to as a dissipative structure; while making cyclic transitions (A → B → C → A) at the molecular level, periodic behaviors are generated as a molecular ensemble, the time scale of which is larger than molecular-level conversion. Although we cannot predict when a chemical reaction happens at the molecular level, we can predict when the macroscopic change occurs by observing the robust temporal pattern. Even if the reaction solution is reduced by half or is frozen temporarily in the middle of the reaction, the autonomous oscillation is robustly maintained. These features are characteristic of dissipative structures.

In a homogeneous system, the variation of the system's enthalpy is dissipated as thermal energy because the system cannot form organized motion. Regarding devices with shapes (such as molecular assemblies), variation in enthalpy can be used to apply mechanical work to the surroundings. This energy conversion function at the multi-molecular scale is the essential attraction of dissipative self-organization with molecular self-assembly.

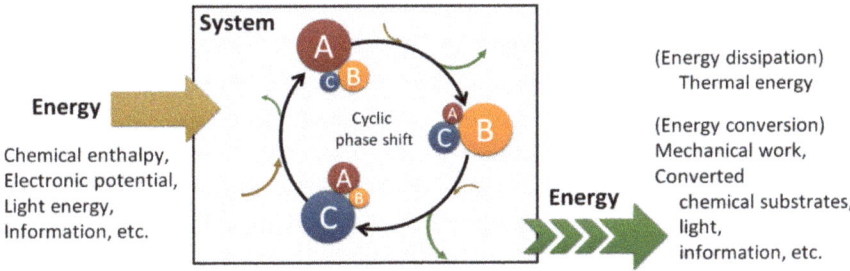

**Scheme 1.** Conceptual scheme for an autonomous system.

## 4. Research Aiming Mechanical Work with Energy Conversion

Molecular machines and motors, which are defined as molecules that are expected to convert supplied energy into mechanical forces and motion, have attracted considerable attention among chemists [9–14]. Molecules that reversibly alter their equilibrium structures, triggered by shifts in external conditions, are known as molecular machines, distinguishing them from molecular motors, a term denoting molecules that maintain their symmetry-broken motion under the continuous supply of energy [12,13]. In contrast to the molecular machines, the symmetry-broken motion of molecular motors may perform autonomous work in molecular-level space if in contact with another object. Feringa's research into nano-cars may provide one example of this, although his team's experiment was conducted using an intermittent electric current rather than continuous light waves [15]. Photo-induced ion or electron transportation across membranes [16,17] provides further instances of this phenomenon [5]. However, to attain more significant autonomous work at the multi-molecular scale, the concept of dissipative self-organization is required. Even if we gather molecular motors and supply energy, the motions are time-averaged and the energy is dissipated as thermal energy. In fact, according to Feringa's report, the motion of a glass-rod on a molecular motor-containing liquid crystal was terminated despite the continuous energy supply [18,19], similar to the system employing molecular machines [20,21]. Therefore, his collaborators are currently directing their efforts to the design of multi-molecular systems where the motors are synchronized to realize mechanical work at the macroscopic level [22].

As mentioned here and elsewhere recently [14,23], the chemical challenge to active work lies in the creation of macroscopic and self-continuous motion. However, one successful example has already been published [24]. This approach, which did not use a chiral molecule, differed from the strategies of other chemists. Molecular-level chirality (e.g., chirality due to an asymmetric form of carbon) is not required for macroscopic objects to move, whereas symmetry-breaking is a critical factor in their motion. Herein, the autonomous flipping and swimming of a submillimeter-size molecular-assembly is introduced [24–26].

The submillimeter-size assembly of an achiral azobenzene derivative (**AZ**) showed an autonomous bending-unbending motion under continuous light irradiation (Figure 3). The photochromic compound repetitively changes its molecular structure between *trans*- and *cis*-isomers under light irradiation. Ordinarily, repetitive photoisomerization yields a mixture of isomers, the composition ratio of which is constant and depends on the photoreaction rates. However, in the study, a phase transition occurred before the ratio achieved the equilibrium value, switching the reaction rates. Coupling the sets of photoisomerization and its induced phase transition led to a periodic transition consisting of an increase in the *cis*-isomer, a morphological change, a decrease in the *cis*-isomer, and then to a morphological recovery [24,27]. The enthalpy also changed periodically with the cyclic process [25]; the cycle enabled autonomous energy conversion from light to mechanical work. At the bending motion, another asymmetry defines the direction of the transformation. According to the single-crystal analysis [25], the space group of the assembly was $P_1$, and it was assumed that the asymmetry concerns the direction of bending. In addition, some break in symmetry is required to produce an autonomous swimming motion: if the bending-and-unbending motion is spatially reciprocal, it is impossible for the assembly to swim in a Newtonian fluid, as the Scallop theorem indicates [28]. The directional swimming of the assembly indicated that the morphological changes occurred in a non-reciprocal manner [26]. Furthermore, the authors also demonstrated signal-dependent pattern formation, a concept that can be applied to molecular-based information processors [25].

Applying stimuli-responsive materials, oscillatory motions, and directional motions has also been reported [29–41]. Stimuli-responsive materials generally terminate their motion when the systems reach equilibrium. The oscillations in those studies were motivated by the coupling of the directionality of external energy-supply and the time-delayed transition of materials [27,39]. Therefore, we can anticipate that the motions are characterized and controllable by the external conditions, but they lack robustness—for example, adjustment of the energy source position seriously affects the mechanical behavior [27]. Some researchers wrongly regard this feature as an instance of taxis, but it actually arises from a lack of self-excitation. On the other hand, dissipative self-organized materials possess the feature of autonomy. The sustainability and directionality of the movements are guaranteed by the internal factors of the materials. The series of studies of BZ-hydrogels, which are hydrogels periodically repeating swelling and unswelling through the Belousov–Zhabotinsky (BZ) reaction, clearly indicated the superiorities of dissipative self-organization for mechanical device applications [42,43].

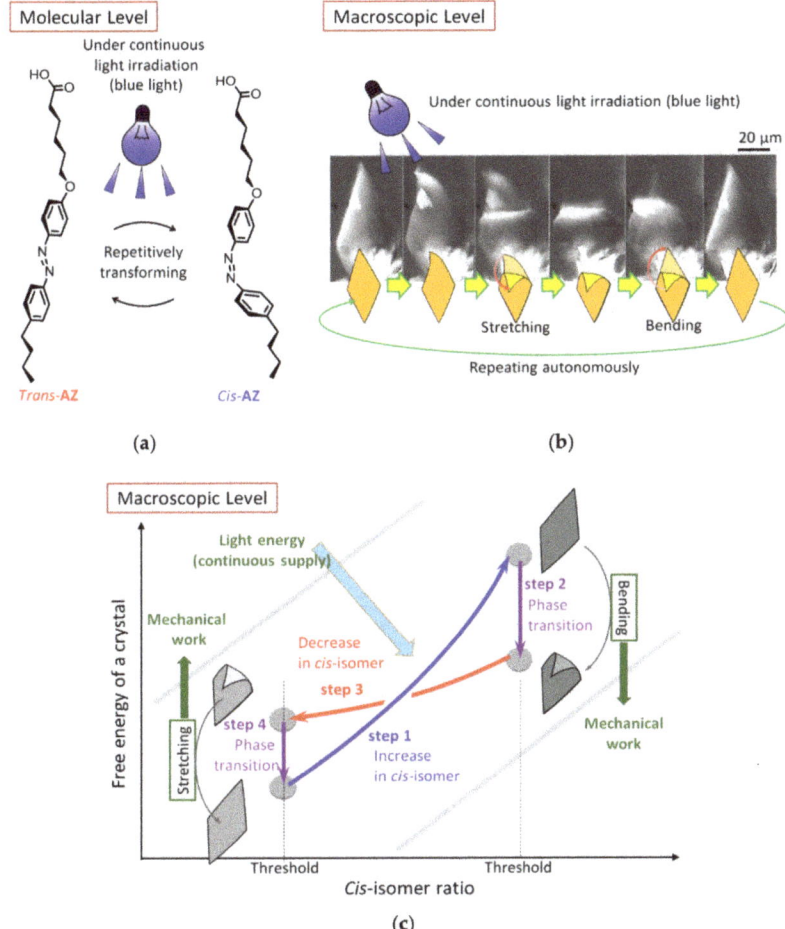

Figure 3. Self-sustaining mechanical motion of azobenzene assembly under continuous blue-light irradiation [24,25]. (a) Molecular-level repetitive transformation of the component molecule (an achiral azobenzene derivative, **AZ**). (b) Micrographs of the macroscopic autonomous flipping motion under light irradiation. (c) Schematic energy diagram for the autonomous cycle; mechanical work was performed at the steps of exergonic morphological change. All figures were reproduced from the original data with permission from Wiley-VCH, 2016.

## 5. Research into Self-Developing Molecular Systems: The Chemical Model of Cell Amplification

Cell amplification is also intrinsic to living things [1,2,44–46]. The pioneering study of the chemical model of cell amplification was reported by Luisi [47,48]. The team monitored the increase in oleate assembly in water following the addition of oleic anhydride and concluded that the autocatalytic hydrolysis of the anhydride occurred in the oleate assembly to form oleic acid [48]. Inspired by the study, Sugawara's group designed a system where a membrane molecule formed from a bipolar amphiphile via hydrolysis, catalyzed by a molecule embedded in the vesicular membrane, and, thereby, captured the growth-and-division dynamics of vesicular assemblies [49,50]. In addition, they succeeded in demonstrating the amplification of DNA via a polymerase chain reaction within the vesicles [51,52]. Furthermore, the group firstly realized autocatalytic vesicular amplification systems, via organic synthesis [53,54]. Following these studies, Devaraj's group [55] also reported the production of an autocatalytic membrane amplification system, in which a molecular catalyst formed spontaneously.

Considering the intrinsic feature of cell amplification, the concept of autocatalysis is fundamental [56,57], and all self-replications occur in an autocatalytic manner [7,58–71]. In the vesicular self-reproduction system without autocatalysis, the generated vesicles were different from the original ones—the catalytic ratio decreased proportionally in each generation, and the system tended to reach equilibrium, even if the precursor for the membrane molecule was continuously supplied (Figure 4a) [49,50,52]. This behavior is the same as inanimate systems. On the other hand, in the autocatalytic system, the catalytic ability is maintained across generations. When the precursor is supplied continuously or an excess amount of precursor is presented, or under both conditions, the individual vesicle repeats its growth and division periodically (Figure 4b) [53–55].

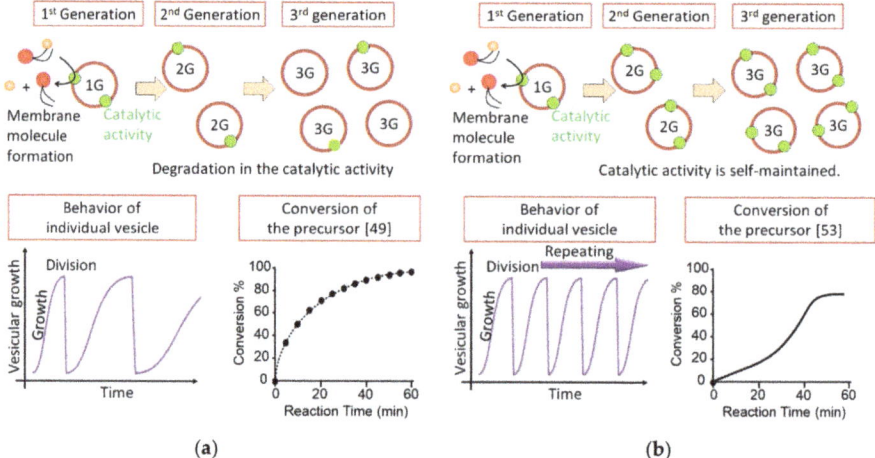

**Figure 4.** The difference between (**a**) non-autocatalytic and (**b**) autocatalytic vesicular self-reproduction.

Besides, as Fletcher comments on his autocatalytic micellar system [72,73], a cyclic reaction with an autocatalytic vesicular formation maintains vesicle numbers in an open system; this scheme is represented in Figure 2a,c. Developing this concept, we would like to consider the existence of one additional autocatalytic reaction: the system where the vesicle becomes prey to a larger self-assembly (the consumption ratio is proportional to the number of the larger assembly) and the larger self-assembly decomposes. The reaction equation of the system is complicated because of the diversity in molecular numbers in each vesicle. Nevertheless, it can be expected that the behavior can be represented by the Lotka–Volterra model (if the decomposition is exhausted to the exterior of the dispersion; Figure 2b) or the similar model shown in Section 3 (if the component is recycled autocatalytically by accepting energy from outside the system; Figure 2f,g). Under such conditions, a spatially and temporally larger oscillation than the vesicular growth-division periodicity will be generated (Figure 5). This predator–prey concept is linked to the robustness of nature in an overall sense.

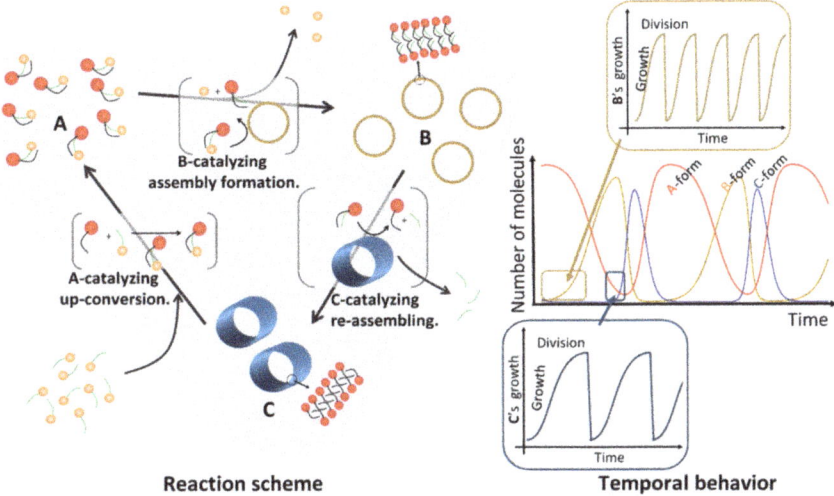

**Figure 5.** Conceptual illustration for a hierarchical system composed of autocatalytic assembling processes. Spatially and temporally larger oscillation generating a molecular assembly that appears and disappears repetitively.

## 6. Conclusions and Perspectives

Dissipative self-organization is the concept of the cooperation of dynamics to form larger dynamic systems [3]. At the nano level, the combination of thermal fluctuation, a symmetry-breaking phenomenon, and a supply of energy results in the formation of a tiny but organized dynamic system. Once organized dynamics are generated, more robustness and more significant organized motion are formed if symmetry-breaking processes are involved. Through iterative self-organization, a highly hierarchical molecular world is constructed. The symmetry-breaking phenomena contribute to making the system differ from the equilibrium structure. Here, a supply of convertible energy is required to realize self-organization, and both the reception and the dissipation of the energy occur autonomously. The concept of dissipative self-organization is directly linked to the chemical energy conversion to achieve autonomous mechanical functions.

We chemists ordinarily encounter self-organized phenomena. If we consider a molecule to be a thermodynamic system composed of electrons and nuclei, the system demonstrates self-organized behavior whether the molecule is termed a machine or not. However, such systems are too small to perform mechanical work, and the assembly of molecules is, thus, proposed to meet this function. However, controlling symmetry-breaking at the multi-molecular level is a complex task; the artificial system of a molecular assembly tends to form its equilibrial structure. The BZ reaction that was applied to mechanical gel [42,43] and the recent azobenzene-assembly system [24–26] provide successful examples. The creation of chemically fueled mechanical work is particularly challenging; if we outdistance the system from the thermal fluctuation level by consuming the dilute energy of chemical compounds, we have to recruit effective ratcheting phenomena.

A recently proposed concept, 'dissipative self-assembly', seems to share the same issues [74–79]. However, from the author's standpoint, the claims of dissipative self-assembly are still unclear and weak in the aspect of the distancing from equilibrium; most of the papers reported consecutive reactions where the intermedium products were assemblies and claimed that the assembled state can be maintained through continuous substance supply [72,73,80–87]. The structure of such nonequilibrium systems is not a dissipative structure, according to Prigogine [3]. In this context, the supplied energy is consumed to sustain the assembly, but not for autonomous tasks [88,89]. To imitate living dynamics clearly, we have to employ one or more symmetry-breaking effects to replace the system from the equilibrium structure at the tentative state to a nonequilibrium resume [90]; according to Chapter 7 in reference 3 and the references therein, at least a cubic nonlinearity in the rate equations is required to destabilize the thermodynamic branch.

However, a negative attitude to the current difficulties in imitating the intrinsic features of living dynamics is misplaced. One of the reasons for the struggle to reproduce autonomous dynamics is that chemists attempt to create mechanical materials using switchable molecules, so-called molecular machines. Such molecules are convenient for chemists because their status is easily controlled. Yet in living things, it is not switching motions but continuous motions that are able to do power- autonomous work. In other words, enzymes work with ratcheting mechanisms to prevent the system from attaining equilibrium. By imitating the function of enzymes [91–93]—molecular-recognition with nonlinear adaptation and repetitive transformation with energy conversion—and by assembling the synthesized products [23], the author believes that we chemists will attain breakthroughs in nanotechnology; indeed, research based on this perspective has recently begun.

**Funding:** This research was funded by Japan society for the promotion of science (JSPS) KAKENHI (Scientific Research on Innovative Areas), Grant Numbers JP18H05423 "Molecular Engine", JP17H05346 "Coordination Asymmetry", and JP20H04622 "Discrete Geometric Analysis for Materials Design".

**Acknowledgments:** The author thanks Professor Sadamu Takeda and Dr. Goro Maruta for their fruitful discussions.

**Conflicts of Interest:** The author declares no conflict of interest.

**Appendix A**

Each graph in Figure 2 was the result of numerical analyses of Equations (A1)–(A7). The symbols of $\chi$s, $k$s, and $t$, denote the molar fractions, kinetic constants, and time, respectively. The term $k_{XY}$ indicates the kinetic constant of the autocatalytic step of X→Y. The calculated timespan was from 0 to 1000 (arbitrary unit). The 'N's denote the relative number of portions. To solve the ordinary differential equation, the ODE45 method was used and operated by Matlab 2020a software provided by The Mathworks Inc., Natick, MA, USA. For Figure 2a the following equations were used:

$$\frac{d}{dt}\chi_B = (k_{AB}\,\chi_B + k_{1F})\,\chi_A - (k_{BA}\,\chi_A + k_{1R})\,\chi_B, \tag{A1}$$

$$\chi_A = 1 - \chi_B. \tag{A2}$$

For the demonstration, 0.02 and 0.01 were assigned for $k_{AB}$ and $k_{BA}$, respectively, $1 \times 10^{-5}$ was assigned for $k_{1F}$ and $k_{1R}$, and 1 was assigned for the initial molar fraction of A. For Figure 2b the following equations were used:

$$\frac{d}{dt}N_B = k_{AB}\,N_B - k_{BC}\,N_B\,N_C, \tag{A3}$$

$$\frac{d}{dt}N_C = k_{BC}\,N_B N_c - k_{CD}\,N_C. \tag{A4}$$

For the visualization, 0.03, 0.02, and 0.1 were assigned for $k_{AB}$, $k_{BC}$, and $k_{CD}$, respectively, and 0.5 was assigned for the initial numbers of B and C. For Figure 2c–g the following equations were used:

$$\frac{d}{dt}\chi_B = (k_{AB}\,\chi_B + k_{1F})\,\chi_A - (k_{BC}\,\chi_C + k_{2F} + k_{BA}\,\chi_B + k_{1R})\,\chi_B + (k_{CB}\,\chi_C + k_{2R})\,\chi_C, \tag{A5}$$

$$\frac{d}{dt}\chi_C = (k_{BC}\,\chi_C + k_{2F})\,\chi_B - (k_{CA}\,\chi_A + k_{3F} + k_{CB}\,\chi_C + k_{2R})\,\chi_C + (k_{AC}\,\chi_A + k_{3R})\,\chi_A, \tag{A6}$$

$$\chi_A = 1 - \chi_B - \chi_C. \tag{A7}$$

Each value for the calculations is indicated in Table A1.

**Table A1.** Values assigned to each parameter for the numerical analysis.

|  | Figure 2c | Figure 2d | Figure 2e | Figure 2f | Figure 2g |
| --- | --- | --- | --- | --- | --- |
| $k_{AB}$ | 0.1 | 0.1 | 0.1 | 0.1 | 0.1 |
| $k_{BA}$ | 0.001 | 0.01 | 0 | 0.005 | 0 |
| $k_{BC}$ | 0 | 0.05 | 0.05 | 0.08 | 0.08 |
| $k_{CB}$ | 0 | 0.005 | 0 | 0.004 | 0 |
| $k_{CA}$ | 0 | 0 | 0 | 0.06 | 0.06 |
| $k_{AC}$ | 0 | 0 | 0 | 0.003 | 0 |
| $k_{2F}$ | 0.01 | 0 | 0 | 0 | 0 |
| $k_{2R}$ | 0.001 | 0 | 0 | 0 | 0 |
| $k_{3F}$ | 0.005 | 0.01 | 0.01 | 0 | 0 |
| $k_{3R}$ | 0.005 | 0.01 | 0 | 0 | 0 |
| $\chi_B$ (initial) | 0.001 | 0.001 | 0.001 | 0.001 | 0.001 |
| $\chi_C$ (initial) | 0.049 | 0.049 | 0.049 | 0.049 | 0.049 |

## References

1. Yewdall, N.A.; Mason, A.F.; van Hest, J.C.M. The hallmarks of living systems: Towards creating artificial cells. *Interface Focus* **2018**, *8*, 20180023. [CrossRef]
2. Ashkenasy, G.; Hermans, T.M.; Otto, S.; Taylor, A.F. Systems chemistry. *Chem. Soc. Rev.* **2017**, *46*, 2543–2554. [CrossRef] [PubMed]
3. Nicolis, G.; Prigogine, I. *Self-Organization in Nonequilibrium Systems*; John Wiley & Sons, Inc.: Toronto, ON, Canada, 1977.
4. Atkins, P.; de Paula, J. Internal Energy. In *Atkins' Physcial Chemistry*, 10th ed.; Chapter 2; Oxford University Press: Oxford, UK, 2014; pp. 64–74.
5. Kageyama, Y. Interplay of Photoisomerization and Phase Transition Events Provide a Working Supramolecular Motor. In *Photosynergetic Responses in Molecules and Molecular Aggregates*; Miyasaka, H., Matsuda, K., Abe, J., Kawai, T., Eds.; Chapter 26; Springer: Singapore, 2020. [CrossRef]
6. The severity of the impact of inverse reactions or coexisting subreactions on molecular collective behavior depends on the spatio-temporal size of the system under consideration. While microscopic reversibility is important for consideration of thermal/mechanical chemical dynamics at the molecular level, there is less necessity to consider reverse processes if the system already has large spatio-temporal structures, such as those which occur in cell amplification. The purpose of the first half of this section is to show that oscillatory phenomena may be generated in an open system due to the involvement of three nonlinear processes. The assumption introduced in the main text is permissible within the context of the objective.
7. Hanopolskyi, A.I.; Smaliak, V.A.; Novichkov, A.I.; Semenov, S.N. Autocatalysis: Kinetics, Mechanisms and Design. *ChemSystemsChem* **2020**. [CrossRef]
8. Haken, H. *Synergetics—An Introduction*; Springer: Berlin/Heidelberg, Germany, 1978. (In Japanese)
9. Kay, E.R.; Leigh, D.A.; Zerbetto, F. Synthetic molecular motors and mechanical machines. *Angew. Chem. Int. Ed.* **2007**, *46*, 72–191. [CrossRef] [PubMed]
10. Silvi, S.; Venturi, M.; Credi, A. Light operated molecular machines. *Chem. Commun.* **2011**, *47*, 2483–2489. [CrossRef]
11. Coskun, A.; Banaszak, M.; Astumian, R.D.; Stoddart, J.F.; Grzybowski, B.A. Great expectations: Can artificial molecular machines deliver on their promise? *Chem. Soc. Rev.* **2012**, *41*, 19–30. [CrossRef]
12. Kassem, S.; van Leeuwen, T.; Lubbe, A.S.; Wilson, M.R.; Feringa, B.L.; Leigh, D.A. Artificial molecular motors. *Chem. Soc. Rev.* **2017**, *46*, 2592–2621. [CrossRef]
13. Dattler, D.; Fuks, G.; Heiser, J.; Moulin, E.; Perrot, A.; Yao, X.; Giuseppone, N. Design of Collective Motions from Synthetic Molecular Switches, Rotors, and Motors. *Chem. Rev.* **2020**, *120*, 310–433. [CrossRef]
14. Aprahamian, I. The Future of Molecular Machines. *ACS Cent. Sci.* **2020**, *6*, 347–358. [CrossRef]
15. Kudernac, T.; Ruangsupapichat, N.; Parschau, M.; Macia, B.; Katsonis, N.; Harutyunyan, S.R.; Ernst, K.H.; Feringa, B.L. Electrically driven directional motion of a four-wheeled molecule on a metal surface. *Nature* **2011**, *479*, 208–211. [CrossRef]

16. Steinberg-Yfrach, G.; Liddell, P.A.; Hung, S.-C.; Moore, A.L.; Gust, D.; Moore, T.A. Conversion of light energy to proton potential in liposomes by artificial photosynthetic reaction centres. *Nature* **1997**, *385*, 239–241. [CrossRef]
17. Mizushima, T.; Yoshida, A.; Harada, A.; Yoneda, Y.; Minatani, T.; Murata, S. Pyrene-sensitized electron transport across vesicle bilayers: Dependence of transport efficiency on pyrene substituents. *Org. Biomol. Chem.* **2006**, *4*, 4336–4344. [CrossRef] [PubMed]
18. Eelkema, R.; Pollard, M.M.; Vicario, J.; Katsonis, N.; Ramon, B.S.; Bastiaansen, C.W.M.; Broer, D.J.; Feringa, B.L. Nanomotor rotates microscale objects. *Nature* **2006**, *440*, 163. [CrossRef] [PubMed]
19. Eelkema, R.; Pollard, M.M.; Katsonis, N.; Vicario, J.; Broer, D.J.; Feringa, B.L. Rotational Reorganization of Doped Cholesteric Liquid Crystalline Films. *J. Am. Chem. Soc.* **2006**, *128*, 14397–14407. [CrossRef]
20. Kim, Y.; Tamaoki, N. Photoresponsive Chiral Dopants: Light-Driven Helicity Manipulation in Cholesteric Liquid Crystals for Optical and Mechanical Functions. *ChemPhotoChem* **2019**, *3*, 284–303. [CrossRef]
21. Thomas, R.; Yoshida, Y.; Akasaka, T.; Tamaoki, N. Influence of a Change in Helical Twisting Power of Photoresponsive Chiral Dopants on Rotational Manipulation of Micro-Objects on the Surface of Chiral Nematic Liquid Crystalline Films. *Chem. Eur. J.* **2012**, *18*, 12337–12348. [CrossRef]
22. Katsonis, N. Cooperative action of molecular motors in mechanically active soft materials. In Proceedings of the 1st International Symposium on Molecular Engine, Chiba University, Chiba, Japan, 8 January 2019.
23. Astumian, R.D. How molecular motors work—Insights from the molecular machinist's toolbox: The Nobel prize in Chemistry 2016. *Chem. Sci.* **2017**, *8*, 840–845. [CrossRef]
24. Ikegami, T.; Kageyama, Y.; Obara, K.; Takeda, S. Dissipative and Autonomous Square-Wave Self-Oscillation of a Macroscopic Hybrid Self-Assembly under Continuous Light Irradiation. *Angew. Chem. Int. Ed.* **2016**, *55*, 8239–8243. [CrossRef]
25. Kageyama, Y.; Ikegami, T.; Satonaga, S.; Obara, K.; Sato, H.; Takeda, S. Light-Driven Flipping of Azobenzene Assemblies-Sparse Crystal Structures and Responsive Behaviour to Polarised Light. *Chem. Eur. J.* **2020**, *26*, 10759–10768. [CrossRef]
26. Obara, K.; Kageyama, Y.; Takeda, S. Swimming motion of micro-sized thin crystals with autonomous and rapid flipping under blue light irradiation. In Proceedings of the 99th Annual Meeting of Chemical Society of Japan, 2I7-50, Konan University, Kobe, Japan, 16–19 March 2019.
27. Kageyama, Y. Light-Powered Self-Sustainable Macroscopic Motion for the Active Locomotion of Materials. *ChemPhotoChem* **2019**, *3*, 327–336. [CrossRef]
28. Purcell, E.M. Life at low Reynolds number. *Am. J. Phys.* **1977**, *45*, 3–11. [CrossRef]
29. Yang, L.; Chang, L.; Hu, Y.; Huang, M.; Ji, Q.; Lu, P.; Liu, J.; Chen, W.; Wu, Y. An Autonomous Soft Actuator with Light-Driven Self-Sustained Wavelike Oscillation for Phototactic Self-Locomotion and Power Generation. *Adv. Funct. Mater* **2020**, *30*, 1908842. [CrossRef]
30. Serak, S.; Tabiryan, N.; Vergara, R.; White, T.J.; Vaia, R.A.; Bunning, T.J. Liquid crystalline polymer cantilever oscillators fueled by light. *Soft Matter* **2010**, *6*, 779–783. [CrossRef]
31. White, T.J.; Tabiryan, N.V.; Serak, S.V.; Hrozhyk, U.A.; Tondiglia, V.P.; Koerner, H.; Vaia, R.A.; Bunning, T.J. A high frequency photodriven polymer oscillator. *Soft Matter* **2008**, *4*, 1796–1798. [CrossRef]
32. Lee, K.M.; Smith, M.L.; Koerner, H.; Tabiryan, N.; Vaia, R.A.; Bunning, T.J.; White, T.J. Photodriven, Flexural-Torsional Oscillation of Glassy Azobenzene Liquid Crystal Polymer Networks. *Adv. Funct. Mater* **2011**, *21*, 2913–2918. [CrossRef]
33. Vantomme, G.; Gelebart, A.H.; Broer, D.J.; Meijer, E.W. A four-blade light-driven plastic mill based on hydrazone liquid-crystal networks. *Tetrahedron* **2017**, *73*, 4963–4967. [CrossRef]
34. Gelebart, A.H.; Vantomme, G.; Meijer, E.W.; Broer, D.J. Mastering the Photothermal Effect in Liquid Crystal Networks: A General Approach for Self-Sustained Mechanical Oscillators. *Adv. Mater.* **2017**, *29*, 1606712. [CrossRef]
35. Gelebart, A.H.; Jan Mulder, D.; Varga, M.; Konya, A.; Vantomme, G.; Meijer, E.W.; Selinger, R.L.B.; Broer, D.J. Making waves in a photoactive polymer film. *Nature* **2017**, *546*, 632–636. [CrossRef]
36. Pilz da Cunha, M.; Peeketi, A.R.; Mehta, K.; Broer, D.J.; Annabattula, R.K.; Schenning, A.; Debije, M.G. A self-sustained soft actuator able to rock and roll. *Chem. Commun.* **2019**, *55*, 11029–11032. [CrossRef]
37. Lu, X.; Guo, S.; Tong, X.; Xia, H.; Zhao, Y. Tunable Photocontrolled Motions Using Stored Strain Energy in Malleable Azobenzene Liquid Crystalline Polymer Actuators. *Adv. Mater.* **2017**, *29*, 1606467. [CrossRef]

38. Ge, F.; Yang, R.; Tong, X.; Camerel, F.; Zhao, Y. A Multifunctional Dye-doped Liquid Crystal Polymer Actuator: Light-Guided Transportation, Turning in Locomotion, and Autonomous Motion. *Angew. Chem. Int. Ed.* **2018**, *57*, 11758–11763. [CrossRef] [PubMed]
39. Zeng, H.; Lahikainen, M.; Liu, L.; Ahmed, Z.; Wani, O.M.; Wang, M.; Yang, H.; Priimagi, A. Light-fuelled freestyle self-oscillators. *Nat. Commun.* **2019**, *10*, 5057. [CrossRef] [PubMed]
40. Uchida, E.; Azumi, R.; Norikane, Y. Light-induced crawling of crystals on a glass surface. *Nat. Commun.* **2015**, *6*, 7310. [CrossRef] [PubMed]
41. Yamada, M.; Kondo, M.; Mamiya, J.; Yu, Y.; Kinoshita, M.; Barrett, C.; Ikeda, T. Photomobile polymer materials: Towards light-driven plastic motors. *Angew. Chem. Int. Ed.* **2008**, *47*, 4986–4988. [CrossRef] [PubMed]
42. Yoshida, R. Development of self-oscillating polymers and gels with autonomous function. *Polym. J.* **2010**, *42*, 777–789. [CrossRef]
43. Yoshida, R.; Ueki, T. Evolution of self-oscillating polymer gels as autonomous polymer systems. *NPG Asia Mater.* **2014**, *6*, e107. [CrossRef]
44. Szostak, J.W.; Bartel, D.P.; Luisi, P.L. Synthesizing life. *Nature* **2001**, *409*, 387–390. [CrossRef]
45. Stano, P.; Luisi, P.L. Achievements and open questions in the self-reproduction of vesicles and synthetic minimal cells. *Chem. Commun.* **2010**, *46*, 3639–3653. [CrossRef]
46. Lancet, D.; Segre, D.; Kahana, A. Twenty Years of "Lipid World": A Fertile Partnership with David Deamer. *Life* **2019**, *9*, 77. [CrossRef]
47. Bachmann, P.A.; Luisi, P.L.; Lang, J. Autocatalytic self-replicating micelles as models for prebiotic structures. *Nature* **1992**, *357*, 57–59. [CrossRef]
48. Walde, P.; Wick, R.; Fresta, M.; Mangone, A.; Luisi, P.L. Autopoietic Self-Reproduction of Fatty Acid Vesicles. *J. Am. Chem. Soc.* **1994**, *116*, 11649–11654. [CrossRef]
49. Takakura, K.; Sugawara, T. Membrane dynamics of a myelin-like giant multilamellar vesicle applicable to a self-reproducing system. *Langmuir* **2004**, *20*, 3832–3834. [CrossRef] [PubMed]
50. Toyota, T.; Takakura, K.; Kageyama, Y.; Kurihara, K.; Maru, N.; Ohnuma, K.; Kaneko, K.; Sugawara, T. Population study of sizes and components of self-reproducing giant multilamellar vesicles. *Langmuir* **2008**, *24*, 3037–3044. [CrossRef] [PubMed]
51. Shohda, K.-I.; Tamura, M.; Kageyama, Y.; Suzuki, K.; Suyama, A.; Sugawara, T. Compartment size dependence of performance of polymerase chain reaction inside giant vesicles. *Soft Matter* **2011**, *7*, 3750. [CrossRef]
52. Kurihara, K.; Tamura, M.; Shohda, K.; Toyota, T.; Suzuki, K.; Sugawara, T. Self-reproduction of supramolecular giant vesicles combined with the amplification of encapsulated DNA. *Nat. Chem.* **2011**, *3*, 775–781. [CrossRef] [PubMed]
53. Takahashi, H.; Kageyama, Y.; Kurihara, K.; Takakura, K.; Murata, S.; Sugawara, T. Autocatalytic membrane-amplification on a pre-existing vesicular surface. *Chem. Commun.* **2010**, *46*, 8791–8793. [CrossRef]
54. Matsuo, M.; Ohyama, S.; Sakurai, K.; Toyota, T.; Suzuki, K.; Sugawara, T. A sustainable self-reproducing liposome consisting of a synthetic phospholipid. *Chem. Phys. Lipids* **2019**, *222*, 1–7. [CrossRef] [PubMed]
55. Hardy, M.D.; Yang, J.; Selimkhanov, J.; Cole, C.M.; Tsimring, L.S.; Devaraj, N.K. Self-reproducing catalyst drives repeated phospholipid synthesis and membrane growth. *Proc. Natl. Acad. Sci. USA* **2015**, *112*, 8187–8192. [CrossRef]
56. Bissette, A.J.; Fletcher, S.P. Mechanisms of autocatalysis. *Angew. Chem. Int. Ed.* **2013**, *52*, 12800–12826. [CrossRef]
57. Kosikova, T.; Philp, D. Exploring the emergence of complexity using synthetic replicators. *Chem. Soc. Rev.* **2017**, *46*, 7274–7305. [CrossRef]
58. Soai, K.; Shibata, T.; Morioka, H.; Choji, K. Asymmetric autocatalysis and amplification of enantiomeric excess of a chiral molecule. *Nature* **1995**, *378*, 767–768. [CrossRef]
59. Tjivikua, T.; Ballester, P.; Rebek, J. Self-replicating system. *J. Am. Chem. Soc.* **1990**, *112*, 1249–1250. [CrossRef]
60. Rotello, V.; Hong, J.I.; Rebek, J. Sigmoidal growth in a self-replicating system. *J. Am. Chem. Soc.* **1991**, *113*, 9422–9423. [CrossRef]
61. Wintner, E.A.; Conn, M.M.; Rebek, J. Self-Replicating Molecules: A Second Generation. *J. Am. Chem. Soc.* **1994**, *116*, 8877–8884. [CrossRef]
62. Chen, J.; Korner, S.; Craig, S.L.; Lin, S.; Rudkevich, D.M.; Rebek, J., Jr. Chemical amplification with encapsulated reagents. *Proc. Natl. Acad. Sci. USA* **2002**, *99*, 2593–2596. [CrossRef] [PubMed]

63. Kamioka, S.; Ajami, D.; Rebek, J., Jr. Autocatalysis and organocatalysis with synthetic structures. *Proc. Natl. Acad. Sci. USA* **2010**, *107*, 541–544. [CrossRef] [PubMed]
64. Zielinski, W.S.; Orgel, L.E. Autocatalytic synthesis of a tetranucleotide analogue. *Nature* **1987**, *327*, 346–347. [CrossRef]
65. Orgel, L.E. Molecular replication. *Nature* **1992**, *358*, 203–209. [CrossRef]
66. Von Kiedrowski, G.; Wlotzka, B.; Helbing, J.; Matzen, M.; Jordan, S. Parabolic Growth of a Self-Replicating Hexadeoxynucleotide Bearing a 3′-5′-Phosphoamidate Linkage. *Angew. Chem. Int. Ed.* **1991**, *30*, 423–426. [CrossRef]
67. Terfort, A.; von Kiedrowski, G. Self-Replication by Condensation of 3-Aminobenzamidines and 2-Formylphenoxyacetic Acids. *Angew. Chem. Int. Ed.* **1992**, *31*, 654–656. [CrossRef]
68. Achilles, T.; von Kiedrowski, G. A Self-Replicating System from Three Starting Materials. *Angew. Chem. Int. Ed.* **1993**, *32*, 1198–1201. [CrossRef]
69. Sadownik, J.W.; Mattia, E.; Nowak, P.; Otto, S. Diversification of self-replicating molecules. *Nat. Chem.* **2016**, *8*, 264–269. [CrossRef] [PubMed]
70. Nowak, P.; Colomb-Delsuc, M.; Otto, S.; Li, J.W. Template-Triggered Emergence of a Self-Replicator from a Dynamic Combinatorial Library. *J. Am. Chem. Soc.* **2015**, *137*, 10965–10969. [CrossRef] [PubMed]
71. Leonetti, G.; Otto, S. Solvent Composition Dictates Emergence in Dynamic Molecular Networks Containing Competing Replicators. *J. Am. Chem. Soc.* **2015**, *137*, 2067–2072. [CrossRef] [PubMed]
72. Colomer, I.; Morrow, S.M.; Fletcher, S.P. A transient self-assembling self-replicator. *Nat. Commun.* **2018**, *9*, 2239. [CrossRef]
73. Morrow, S.M.; Colomer, I.; Fletcher, S.P. A chemically fuelled self-replicator. *Nat. Commun.* **2019**, *10*, 1011. [CrossRef]
74. Della Sala, F.; Neri, S.; Maiti, S.; Chen, J.L.Y.; Prins, L.J. Transient self-assembly of molecular nanostructures driven by chemical fuels. *Curr. Opin. Biotechnol.* **2017**, *46*, 27–33. [CrossRef]
75. Van Rossum, S.A.P.; Tena-Solsona, M.; van Esch, J.H.; Eelkema, R.; Boekhoven, J. Dissipative out-of-equilibrium assembly of man-made supramolecular materials. *Chem. Soc. Rev.* **2017**, *46*, 5519–5535. [CrossRef]
76. De, S.; Klajn, R. Dissipative Self-Assembly Driven by the Consumption of Chemical Fuels. *Adv. Mater.* **2018**, *30*, e1706750. [CrossRef]
77. Ragazzon, G.; Prins, L.J. Energy consumption in chemical fuel-driven self-assembly. *Nat. Nanotechnol.* **2018**, *13*, 882–889. [CrossRef]
78. Leng, Z.; Peng, F.; Hao, X. Chemical-Fuel-Driven Assembly in Macromolecular Science: Recent Advances and Challenges. *ChemPlusChem* **2020**, *85*, 1190–1199. [CrossRef] [PubMed]
79. Fialkowski, M.; Bishop, K.J.; Klajn, R.; Smoukov, S.K.; Campbell, C.J.; Grzybowski, B.A. Principles and implementations of dissipative (dynamic) self-assembly. *J. Phys. Chem. B* **2006**, *110*, 2482–2496. [CrossRef]
80. Boekhoven, J.; Brizard, A.M.; Kowlgi, K.N.; Koper, G.J.; Eelkema, R.; van Esch, J.H. Dissipative self-assembly of a molecular gelator by using a chemical fuel. *Angew. Chem. Int. Ed.* **2010**, *49*, 4825–4828. [CrossRef] [PubMed]
81. Maiti, S.; Fortunati, I.; Ferrante, C.; Scrimin, P.; Prins, L.J. Dissipative self-assembly of vesicular nanoreactors. *Nat. Chem.* **2016**, *8*, 725–731. [CrossRef]
82. Hao, X.; Sang, W.; Hu, J.; Yan, Q. Pulsating Polymer Micelles via ATP-Fueled Dissipative Self-Assembly. *ACS Macro. Lett.* **2017**, *6*, 1151–1155. [CrossRef]
83. Riess, B.; Boekhoven, J. Applications of Dissipative Supramolecular Materials with a Tunable Lifetime. *ChemNanoMat* **2018**, *4*, 710–719. [CrossRef]
84. Bal, S.; Das, K.; Ahmed, S.; Das, D. Chemically Fueled Dissipative Self-Assembly that Exploits Cooperative Catalysis. *Angew. Chem. Int. Ed.* **2019**, *58*, 244–247. [CrossRef]
85. Wanzke, C.; Jussupow, A.; Kohler, F.; Dietz, H.; Kaila, V.R.I.; Boekhoven, J. Dynamic Vesicles Formed By Dissipative Self-Assembly. *ChemSystemsChem* **2019**, *2*, e1900044. [CrossRef]
86. Post, E.A.J.; Fletcher, S.P. Controlling the Kinetics of Self-Reproducing Micelles by Catalyst Compartmentalization in a Biphasic System. *J. Org. Chem.* **2019**, *84*, 2741–2755. [CrossRef]
87. Post, E.A.J.; Fletcher, S.P. Dissipative self-assembly, competition and inhibition in a self-reproducing protocell model. *Chem. Sci.* **2020**, *11*, 9434–9442. [CrossRef]

88. Penocchio, E.; Rao, R.; Esposito, M. Thermodynamic efficiency in dissipative chemistry. *Nat. Commun.* **2019**, *10*, 3865. [CrossRef] [PubMed]
89. Koper, G.J.M.; Boekhoven, J.; Hendriksen, W.E.; van Esch, J.H.; Eelkema, R.; Pagonabarraga, I.; Rubí, J.M.; Bedeaux, D. The Lost Work in Dissipative Self-Assembly. *Int. J. Thermophys.* **2013**, *34*, 1229–1238. [CrossRef]
90. The author does not exclude the possibility that the reported systems form a dissipative structure, because there may be additional symmetry-breaking phenomena that are not described in the papers.
91. Ragazzon, G.; Baroncini, M.; Silvi, S.; Venturi, M.; Credi, A. Light-powered autonomous and directional molecular motion of a dissipative self-assembling system. *Nat. Nanotechnol.* **2014**, *10*, 70–75. [CrossRef] [PubMed]
92. Wilson, M.R.; Solà, J.; Carlone, A.; Goldup, S.M.; Lebrasseur, N.; Leigh, D.A. An autonomous chemically fuelled small-molecule motor. *Nature* **2016**, *534*, 235–240. [CrossRef] [PubMed]
93. Fletcher, S.P.; Dumur, F.; Pollard, M.M.; Feringa, B.L. A Reversible, Unidirectional Molecular Rotary Motor Driven by Chemical Energy. *Science* **2005**, *310*, 80–82. [CrossRef]

**Publisher's Note:** MDPI stays neutral with regard to jurisdictional claims in published maps and institutional affiliations.

© 2020 by the author. Licensee MDPI, Basel, Switzerland. This article is an open access article distributed under the terms and conditions of the Creative Commons Attribution (CC BY) license (http://creativecommons.org/licenses/by/4.0/).

Article

# Off-Shell Quantum Fields to Connect Dressed Photons with Cosmology

Hirofumi Sakuma [1,*,†], Izumi Ojima [1,†], Motoichi Ohtsu [1,†] and Hiroyuki Ochiai [2]

1. Research Origin for Dressed Photon, Yokohama-shi, Kanagawa 221-0022, Japan; ojima@gaia.eonet.ne.jp (I.O.); ohtsu@rodrep.or.jp (M.O.)
2. Institute of Mathematics for Industry, Kyushu University, Fukuoka-shi, Fukuoka 819-0395, Japan; ochiai@imi.kyushu-u.ac.jp
* Correspondence: sakuma@rodrep.or.jp
† These authors contributed equally to this work.

Received: 22 April 2020; Accepted: 21 July 2020; Published: 28 July 2020

**Abstract:** The anomalous nanoscale electromagnetic field arising from light–matter interactions in a nanometric space is called a dressed photon. While the generic technology realized by utilizing dressed photons has demolished the conventional wisdom of optics, for example, the unexpectedly high-power light emission from indirect-transition type semiconductors, dressed photons are still considered to be too elusive to justify because conventional optical theory has never explained the mechanism causing them. The situation seems to be quite similar to that of the dark energy/matter issue in cosmology. Regarding these riddles in different disciplines, we find a common important clue for their resolution in the form of the relevance of space-like momentum support, without which quantum fields cannot interact with each other according to a mathematical result of axiomatic quantum field theory. Here, we show that a dressed photon, as well as dark energy, can be explained in terms of newly identified space-like momenta of the electromagnetic field and dark matter can be explained as the off-shell energy of the Weyl tensor field.

**Keywords:** off-shell quantum field; space-like momentum; dressed photon; micro-macro duality; Clebsch dual field; Majorana fermion; the cosmological term; Weyl tensor; dark energy; dark matter

---

## 1. Introductory Review of Dressed Photon Technology

### 1.1. Broad Overview

Suppose that a nanometer-sized material (NM) is illuminated by propagating light whose diffraction-limited size is much larger than the size of the NM. Then, an anomalous non-propagating localized light field is generated around the NM, contrary to the accepted knowledge of optics. This non-propagating light field is called the optical near field [1], which is undetectable by a separately placed conventional photodetector. The studies on the optical near field initiated practically in the late 20th century have led, through trial-and-error approaches, to the novel concept of a quasi-particle created as a result of light–matter interaction in a nanometric space. This quasi-particle is figuratively called the dressed photon (DP), that is, a metaphoric expression of photon energy partly fused with the energies of the material involved in the interaction. Figure 1 shows typical experimental setups for creating a DP. Studies on DPs are now rapidly progressing, yielding innovative generic technologies [2] of "small light", which accomplish the impossible in a variety of application fields.

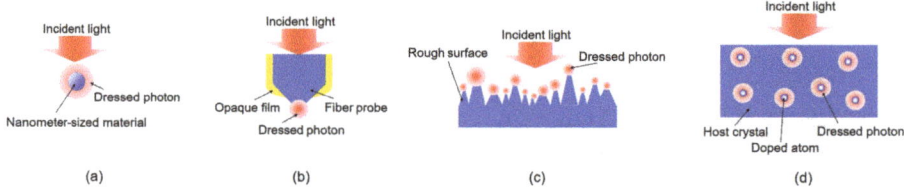

**Figure 1.** Typical experimental setups for creating a DP. (**a**) on a nanoparticle; (**b**) on the tip of a fiber probe; (**c**) on bumps of a rough material surface; (**d**) on doped atoms in a host crystal.

One should bear in mind, however, that the concept of a DP first proposed by one of the authors (M.O.) still seems to be either ignored or tacitly understood differently by the mainstream researchers in optical sciences because of the conceptual difficulty of dealing with off-shell quantities in the midst of field interactions. We think that this kind of refusal or confusion about DPs stems from a certain degree of ambiguity in using such an abstract expression as light–matter field interactions in a nanometric space; the main purpose of the above remark is not to criticize the incorrect usage of DPs in the literature but rather to promote renewed awareness that the issue of DP phenomena addressed here is a remarkable one that cannot be understood within the conventional framework of optical theory.

To elucidate the essence of the DP problem, we start by listing five conventional views in optics and briefly show how a DP violates them.

**Five conventional common views in optics**

I. Light is a propagating wave that fills a space. Its spatial extent (size) is much larger than its wavelength.
II. Light cannot be used for imaging and fabrication of sub-wavelength-sized materials. Furthermore, light cannot be used for assembling and operating sub-wavelength-sized optical devices.
III. For optical excitation of an electron, the photon energy must be equal to or higher than the energy difference between the relevant two electronic energy levels.
IV. An electron cannot be optically excited if the transition between the two electric energy levels is electric dipole forbidden.
V. Crystalline silicon has a very low light emission efficiency and is thus unsuitable for use as an active medium in light-emitting devices.

Contrary to I–V above, the intrinsic natures of DPs have enabled the advent of innovative technologies such as the following: (1) Nanometer-sized optical devices. These devices are operated on the basis of the spatially localized nature of DPs created on an NM (Figure 1a) and the autonomous DP energy transfer between NMs. These devices are based on the intrinsic nature of a DP that is contrary to I, II and IV. Integrated 2D arrays of NOT- and AND-logic gates operating at room temperature have been fabricated using InAs NMs [3]. (2) Nanofabrication technology, such as chemical vapor deposition and autonomous smoothing of a material surface (Figure 1b,c). These technologies are based on the intrinsic nature of a DP that is contrary to I–IV. Their details will be described in subsection 1.2 since the information on the maximum size of a DP will be used in Section 4 on cosmology. (3) Light-emitting devices using indirect-transition-type semiconductors. These devices are based on the intrinsic nature of a DP that is contrary to V. Infrared light-emitting diodes using crystalline silicon (Si) have been realized. For their fabrication, a method of DP-assisted annealing has been invented to autonomously control the spatial distribution of the dopant atoms on which DPs are created and localized (Figure 1d). Their output optical powers are as high as 2 W [4]. Infrared Si lasers have also been developed whose CW output optical power is as high as 100 W, and the threshold current density is as low as 60 A/cm$^2$ [2] at room temperature [5]. Their high power and low energy consumption factors are $10^4$ and 0.05, respectively, relative to those of the conventional single-stripe double heterojunction-structured semiconductor lasers fabricated using the direct-transition-type compound InGaAsP. Furthermore, novel polarization rotators have been developed using crystalline SiC that exhibit a gigantic ferromagnetic magneto-optical effect [6].

*1.2. Nanofabrication Technology and the Size of a DP*

This subsection describes two examples of nanofabrication technology that provide key experimental data for theoretical discussions in Sections 2–4. In particular, the maximum size of a DP was determined by analyzing a large number of experimental results.

1.2.1. Photochemical Vapor Deposition

In this method, a material is grown by depositing atoms on a substrate. Gaseous molecules are dissociated when a DP is created, for example, on the fiber probe tip of Figure 1b. Atoms created by this dissociation are deposited on the substrate installed below the fiber probe tip. Since the size of a DP is equivalent to that of the fiber probe tip, a sub-wavelength-sized NM can be grown, which is contrary to common views I and II. Furthermore, the photon energy $h\nu_{in}$ of the light incident on the end of the fiber probe can be lower than the excitation energy $E_{excite}$ of the electrons in the molecule, which is contrary to common view III. This phenomenon occurs because the energy $h\nu_{DP}$ of the created DP is given by the sum of $h\nu_{in}$, the energies of excitons and phonons in the fiber probe, and thus $h\nu_{DP} \geq E_{excite}$. For example, DP dissociated gaseous $Zn(C_2H_5)_2$ molecules (the wavelength $\lambda_{excite}$ of light whose energy corresponds to $E_{excite}$ was as short as 270 nm), thus depositing an NM composed of Zn atoms on a sapphire substrate. In contrast to common view III, blue incident light (wavelength $\lambda_{in} = 488$ nm $> \lambda_{excite}$) was used. Figure 2a shows 3D atomic force microscopic (AFM) images of the grown NMs [7]. Furthermore, the electric dipole-forbidden transition of electrons could be used for dissociation, which is contrary to common view IV. This approach was possible because the conventional long-wave approximation is not valid in the case of a DP due to its sub-wavelength size. For example, DPs dissociated optically inactive $Zn(acac)_2$ molecules, thus depositing Zn atoms, using visible incident light (457 nm wavelength) (Figure 2b) [8,9].

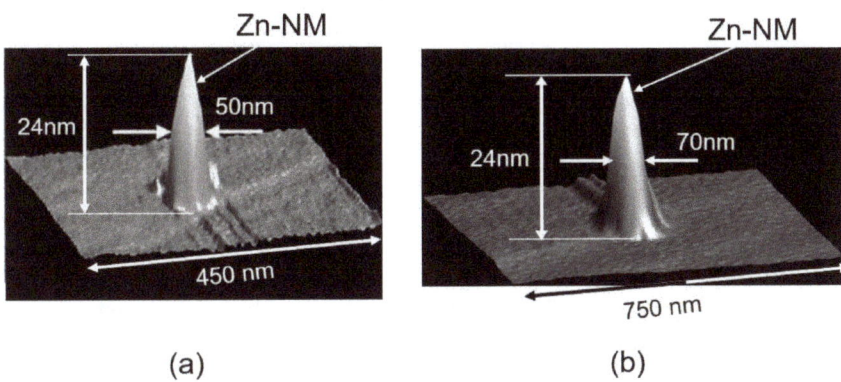

**Figure 2.** AFM images of Zn-NMs formed on a sapphire substrate. Dissociated molecules are (**a**) [7] $Zn(C_2H_5)_2$ and (**b**) [8] $Zn(acac)_2$. The values of the height and FWHM are given in each figure.

The maximum size of a DP was estimated by measuring temporal variations in the deposition rate of the number of Zn atoms and the full-width at half-maximum (FWHM) of the 3D image [10]. The results showed that, in the initial stage of deposition, the deposition rate and the FWHM increased monotonically with time. When the FWHM reached the size of the fiber probe tip, the deposition rate was the maximum, which is the phenomenon known as size-dependent resonance between the fiber probe tip and the size of the NM [11]. Then, the deposition rate monotonically decreased and approached a constant value. This result indicated that the size of the NM asymptotically approached a certain value. Since the height of the NM monotonically increased with the deposition time, this approach indicated that the value of the FWHM approached a certain value. To confirm this indication, the maximum value of the FWHM was evaluated from the 3D images of the saturated-sized NMs. As a result, a value of 50–70 nm was obtained after compensating for the errors and inaccuracies in the experimental data. Figure 2a,b just show the images of the NMs with the maximum value of the FWHM. This maximum value also indicated that the maximum size of a DP is 50–70 nm because the size of the DP transferred from the fiber probe tip to the NM corresponds to the size of the NM. A series of experiments confirmed that this value was independent of the molecular species, the wavelength and power of the incident light, the conformation, and the size of the fiber probe tip.

1.2.2. Smoothing Material Surfaces

Galilei chemical-mechanically polished the lens surfaces of his telescope as early as the 17th century. Although this method is popularly used even now in industry, polishing 3D or microsurfaces using this method is difficult because it is a contact method that employs a polishing pad. Furthermore, small scratches are created on the surface during the polishing process. To solve these problems, a non-contact dry-etching method was invented using DPs [12]. Its principle is nearly the same as that of the photochemical vapor deposition above. That is, DPs are created on small bumps of the rough surface by light irradiation (Figure 1c). A gaseous $Cl_2$ molecule, as an example, is dissociated if it jumps into the DP field. Since the created Cl atom is chemically active, it etches the bump without using any devices such as a fiber probe. Thus, etching autonomously starts upon light irradiation, varying the conformation and size of the bumps, and stops when the surface becomes flat, i.e., when DPs are no longer created. A variety of 3D surfaces, such as convex surfaces, concave surfaces, and the side walls and inner wall of a cylinder, have been smoothed. The microsized side walls of the corrugations of a diffraction grating were also smoothed. This method has been applied to a variety of materials, such as glasses, crystals, ceramics, and plastics, to decrease the roughness to sub-nanometer. It has been employed in industry to increase the optical damage threshold of high-power UV laser mirrors [13], repair the surface of photomasks for UV lithography [14], and so on.

The maximum size of a DP was also evaluated by this method: Figure 3 shows the experimental results of polishing a plastic PMMA surface by dissociating $O_2$ molecules by DPs [15]. The wavelength $\lambda_{excite}$ of light corresponding to the $E_{excite}$ of the $O_2$ molecule was 242 nm. However, the wavelength $\lambda_{in}$ of the incident light was as long as 325 nm ($>\lambda_{excite}$), contrary to common view III. The horizontal axis represents the period $l$ of the surface roughness. The vertical axis is the standard deviation $\sigma$ of the roughness acquired from the AFM images. Here, the ratio $\sigma_{after}/\sigma_{before}$ between the values before ($\sigma_{before}$) and after ($\sigma_{after}$) the etching is plotted on a logarithmic scale. This figure shows that $\sigma_{after}/\sigma_{before}$ was less than 1 in the range of $l < 50$–$70$ nm, from which the maximum size of the DP is again confirmed to be 50–70 nm, as displayed by the grey band in this figure.

For comparison, in the case when $\lambda_{in} = 213$ nm ($<\lambda_{excite}$), which follows common view III, the value of $\sigma_{after}/\sigma_{before}$ was less than unity only in the range of $l > \lambda_{in}$. In contrast, $\sigma_{after}/\sigma_{before} > 1$ in the range of $l < \lambda_{in}$ was obtained. By comparing these results, etching by DPs is confirmed to be effective for selectively removing fine bumps of sub-wavelength size.

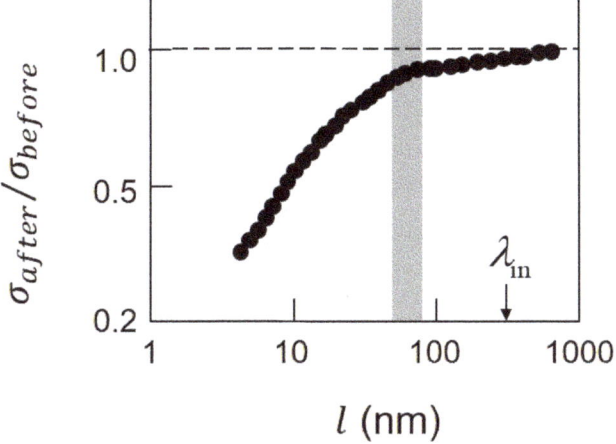

**Figure 3.** Ratio of the standard deviation of the roughness of a plastic PMMA surface before and after etching. The downward arrow represents the value of $l$ that is equal to $\lambda_{in}$. The width of the grey band corresponds to the maximum size of the DP. The ratio $\sigma_{after}/\sigma_{before}$ was derived from the values of $\sigma_{before}$ and $\sigma_{after}$ given in Figure 4 of Ref. [15].

## 2. New Theory for Dressed Photon

### 2.1. Missing Aspect of Quantum Field Interaction Theory

In the usual quantum field theory (QFT), scattering processes are described by the LSZ reduction formulae [16], which determine the S-matrix elements $S_{\beta\alpha}$ connecting the in(-coming) and out(-going) scattering states, $|\alpha, in\rangle$ and $|\beta, out\rangle$, respectively, as on-shell projections of the time-ordered Green functions. This way of description is suitable for experimental situations satisfying the asymptotic completeness which means that interactions among fields can be reduced to scattering processes. In such situations, in-states $|\alpha, in\rangle \in \mathfrak{H}^{in}$ describe the states of on-shell particles in the Heisenberg picture in Hilbert space $\mathfrak{H}$ before the interaction at $t = -\infty$, and out-states $|\beta, out\rangle \in \mathfrak{H}^{out}$ representing the final states of on-shell particles after the interaction at $t = +\infty$, which can be determined by modeled scattering processes under the assumption $\mathfrak{H} = \mathfrak{H}^{in} = \mathfrak{H}^{out}$ of asymptotic completeness.

In the situations with asymptotic completeness being valid, all the discussions can safely be focused on the on-shell aspects in terms of the S-matrix for which the LSZ formulae in QFT are used commonly among particle physicists. In this case, however, an important issue has been forgotten regarding the roles played by off-shell Heisenberg fields at the center of given field interactions. We note here that the Greenberg and Robinson theorem [17,18] proved in axiomatic quantum field theory shows that an interaction among quantum fields must inevitably accompany space-like momentum supports whenever this interaction can non-trivially transform in-states of asymptotic field $\phi^{in}$ into out-states of asymptotic field $\phi^{out}$ describing particles with time-like momentum support. Note that space-like momentum here does not mean the presence of a tachyonic field [19] carrying unstable particles with space-like momenta.

Thus, the consequence of the axiomatic theory claiming the existence of space-like momentum support is a remarkable feature in sharp contrast to the conventional perturbative expansion method for field interactions, in which only on-shell particles with time-like or light-like momentum support are considered physical. This important result has been totally neglected thus far, perhaps owing to such prejudice that abstract consequences in mathematical theorems are irrelevant to specific physical aspects of interacting fields. In the following subsection, however, our discussion on DPs will exhibit the existence of space-like momentum supports in such a form linked to the existence of well-known $U(1)$ gauge bosons mediating electromagnetic interactions as virtual photons. Since the notion of virtual photons is closely-linked to perturbative expansion methods not necessarily related to space-like momenta of the field under consideration, the term of virtual photons mentioned above is used in a loose sense. In what follows, we are going to reexamine the problem of electromagnetic interaction from the viewpoint of Micro-Macro duality theory to be touched upon shortly below in which not only microscopic "particle modes" but also macroscopic "non-particle condensates" play key roles to attain complete description of given electromagnetic fields.

## 2.2. Augmented Maxwell's Equation

In view of the unfamiliarity in the science community at large with the relevant subjects, we recapitulate the important points to make this article self-contained, on the basis of [5], the latest tutorial paper on DPs, which summarizes most of the results reported in a series of works [5,20–22]. In our new theory on DPs, we have introduced a new mathematical formulation called the Clebsch dual field, and some of the important outcomes derived from the formulation will be used in our arguments without detailed explanation; hence, we reserve the Method section until the end to give a revised concise explanation as background information for interested readers.

To identify precisely the nature of the problem under consideration, we emphasize first that quantum fields with infinite degrees of freedom are accompanied in general by disjoint representations [23], which are mutually separated by the absence of intertwiners, as the stronger, refined, and clear-cut version of unitary non-equivalence. For those who are familiar only with quantum mechanics with finite degrees of freedom, the existence of such disjoint representations may look like a pathology in the system with infinite degrees of freedom. However, the familiar situation encountered in the systems with finite degrees of freedom is, actually, an exceptional one specific to the finite system. The emergence of characteristic structures where "invisible" microscopic levels become "visible" to us is due to the sector structure arising from the spectral decomposition of the center of the observable algebra. Each sector labeled by macroscopic order parameters is mutually disjoint owing to the absence of intertwiners between different sectors at the microscopic level. This fact is the crux of the mathematically reformulated "quantum-classical" correspondence explained in the Micro-Macro duality proposed in [24,25] by one (I. O.) of the present authors.

Recall that, in the relativistically covariant formulation of the electromagnetic field, only transverse modes are considered physical and the longitudinal mode is eliminated as unphysical because of the indefiniteness of the metric of the longitudinal mode. At the classical macroscopic level, however, the Coulomb mode corresponding to the unphysical longitudinal mode plays the dominant role in electromagnetic interactions. The clue to resolving this contradiction related to the gap between microscopic and macroscopic worlds must lie in the disjointness of representations at the microscopic level and in the presence of space-like momentum support related to the former.

By one (I. O.) [26] of the authors, the important role played by macroscopic non-particle condensates (touched upon at the end of the preceding subsection) has first been discussed in electromagnetic theory in the attempt to reexamine the essence of Nakanish–Lautrup formalism [27] of abelian gauge theory: one of the remarkable points important for the present discussion on the classical Clebsch dual field is concerning the contrast between gauge invariance (in algebraic sense) and physicality of specific modes (changing from a representation to another, dependent on the choice of physical situations): in the usual treatment of gauge theories, it is believed that a physical quantity must be gauge invariant, on the basis of such algebraic judgment as whether $\tau_\Lambda(A) = A$ or not, in terms of the algebraic gauge transformation $\tau_\Lambda$. In the actual situations, however, such gauge non-invariant quantities as the longitudinal Coulomb tail $A^c$ and/or the Cooper pairs $\chi^c$ are to be treated as physical modes in spite of their gauge non-invariance! In order to treat correctly such gauge non-invariant physical modes, we need to introduce such viewpoint that a quantity $A$ is physical or not in a given situation should be judged by means of the gauge transformation represented by a commutation relation at the operator level: $[Q_\Lambda, A] = 0$ or not, in each representation of physical relevance, where $Q_\Lambda$ denotes a conserved charge defined in the Nakanishi–Lautrup B-field formalism which generates an infinite-dimensional abelian Lie group of local gauge transformation. Thus, in spite of their gauge dependence, the Coulomb tail $A^c$ or the Cooper pairs $\chi^c$ as c-number condensates become physical quantities owing to this commutativity. One should bear in mind that this point is helpful for reading Section 5 of Method on the formulation of the Clebsch dual field.

As to the electromagnetic 4-vector potential $A_\mu$, one should also pay attention to the fact that it possesses a nonlocal off-shell (out-of-light-cone) characteristic in the sense that an observable quantity $\oint_\gamma A_\nu dl^\nu$ in the Aharonov–Bohm (AB) effect [28] does not correspond to the value of $A_\mu$ at a certain point in spacetime but to the integrated value along the Wilson loop $\gamma$ is worth mentioning. Thus, motivated by the above-mentioned concept of disjoint representations and space-like momentum support, which seem to be closely linked to the nonlocal characteristic of $A_\mu$, let us see how Maxwell's equation (1), represented in terms of vector potential $A^\mu$, whose Helmholtz decomposition is given by (2), and the mixed form of energy-momentum (EM) tensor $T_\mu^\nu$ given in (3),

$$\partial_\nu F^{\mu\nu} = \partial_\nu(\partial^\mu A^\nu - \partial^\nu A^\mu) = [-\partial^\nu \partial_\nu A^\mu + \partial^\mu(\partial_\nu A^\nu)] = j^\mu, \tag{1}$$

$$A^\mu = \alpha^\mu + \partial^\mu \chi, \quad (\partial_\nu \alpha^\nu = 0, \quad \partial_\nu A^\nu = \partial_\nu \partial^\nu \chi), \tag{2}$$

$$T_\mu^\nu = -F_{\mu\sigma} F^{\nu\sigma} + \frac{1}{4} \eta_\mu^\nu F_{\sigma\tau} F^{\sigma\tau}, \tag{3}$$

can be extended into a thus far unknown space-like 4-momentum sector of the electromagnetic field, where the notations are conventional and the sign convention of the Lorentzian metric ($\eta_{\mu\nu}$) signature $(+---)$ is employed.

In the Clebsch dual formulation, the 4-vector potential in the space-like sector is denoted by $U_\mu$, and for the light-like case of $U^\nu(U_\nu)^* = 0$, where $*$ denotes a complex conjugate, this potential is parametrized in terms of a couple of Clebsch parameters $\lambda$ and $\phi$ satisfying

$$U_\mu := \lambda \partial_\mu \phi, \quad \partial^\nu \partial_\nu \lambda - (\kappa_0)^2 \lambda = 0, \quad \partial^\nu \partial_\nu \phi = 0, \tag{4}$$

$$C^\nu L_\nu = 0, \quad (C^\nu := \partial^\nu \phi, \quad L_\nu := \partial_\nu \lambda), \tag{5}$$

where $\kappa_0$ is an important constant to be determined in Section 3. Our goal is to show that, as a dual of the Proca equation of the form $\partial^\nu \partial_\nu A^\mu + m^2 A^\mu = 0$, the newly identified vector potential $U^\mu$, called the Clebsch dual (electromagnetic wave) field, given by

$$\partial^\nu \partial_\nu U^\mu - (\kappa_0)^2 U^\mu = 0, \iff (\partial^\nu \partial_\nu A^\mu + m^2 A^\mu = 0) \tag{6}$$

can satisfy "Maxwell's equation" in the space-like momentum sector and behaves like a classical version of a longitudinal virtual photon, which is shown in the Method section. While the space-like Klein Gordon (KG) equation in (4) is necessarily related to negative energy, this equation has been forgotten in the predominant arguments in the state vector space involving the Fock space structure equipped with the vacuum vector $|0\rangle$ characterized by $a|0\rangle = 0$ in terms of the annihilation operator $a$. However, this is not the whole story. Interestingly, if we move from the vacuum situation to thermal one, then we find that the modular inversion symmetry in the Tomita–Takesaki extension [29] of the thermal equilibrium, which can physically be interpreted as the right/left symmetry of the state vector of the Gibbs state, implies the existence of stable states with two-sided (positive and negative) energy spectra. Thus, one should not neglect the possibility that the two-sided "energy" as in (6) satisfies the stability of the Fock space structure.

As shown in the Method section, one of the important characteristics of the Clebsch dual field is that the field strength $S_{\mu\nu} := \partial_\mu U_\nu - \partial_\nu U_\mu$ corresponding to $F_{\mu\nu}$ is given by a simple bivector of

$$S_{\mu\nu} = L_\mu C_\nu - L_\nu C_\mu, \quad C^\nu L_\nu = 0. \tag{7}$$

In addition, it is shown in the Method section that the light-like ($U^\nu(U_\nu)^* = 0$) Clebsch dual field corresponding to the classical version of the $U(1)$ gauge boson can be extended to cover the gauge symmetry broken space-like $U^\nu(U_\nu)^* < 0$ case, in which both $\lambda$ and $\phi$ satisfy the same space-like KG equation of (4). By this extension, the form of the EM tensor of the Clebsch dual field changes from (8) to the first equation in (9),

$$\hat{T}_\mu^\nu := S_{\mu\sigma}S^{\nu\sigma} = \rho C_\mu C^\nu, \quad \rho := L^\nu L_\nu, \tag{8}$$

$$\hat{T}_\mu^\nu := S_{\mu\sigma}S^{\nu\sigma} - S_{\sigma\tau}S^{\sigma\tau}\eta_\mu^\nu/2, \iff G_\mu^\nu := R_\mu^\nu - R g_\mu^\nu/2. \tag{9}$$

$\hat{T}_\mu^\nu$ in (9) becomes isomorphic to the Einstein tensor $G_\mu^\nu$ given in the second equation of (9) where $R_\mu^\nu$ denotes Ricci tensor defined as the contraction of Riemann curvature tensor of the form: $R_\mu^\nu := R_{\sigma\mu}{}^{\nu\sigma}$ and $R$ is the scalar curvature defined as $R := R_\nu^\nu$. Riemann curvature tensor $R_{\alpha\beta\gamma\delta}$ satisfies the following properties:

$$R_{\beta\alpha\gamma\delta} = -R_{\alpha\beta\gamma\delta}, \quad R_{\alpha\beta\delta\gamma} = -R_{\alpha\beta\gamma\delta}, \quad R_{\gamma\delta\alpha\beta} = R_{\alpha\beta\gamma\delta}, \tag{10}$$

$$R_{\alpha\beta\gamma\delta} + R_{\alpha\gamma\delta\beta} + R_{\alpha\delta\beta\gamma} = 0, \quad \nabla_\nu G_\mu^\nu = 0. \tag{11}$$

where $\nabla_\nu$ denotes a covariant derivative defined on a curved spacetime. Note that, if we define $\hat{R}_{\alpha\beta\gamma\delta}$ as $\hat{R}_{\alpha\beta\gamma\delta} := -S_{\alpha\beta}S_{\gamma\delta}$, then we readily see that it satisfies exactly the same equations as those in (10). The fact that $\hat{R}_{\alpha\beta\gamma\delta}$ also satisfies the first equation in (11) can be directly shown from (7), namely, $S_{\alpha\beta}$ is a simple bivector field. Since "Ricci tensor" $\hat{R}_\mu^\nu$ in this case is defined as $\hat{R}_\mu^\nu = \hat{R}_{\sigma\mu}{}^{\nu\sigma} = -S_{\sigma\mu}S^{\nu\sigma} = S_{\mu\sigma}S^{\nu\sigma}$ and "scalar curvature" $\hat{R}$ becomes $\hat{R}_\nu^\nu = S_{\sigma\tau}S^{\sigma\tau}$, we see that the first equation in (9) is rewritten as $\hat{T}_\mu^\nu = \hat{R}_\mu^\nu - \hat{R}\eta_\mu^\nu/2$ and is isomorphic to the second one. In addition, the divergence free condition $\partial_\nu \hat{T}_\mu^\nu = 0$ which qualifies $\hat{T}_\mu^\nu$ as the energy-momentum tensor of the Clebsch dual field corresponds to the second equation in (11). We conjecture that the above isomorphism (9) is a sort of "conjugated" manifestation of the isomorphism between the Coulomb force and the universal gravitation, since, as we already explained, the Clebsch dual field represents the longitudinal Coulomb modes of electromagnetic field. In addition, it also implies an intriguing possibility that *the quantization of the Clebsch dual field to be discussed in the following Section 3 is also closely related to that of spacetime.*

For a space-like case, when the $\lambda$ and $\phi$ fields are given by plane waves of $\psi = \hat{\psi}_c \exp[i(k_\nu x^\nu)]$ satisfying (4), together with $\partial^\nu \psi \partial_\nu \psi^* = -(\kappa_0)^2(\hat{\psi}_c \hat{\psi}_c^*)$, we obtain $\hat{T}_\nu^\nu = -S_{\sigma\tau}S^{\sigma\tau}$ directly from (7) and (9), leading to showing that the trace of $\hat{T}_\nu^\nu$ defined as the norm of $\hat{T}_\nu^\nu$ ($||\hat{T}_\nu^\nu||$) is negative:

$$||\hat{T}_\nu^\nu|| := -S^{\sigma\tau}(S_{\sigma\tau})^* = 4(\kappa_0)^2[U^\nu(U_\nu)^*] = -2(\kappa_0)^4[\hat{\lambda}_c(\hat{\lambda}_c)^*][\hat{\phi}_c(\hat{\phi}_c)^*] < 0, \tag{12}$$

which will be used in Section 4 on cosmology.

### 3. Quantization of the Clebsch Dual Field and DP Model

Using the plane wave form mentioned above, $L^\mu$ derived from (4) satisfies

$$L^\nu L_\nu^* = -(\kappa_0)^2(\hat{\lambda}_c \hat{\lambda}_c^*) = const. < 0, \tag{13}$$

which shows that "momentum-like vector" $L^\mu = \partial^\mu \lambda$ lies in a submanifold of the Lorentzian manifold called de Sitter space in cosmology, which is a pseudo-hypersphere with radius $(\sqrt{\Lambda_{dS}})^{-1/2}$ embedded in $R^5$. The importance of this space in the context of spacetime quantization was first noted by Snyder [30], who proposed a quantization scheme with Planck length and the built-in Lorentz invariance based on the assumption that hypothetical momentum 5-vector $\hat{p}^\mu$ ($0 \leq a \leq 4$) in $R^5$ is constrained to lie on de Sitter space, i.e., $\hat{p}^\nu \hat{p}_\nu^* = -\Lambda_{dS}$. The similarity between (13) and the de Sitter space structure of $\hat{p}^\nu \hat{p}_\nu^* = -\Lambda_{dS}$ seems to imply that the isomorphism (9) between $\hat{T}_\mu^\nu$ and $G_\mu^\nu$ derived for the classical field equation is valid also for quantized fields, which is surely an important issue to be investigated.

A particularly interesting point concerning this similarity is the following contrast: in Snyder's quantization scheme, the parameter $\Lambda_{dS}$ does not explicitly appear although a Planck scale is introduced independently. On the contrary, $(\kappa_0)^2$ plays a key role in the Clebsch dual field. This observation suggests that conformal symmetry breaking related to (4) may be closely related to the dynamic origin of the cosmological constant $\Lambda$, which Snyder did not discuss. In this section, firstly, we will show that the introduction of $\kappa_0$ can be justified only when we consider the quantization of the Clebsch dual field. In addition, its physical implication for cosmology will be discussed in Section 4 from the viewpoint of simultaneous conformal symmetry breaking of electromagnetic and gravitational fields.

Note that $\hat{T}_\mu^{\ \nu} = \rho C_\mu C^\nu$ in (8) is isomorphic to the EM tensor of freely moving fluid particles, so the kinetic theory of molecules suggests that the $\rho$ field can be quantized. Since the physical dimensions of $\rho C_\mu C^\nu = \rho \partial_\mu \phi \partial^\nu \phi$ and $\phi$ are the same as those of $F_{\mu\sigma} F^{\nu\sigma}$ and $F_{\mu\nu}$, respectively, using $\rho = L^\nu L_\nu$ given in (8), we see that $L^\mu$ has the dimension of length. Therefore, the quantization of $\rho$ means that there exists a certain quantized length of which the inverse is $\kappa_0$. Now, let us consider the Dirac equation of the form

$$(i\gamma^\nu \partial_\nu + m)\Psi = 0, \tag{14}$$

which can be regarded as the "square root" of the time-like KG equation: $(\partial^\nu \partial_\nu + m^2)\Psi = 0$. Therefore, the Dirac equation for $(\partial^\nu \partial_\nu - (\kappa_0)^2)\Psi = 0$ must be $i(\gamma^\nu \partial_\nu + \kappa_0)\Psi = 0$. On the other hand, an electrically neutral Majorana representation exists for (14), in which all the values of the $\gamma$ matrix become purely imaginary numbers such that this matrix has the form of $(\gamma^\nu_{(M)} \partial_\nu + m)\Psi = 0$, which is identical to the Dirac equation for the above space-like KG equation. The reason why we have introduced the Clebsch dual field as the space-like extension of the electrically neutral electromagnetic wave field is because the Greenberg and Robinson theorem mentioned in in Section 2.1 requires such a field for quantum field interactions, so that the above arguments suggest that Majorana field must be such a quantum field.

The Majorana field is fermionic with a half-integer spin 1/2, so the same state cannot be occupied by two fields according to Pauli's exclusion principle. A possible configuration of a couple of Majorana fields corresponding to the Clebsch dual field that behaves like a boson with spin 1 can be identified with the help of Pauli–Lubanski vector $W_\mu$ describing the spin states of moving particles. $W_\mu$ has the form of $W_\mu = M_{\mu\nu} p^\nu$, where $M_{\mu\nu}$ and $p^\nu$ are the angular and linear momenta of the Majorana field, respectively. Note that the two fields $M_{\mu\nu}$ and $N_{\mu\nu}$ can share the same $W_\mu$ such that

$$M_{\mu\nu} p^\nu = N_{\mu\nu} q^\nu = W_\mu \tag{15}$$

when their linear momenta $p^\mu$ and $q^\mu$ are orthogonal, i.e., $p^\nu q_\nu = 0$. Two Majorana fields satisfying this orthogonality condition can be combined, as in the case of a Cooper pair in the superconducting phenomenon, to form a vector boson with spin 1, which can be identified as the quantized Clebsch dual field satisfying the orthogonality condition in (5).

Now, we are ready to consider the mechanism through which a DP emerges. As a mathematically simple situation, let us consider the case in which the space-like KG Equation (4) is perturbed by the interaction with a point source $\delta(x^0)\delta(r)$, where $r$ denotes the radial coordinate of a spherical coordinate system. The essential causal aspects of this problem were already investigated by Aharonov et al. [31], who showed that the resulting time-dependent behavior of the solution can be expressed by the superposition of a superluminal (space-like) stable oscillatory mode and a time-like linearly unstable mode whose combined amplitude spreads with a speed slower than the light velocity. A time-like unstable mode of the solution to (4) expressed in a polar coordinate system with spherical symmetry has the form of $\lambda(x^0, r) = \exp(\pm k_0 x^0) R(r)$, where $R(r)$ satisfies

$$R'' + \frac{2}{r} R' - (\hat{\kappa}_r)^2 R = 0, \quad (\hat{\kappa}_r)^2 := (k_0)^2 - (\kappa_0)^2 > 0, \tag{16}$$

whose solution becomes the Yukawa potential: $R(r) = \exp(-\hat{\kappa}_r r)/r$, which rapidly falls off as $r$ increases. The nonzero component of the *deformed* Clebsch dual bivector field $^dS_{ab}$ derived by the combined use of (16) and (7) is $^dS_{0r}$, namely, $^dS_{0r}^\dagger := k_0 R' \exp(k_0 x^0)$ and $^dS_{0r} := -k_0 R' \exp(-k_0 x^0)$, which are, in the classical interpretation, growing and damping solutions. However, quantum mechanically, these two can be interpreted as follows. The transmutation from a space-like mode to a pair of these two time-like modes through the interaction with a point source can be regarded as a pair creation of Majorana particles: one going forward in time and the other antiparticle going backward in time. This pair creation is possible because the Clebsch dual field consists of a pair of Majorana fields. Since these modes are non-propagating, they are superimposed to yield a non-propagating light field called a DP that can be regarded as a pair annihilation. The energy density of the DP generated by these processes is given by $-(^dS_{0r}^\dagger)(^dS_{0r}) = (k_0 R')^2$. If we use a natural unit system, then $\kappa_0$ possessing the dimension of $m^{-1}$ may be regarded as an elemental block of DP energy. In subsection 1.2, we have observed that the maximum size of a DP is approximately 50 nm. Since this size can naturally be assumed to correspond to the minimum energy of the DP, we have $\text{Min}[\hat{\kappa}_r] \approx \kappa_0 \approx (50\,\text{nm})^{-1}$ using (16).

## 4. Connection with Cosmology

Since the spatial dimension of our physical spacetime is three, the maximum number of momentum vectors satisfying the orthogonality condition (15) is also three, that is, $M_{\mu\nu}p^\nu = N_{\mu\nu}q^\nu = L_{\mu\nu}r^\nu = W_\mu$, which indicates the existence of a compound state of Majorana fermions with spin 3/2 denoted by $|M3\rangle_g$. Note that this state can play the role of "the ground state" of the Clebsch dual field in the sense that Clebsch dual fields as extended virtual photons can be excited from any of the three different configurations of the "Clebsch dual structure" (15) embedded in $|M3\rangle_g$. Electromagnetic interactions are ubiquitous phenomena such that incessant occurrence of excitation–deexcitation cycles between "the ground" and non-ground states makes the former a fully occupied state from the viewpoint of a macroscopic time scale. In such a situation, $|M3\rangle_g$ would exist not as an extremely ephemeral virtual state but as a stable unseen off-shell state.

In order to apply our new idea on the Clebsch dual field to cosmological problems, we first point out that the formulation of it derived for Minkowski space in Sections 2 and 3 is readily generalized to cover the case of a curved spacetime for which the partial derivative $\partial_\mu$ of a given field defined on the former must be replaced by the covariant derivative $\nabla_\mu$ of the field defined on the latter. At the end of Section 2, we have shown the isomorphsm between the energy-momentum tensor of Clebsch dual field and Einstein's field equation by utilizing $\hat{R}_{\mu\nu\rho\sigma} = -S_{\mu\nu}S_{\rho\sigma}$. It is clear that a curved spacetime does not create any problem for defining the skew-symmetric simple bivector field $S_{\mu\nu}$ and hence $\hat{R}_{\mu\nu\rho\sigma} = -S_{\mu\nu}S_{\rho\sigma}$. One of the notable problems we have in the case of dealing with a curved spacetime is that differential operators do not commute in general. For a given vector field $V_\mu$ on Minkowski space, we have $\partial^2_{\nu\rho}V_\mu = \partial^2_{\rho\nu}V_\mu$. On a curved spacetime, however, we have $\nabla_{\nu\rho}V_\mu = \nabla_{\rho\nu}V_\mu + V_\sigma R^\sigma_{\ \mu\nu\rho}$ where $R^\sigma_{\ \mu\nu\rho}$ denotes Riemann curvature tensor, so that the order of differentiation matters. The sole exception for this non-commuting rule is the case where a vector field $V_\mu$ is replaced by a scalar field $S$, for which we have $\nabla_\nu S = \partial_\nu S$ and $\nabla_{\nu\rho}S = \partial^2_{\nu\rho}S - \Gamma^\sigma_{\ \nu\rho}\partial_\sigma S = \nabla_{\rho\nu}S$ because the affin connection $\Gamma^\sigma_{\ \nu\rho}$ is symmetric with respect to the subscripts $\nu$ and $\rho$. Notice again that the skew-symmetric Clebsch dual field $S_{\mu\nu}$ given in (7) is a bivector field represented in terms of the exterior product of a couple of gradient vector $L_\mu = \partial_\mu \lambda = \nabla_\mu \lambda$ and $C_\mu = \partial_\mu \phi = \nabla_\mu \phi$. Therefore, while $S_{\mu\nu}$ only contains the first derivatives of scalar fields $\phi$ and $\lambda$, the entire formulation of the Clebsch dual field covering, for instance, $\nabla_\nu \hat{T}^\nu_{\ \mu}$ involves the first and second derivatives of them, for the latter of which the order of differentiation does not matter. We mentioned already that the simple bivector property of $S_{\mu\nu}$ is a crucial element for deriving the first equation in (11). In reference [5], we show that, not only for (11) but also for the other parts of the Clebsch dual formulation, the simple bivector property of $S_{\mu\nu}$ and the commutativity of the second derivatives of scalar fields $\lambda$ and $\phi$ are essential elements. By using those properties, we can prove $\nabla_\nu \hat{T}^\nu_{\ \mu} = 0$ since, as far as the mathematical manipulations are concerned, those in a curved spacetime are essentially similar to those in Minkowsky space. Thus, we show that the isomorphism (9) can be extended to that in a curved spacetime.

Having stated this, we now move on to the well-known isotropic spacetime structure employed in cosmological arguments:

$$ds^2 = (cdt)^2 - (R(t))^2 \left[ \frac{dr^2}{1 - \zeta r^2} + r^2(d\theta^2 + \sin^2\theta d\phi^2) \right], \tag{17}$$

where $\zeta$ denotes the curvature parameter taking one of the triadic values of $(0, +1, -1)$ and the other notations are conventional. The coordinate system employed in (17) is a unique co-moving (co-moving with matter) one singled out by Weyl's hypothesis on the cosmological principle with which the energy-momentum tensor $T_\mu^\nu$ of the universe becomes identical in form to the following one of the hydrodynamics:

$$T_\mu^\nu = \begin{pmatrix} \rho c^2 & 0 & 0 & 0 \\ 0 & -p & 0 & 0 \\ 0 & 0 & -p & 0 \\ 0 & 0 & 0 & -p \end{pmatrix}. \tag{18}$$

In addition, corresponding to (18), the components of metric tensor $g_{\mu\nu}$ can be chosen in such that off-diagonal elements of Einstein tensor $G_\mu^\nu$ are also zeros. A caveat in using this coordinate system for our Clebsch dual field is that, due to its space-like property, the energy-momentum tensor $\hat{T}_\mu^\nu$ of the Clebsch dual field to be given by (23) cannot be diagonalized as in the case of (18) since the field resides outside the familiar time-like universe. In spite of that, the above coordinates system introduced by Weyl is a quite informative one from the viewpoint of cosmological observations, so that we think one of the meaningful approaches to estimate the impact of $\hat{T}_\mu^\nu$ on our time-like universe would be to focus solely on its diagonal components, especially the trace $\hat{T}_\nu^\nu$ as the sum of them whose justification will be given shortly, projected on the four-dimensional "screen" spanned by the set of basis vectors of the Weyl coordinates.

In what follows, we are going to derive the energy-momentum tensor ((23) or (27)) directly related to a compound state of Majorana fermions $|M3\rangle_g$ referred to at the beginning of this section. To avoid misunderstanding of the characters of this tensor, the following remark on fermionic fields is important to be made in advance: in quantum theory, the time change of a state is described by the dynamics acting on the (C*-)algebra of observables. The non-commutativity inherent to quantum theory requires the notions of quantum "observables" and "states" of a given system to be distinguished more clearly than in the classical case. Even in the classical Einstein field equation, it is true that "observables" or "physical quantities" (represented typically by the energy-momentum) and "states" (represented by the curvature of spacetime) are seen to occupy different places in a way that the former and the latter appear in the right and the left hand sides of the equation, respectively. In regard to fermionic fields, we can say that, though state changes of fermionic fields are visible, the physical quantities satisfying Fermi statistics with anti-commutation relations cannot be visible. In the conventional quantum field theory, such invisible entities as fermionic fields were introduced as an ad hoc fashion and it is not until the advent of Doplicher-Haag-Roberts theory [32] that their existence was justified through a process of reconstructing all the members of a standard formulation of the theory involving fermionic entities, just starting from the formalism consisting of only observable data structure in the context of Galois theory.

According to these arguments, the physical quantities associated with ((23) or (27)) derived from the spacelike Majorana fermionic field explained in Section 3 should be invisible in nature. The reason is as follows: the Clebsch dual field can be manipulated mathematically as if it is a classical field, similarly to the case of Schroedinger's wave equation. As far as the invisible nature of a spacelike 4 momentum vector is concerned, however, we have to take the above-mentioned property of Fermi statistics into consideration. (The close relation between the quantization of spacelike 4 momentum and Fermi statistics was pointed out first by Feinberg [33].) The key question in our analysis on dark energy is, therefore, whether we can find observable quantities or not. Since the relevant criterion for singling out such quantities may change depending on the choices of situations and aspects, however, we have no choice but to make a good guess. The fact which seems to work as "the guiding principle" is that, within the framework of relativistic quantum field theory, any observable without exception associated with the given internal symmetry is the invariant under the action of transformation group materializing the symmetry under consideration. By extending this knowledge on the internal symmetry to the external (spacetime) one, we assume that the trace $\hat{T}_\nu^\nu$ as the invariant of general coordinate transformation is observable since it is directly related to the actual observable quantity of the expansion rate of the universe through the isomorphism (9) which has been shown to be valid for a curved spacetime through the arguments in the second paragraph in this section.

To implement our analyses on dark energy, for the sake of simplicity, we take two-stage approach I and II. In the first stage I, we confine the scope of our argument to sub-Hubble scales in which the spacetime of the isotropic universe can be regarded as Minkowski space in an approximate sense. Then, in the second stage II, we smoothly extend our argument beyond those limits to cover the entire curved spacetime.

**Stage I analyses**

Firstly, to incorporate the fundamental quantum condition of $E = h\nu$ into the Clebsch dual field, let us consider the light-like case given by (8), where we have $\hat{T}_\mu^\nu = \rho C_\mu C^\nu = (\partial_\sigma \lambda \partial^\sigma \lambda) \partial_\mu \phi \partial^\nu \phi$. Using plane wave expressions of

$$\phi = \hat{\phi}_c \exp(ik_\nu x^\nu), k_\nu k^\nu = 0; \ \lambda = iN_\lambda \hat{\lambda}_0 \exp(il_\nu x^\nu), l_\nu l^\nu = -(\kappa_0)^2, \qquad (19)$$

where $i$, $\hat{\lambda}_0$ and $N_\lambda$ denote the imaginary unit, the quantized elemental amplitude and the number of such an elemental mode, we obtain

$$(C_\mu)^* C^\nu = k_\mu k^\nu \hat{\phi}_c (\hat{\phi}_c)^*, \ \rho = (iN_\lambda)^2 (\kappa_0)^{-2}. \qquad (20)$$

In deriving the second equation of (20), $\hat{\lambda}_0 (\hat{\lambda}_0)^* = (\kappa_0)^{-4}$ has been used since the dimension of $\hat{\lambda}_0$ is length squared. Now, we introduce Cartesian coordinates $x^1, x^2$, and $x^3$ such that the $k$ vector for $\phi$ is parallel to the $x^1$ direction and consider a rectangular parallelepiped $V$ spanned by the length vector $(1/k_1, 1, 1)$. Using (20) and $k_0 = \nu_0/c$ where $c$ denotes the light velocity, the volume integration of $\hat{T}_0^0 / (iN_\lambda)^2$ over $V$ as the energy per quantum becomes

$$\frac{1}{(iN_\lambda)^2} \int_V \hat{T}_0^0 dx^1 dx^2 dx^3 = (\kappa_0)^{-2} \epsilon [\hat{\phi}_c (\hat{\phi}_c)^*] \frac{\nu_0}{c} \ \to \ h = \frac{1}{c}(\kappa_0)^{-2} \epsilon [\hat{\phi}_c (\hat{\phi}_c)^*], \qquad (21)$$

from which the condition corresponding to $E = h\nu$ is identified as the second equation in (21), where $\epsilon$ denotes a unit square meter. For the non-light-like case of $U^\nu (U_\nu)^* < 0$, using (12), since we have $||\hat{T}_\nu^\nu|| = -S^{\mu\nu}(S_{\mu\nu})^* = 2(iN_\lambda)^2 [\hat{\phi}_c (\hat{\phi}_c)^*]$, $||\hat{T}_\nu^\nu||_1 := -[S^{\mu\nu}(S_{\mu\nu})^*]_1$ defined as that for $(N_\lambda)^2 = 1$ becomes

$$||\hat{T}_\nu^\nu||_1 = -[S^{\mu\nu}(S_{\mu\nu})^*]_1 = -2[\hat{\phi}_c (\hat{\phi}_c)^*]. \qquad (22)$$

Since the Clebsch dual wave field, as in the case of an electromagnetic wave, has a propagating direction, to have isotropic radiation, we need three fields, any pair of which is mutually orthogonal. Such three fields are given, for instance, by $(S_{23}, S_{02})$, $(S_{31}, S_{03})$ and $(S_{12}, S_{01})$. $\hat{T}_\mu^\nu(3)$ derived by the superposition of these fields with $S_{23} = S_{31} = S_{12} = \sigma$ and $S_{01} = S_{02} = S_{03} = \tau$ turns out to be

$$\hat{T}_\mu^\nu(3) = \begin{pmatrix} -3\sigma^2 & -\tau\sigma & -\tau\sigma & -\tau\sigma \\ \tau\sigma & 2\tau^2 - \sigma^2 & 0 & 0 \\ \tau\sigma & 0 & 2\tau^2 - \sigma^2 & 0 \\ \tau\sigma & 0 & 0 & 2\tau^2 - \sigma^2, \end{pmatrix}, \quad (23)$$

which is the energy-momentum tensor of the anti dark energy (dark energy with negative energy density, that is, $\hat{T}_0^0(3) = -3\sigma^2 < 0$) we propose in this paper. As we will see shortly, the dark energy (with positive energy density) $^*\hat{T}_\mu^\nu(3)$ given by (27) having exactly the same trace as that of the anti dark energy (23) can be introduced accordingly. Here, a remark must be made to clear the following point concerning different types of dark energy. Although the cosmological term $\lambda g_{\mu\nu}$ with $\lambda > 0$ is well-known and presumably the simplest candidate model of the dark energy, the up-to-date notion of dark energy includes presently-unknown entities other than $\lambda g_{\mu\nu}$. The present model $^*\hat{T}_\mu^\nu(3)$ now we are considering belongs to the latter type.

**Stage II analyses**

The above analyses in I shows that $\hat{T}_\nu^\nu(3) = -6\sigma^2 + 6\tau^2 = -6[\hat{\phi}_c(\hat{\phi}_c)^*]$. As we already pointed out, the isomorphism between $\hat{T}_\mu^\nu$ and $G_\mu^\nu$ in (9) can be extended to the one in a curved space-time. Using this relation, we can say that the existence of $\hat{T}_\nu^\nu(3)$ induces a constant negative scalar curvature in the universe. The configuration of such a universe is described as a four-dimensional hyper pseudo-sphere with a certain "radius" $3/\Lambda$ embedded in a fifth dimensional Minkowski space. This universe is known as de Sitter space whose metric invariant $ds^2$ can be rewritten with polar coordinates $(r, \theta, \phi)$ as

$$ds^2 = (cdt)^2 - (R_0)^2 \exp\left(2\sqrt{\frac{\Lambda}{3}}ct\right)\{dr^2 + r^2(d\theta^2 + \sin^2\theta)d\phi^2\}, \quad (24)$$

where $R_0$ denotes a constant initial radius of the universe. By comparing (24) with (17), we see that the curvatue parameter $\zeta$ of de Sitter space is zero, which shows that the analyses in the first stage I can be extended smoothly to the second stage II. Since de Sitter space is a unique solution of the Einstein field equation for the cosmological term of $\Lambda g_{\mu\nu}$, we see that the impact of $\hat{T}_\mu^\nu(3)$ can be observed in a form of cosmological constant.

To the best of authors' knowledge, the observational data available to us on our expanding universe is the cosmological constant $\lambda_{obs}$ derived on the assumption that the dark energy may be modeled by the cosmological term $\lambda g_{\mu\nu}$. If the dark energy is modeled by $\lambda_{obs} g_{\mu\nu}$, then the Einstein field equation with the sign convention of $R_{\mu\nu} = R^\sigma_{\mu\nu\sigma}$ becomes the first equation in (25), and if it is modeled by $^*\hat{T}_\mu^\nu(3)$, then the Einstein field equation becomes the second one in (25):

$$R_\mu^\nu - \frac{R}{2}g_\mu^\nu + \lambda_{obs}g_\mu^\nu = -\frac{8\pi G}{c^4}T_\mu^\nu, \quad R_\mu^\nu - \frac{R}{2}g_\mu^\nu = -\frac{8\pi G}{c^4}(T_\mu^\nu + ^*\hat{T}_\mu^\nu(3)), \quad (25)$$

which suggests that one of the meaningful observational validations of our dark energy candidate model $^*\hat{T}_\mu^\nu(3)$ would be to compare the traces of $\lambda_{obs}g_\mu^\nu$ and $(-8\pi G/c^4)^*\hat{T}_\mu^\nu(3)$. Since the trace of $^*\hat{T}_\mu^\nu(3)$ is the same as that of $\hat{T}_\mu^\nu(3)$, we see that, using (22), the magnitude of $\hat{T}_\nu^\nu(3)$ corresponding to the above-mentioned isotropic radiation is evaluated as $-3 \times 2[\hat{\phi}_c(\hat{\phi}_c)^*]$, whose numerical value can be derived by the use of (21), and the experimentally determined value of $\kappa_0$. Using $(\kappa_0)^{-1} \approx 50$ nm, we get $\lambda_{DP} := (-8\pi G/c^4)\hat{T}_\nu^\nu(3)/g_\nu^\nu \approx 2.47 \times 10^{-53} m^{-2}$, which may be regarded as the "reduced cosmological constant" of $^*\hat{T}_\mu^\nu(3)$, while the value of $\lambda_{obs}$ derived by Planck satellite observations [34] is $\lambda_{obs} \approx 3.7 \times 10^{-53} m^{-2}$. Thus, we can say that $|M3\rangle_g$ is a promising candidate for dark energy.

Note that the energy density $\hat{T}_0^0(3)$ in (23) is negative. In order to figure out the meaning of $\hat{T}_0^0(3)$, let us consider a simple case of the on-shell condition of a real-valued 4-momentum vector $p_\mu = (p_0, p_1, p_2, p_3)$. Without the loss of generality, we can choose a coordinate system in which $p_2$ and $p_3$ vanish, so that we have

$$p^\nu p_\nu = (p_0)^2 - (p_1)^2 = \Pi = \text{const.} \tag{26}$$

Clearly, $(-p_0, -p_1)$ satisfies (26) when $(p_0, p_1)$ is a solution to it. Since energy and time are canonically conjugate variables, the time evolution of a given dynamical system with negative energy (Hamiltonian) can be reinterpreted as the backward time evolution of the counterpart system with positive energy. We often encounter such reinterpretations in Feynman diagrams to distinguish the anti-particle arising from a pair creation, so that, at the microscopic quantum level, the emergence of negative energy does not create any fundamental problem, as we already referred to the two-sided energy spectra of the Tomita–Takesaki extension of the thermal equilibrium. At the macroscopic classical level, however, there is no hint of the existence of anti-matter in abundance. To explain it, the idea of a twin universe as the cosmic version of a pair creation was proposed by Petit [35], though the issue remains unsettled yet. Whatever the reason may be, the weak energy condition (positivity of the energy) in the classical general theory of relativity related to the stability of a given dynamical system under consideration must be tied to the matter (with positive energy) dominated property of our universe.

The simple argument on (26) suggests that the classically unfavorable negative property of $\hat{T}_0^{\ 0}(3)$ can be circumvented as follows. In (26), if we formally replace $p_0$ by $ip_1$ and $p_1$ by $ip_0$, then we readily see that (26) remains the same. This procedure can be applied to transform (23) into the following trace invariant (27). Notice that, with the Hodge dual exchanging between $(\sigma, \tau)$ and $(i\tau, i\sigma)$ in (23), which corresponds qualitatively to the above exchange between $(p_0, p_1)$ and $(ip_1, ip_0)$ because electric and magnetic fields respectively bear temporal and spatial attributes from the Lorentz group theoretical viewpoint, $\hat{T}_\mu^{\ \nu}(3)$ turns into the following ${}^*\hat{T}_\mu^{\ \nu}(3)$

$$ {}^*\hat{T}_\mu^{\ \nu}(3) = \begin{pmatrix} 3\tau^2 & \tau\sigma & \tau\sigma & \tau\sigma \\ -\tau\sigma & -2\sigma^2 + \tau^2 & 0 & 0 \\ -\tau\sigma & 0 & -2\sigma^2 + \tau^2 & 0 \\ -\tau\sigma & 0 & 0 & -2\sigma^2 + \tau^2 \end{pmatrix}, \tag{27}$$

in which the transformed 4-momentum vector density in the first row (in comparison to that in (23)), which changes the sign while the trace of it remains exactly the same as that of $\hat{T}_\mu^{\ \nu}(3)$ in (23). The sign change for the spatial components in the first row occurs in exactly the same manner as the one in (26), though the sign change for the temporal component differs from it. This is because, as we already pointed out, electric and magnetic field respectively bear temporal and spatial attributes, so that the appearance of $\tau$ in (27) is a consistent change in this respect. Thus, the physical meaning of the dual existence of (23) and (27) is that the notion of matter-antimatter duality can be extended to the dark energy model based on the Clebsch dual field. Notice that the diagonal components of ${}^*\hat{T}_\mu^{\ \nu}(3)$ resemble the artificial partition of the diagonal components of $\lambda g_\mu^\nu$ into $\rho_\lambda = \lambda c^4/(8\pi G)$ and $p_\lambda = -\lambda c^4/(8\pi G)$ (cf. (18)) already employed as the hypothetical equation of state of dark energy in the conventional cosmology.

In considering the problem of dark matter from the viewpoint of conformal symmetry breaking mentioned at the beginning of Section 3, we cast a spotlight on the Bel–Robinson tensor [36] $B_{\alpha\beta\gamma\delta}$ satisfying $\nabla^\alpha B_{\alpha\beta\gamma\delta} = 0$, where $\nabla^\alpha$ denotes the covariant derivative. We can readily show that

$$\frac{1}{2} B_{\mu\nu\sigma}^{\ \ \ \sigma} = W_{\mu\alpha\beta\gamma} W_\nu^{\ \alpha\beta\gamma} - \frac{1}{4} g_{\mu\nu} W^2, \quad W^2 := W_{\alpha\beta\gamma\delta} W^{\alpha\beta\gamma\delta}, \tag{28}$$

where $W_{\alpha\beta\gamma\delta}$ denotes the Weyl tensor. A lengthy but straightforward calculation [37] shows that $B_{\mu\nu\sigma}^{\ \ \ \sigma}$ vanishes identically, which indicates that

$$g_{\mu\nu} = \frac{4 W_{\mu\alpha\beta\gamma} W_\nu^{\ \alpha\beta\gamma}}{W^2}, \quad \text{if } W^2 \neq 0. \tag{29}$$

Since the magnitude of $W^2$ in the well-known Schwarzschild outer solution of a given star decreases monotonously along radius direction, for discussions on cosmological phenomena for which mass distributions can be approximated as that of continuous medium, we would have no need to worry about the singular point of $W^2 = 0$. Notice that (29) shows an intriguing possibility that we can figure out the physical meaning of the cosmological term $\lambda g_{\mu\nu}$ which remains a unsettled issue ever since the time of Einstein, though it is tentatively used as a dark energy model. The unique property of (29) that should be distinguished from the one of usual $g_{\mu\nu}$ as a metric tensor is the fact that the former can be defined in the spacetime whose dimension is larger than or equal to 4 because the Weyl tensor does not exist in the lower dimension and that it is directly related to gravitational field. Such being the case, we introduce a new notation $\hat{g}_{\mu\nu}$ to represent the right-hand side of (29).

In our preceding arguments on dark energy, we have shown a possibility that dark energy may be explained by a new model different from the cosmological term $\lambda g_{\mu\nu}$. If that is the case, then $\lambda g_{\mu\nu}$ must represent another phenomenon. Note that the magnitude $W^2$ measures the deviation of spacetime from the conformally flat FRW metric for the isotropic universe. Thus, a field whose energy-momentum tensor $\tilde{T}_{\mu\nu}$ having the following form:

$$\tilde{T}_{\mu\nu} = -\lambda \hat{g}_{\mu\nu}, \quad \lambda > 0 \tag{30}$$

would behave like a field with an attractive nature of gravity, that is to say, that it must work as the seed of galaxy formations, which suggests us to look into a possibility that $-\lambda \hat{g}_{\mu\nu}$ is one of the candidates of the dark matter model. One of the intriguing properties of $-\lambda \hat{g}_{\mu\nu}$ is that its form remains the same irrespective of the magnitude of $W^2$. Considering its attractive nature of gravity, the initial quite small magnitude $(W_{init})^2$ which seems to be relating to the observed slight density variations in the early universe identified by COBE mission would grow monotonously. Thus, $(W_{init})^2$ is a parameter playing a similar role as $R_0$ in (24) and the existence of $-\lambda \hat{g}_{\mu\nu}$ may be regarded as a major dynamical cause for monotonously increasing $W^2$ field.

The important question in fixing the dark matter model $-\lambda \hat{g}_{\mu\nu}$ is the determination of $\lambda$. For this problem, we think that the isomorphism between conformally broken space-like electromagnetic field (Clebsch dual field) and gravitational one (9) must play a key role. At the end of Section 2, we show that $||\hat{T}_\nu^\nu||$ in (12) is an elemental contribution of the former to the scalar curvature of spacetime. As we have already shown, the magnitude of this elemental contribution corresponds in the converted unit of cosmological constant to $\lambda_{DP}/3$ where $\lambda_{DP}$ is the reduced cosmological constant of our dark energy model defined in the 6th line from Equation (25). Since (9) is the isomorphism between Clebsch dual field and Ricci part of gravitational field, it would be natural to assume that $\lambda$ in $-\lambda \hat{g}_{\mu\nu}$ as a conformally broken scale parameter associated with Weyl part is equal to $\lambda_{DP}/3$, which we call simultaneous conformal symmetry breaking of electromagnetic and gravitational fields. As a partial justification of this hypothesis, we point out that the consensus ranges of the estimated percentage of dark energy and matter are $(68\% - 76\%; mean = 72\%)$ and $(20\% - -28\%; mean = 24\%)$, so that the coefficient $1/3$ of $\lambda_{DP}/3$ is consistent with the mean values of these ranges. In the limit of $W^2 \to 0$, where $\hat{g}_{\mu\nu} \to g_{\mu\nu}^{(FRW)}$, $-\lambda_{DP} \hat{g}_{\mu\nu}/3$ asymptotically approaches to the anti-de Sitter space extensively studied in the Maldacena duality [38]. Thus, if $-\lambda_{DP} \hat{g}_{\mu\nu}/3$ actually exists, then we can say that the anti-de Sitter space existed in the early universe.

## 5. Methods: Formulation of the Clebsch Dual Field

The quantization of the electromagnetic field cannot be performed without gauge fixing of some sort, which suggests that $\partial_\nu A^\nu$ can be specified in a physically meaningful fashion. We next discuss that the Feynman gauge first introduced by Fermi in the Lagrangian density $L_{GF}$, containing a gauge fixing term $-(\partial_\nu A^\nu)^2/2$ whose variation with respect to $A_\mu$ is the second equation in (31),

$$L_{GF} := -\frac{1}{4} F_{\mu\nu} F^{\mu\nu} - \frac{1}{2}(\partial_\nu A^\nu)^2 \rightarrow [\partial_\nu F^{\nu\mu} + \partial^\mu (\partial_\nu A^\nu)]\delta A_\mu = 0, \tag{31}$$

which is exactly such a gauge specification. Combining (1) and (31) with the well-documented equation $\partial_\nu T_\mu^\nu = F_{\mu\nu} j^\nu$ on the divergence of the EM tensor $T_\mu^\nu$ given by (3), we obtain

$$\partial^\nu \partial_\nu A^\mu = 0, \quad \text{and} \quad \partial_\nu T_\mu^\nu = F_{\mu\nu}(\partial^\nu \phi), \ (\phi := \partial_\nu A^\nu), \tag{32}$$

of which the second equation shows that the EM conservation $\partial_\nu T_\mu^\nu = 0$ holds well, even in the case of $\partial^\nu \phi \neq 0$, as long as the vector $\partial^\nu \phi$ is perpendicular to $F_{\mu\nu}$. In addition, directly from the second equation in (31), using the antisymmetry of $F^{\mu\nu}$, we have

$$\partial^\nu \partial_\nu \phi = 0. \tag{33}$$

Using Nakanishi–Lautrup (NL) B-field formalism mentioned in Section 2.2, we can show that (33) is the gauge-fixing condition we want to obtain. NL formalism realizes manifestly-covariant quantization of electromagnetic field in which the Lorentz gauge condition ($\partial_\nu A^\nu = 0$) can be *generalized* to the covariant linear gauges of the form:

$$L_B = B \partial_\nu A^\nu + \frac{\alpha}{2} B^2, \tag{34}$$

where $L_B$, $B$ and $\alpha$ respectively denote a gauge-fixing Lagrangian density to be added to the gauge-invariant Lagrangian density $-F_{\mu\nu} F^{\mu\nu}/4$, NL B-field to be defined below and a real parameter. The gauge-fixing condition and B-field are given by

$$\partial_\nu A^\nu + \alpha B = 0, \quad \partial^\nu \partial_\nu B = 0. \tag{35}$$

In particular, the gauge-fixing condition with $\alpha = 1$ is known as the Feynman gauge and we readily show that the total Lagrangian density $L_{GF}$ with this gauge becomes equal to the first equation in (31). The second equation in (35) is called a subsidiary condition necessary to identify the physically meaningful sector in which quantized transverse modes reside. Quantum mechanically, B-field is shown to be a physical quantity in the sense that it is "non-ghost" field though it is invisible.

Notice that the subsidiary condition on $B$ given in (35) is identical to (33) on $\phi$ defined in (32) and the Feynman gauge shows that $\phi = \partial_\nu A^\nu = -B$. Since the classical physicality of $\phi$ in the sense of $\partial_\nu T_\mu^\nu = 0$ is assured by the orthogonality condition of $F_{\mu\nu} \perp \partial^\nu \phi$, we are going to look into this condition further. Using (2), the first equation in (32) can be regarded as a partial differential equation on $\alpha^\mu$ given the above result of (33) specifying $\chi$, namely,

$$\partial^\nu \partial_\nu \alpha^\mu_{(h)} = 0, \quad \partial^\nu \partial_\nu \alpha^\mu_{(i)} + \partial^\nu \partial_\nu (\partial^\mu \chi) = 0, \tag{36}$$

where $\alpha^\mu_{(h)}$ and $\alpha^\mu_{(i)}$ denote homogeneous and inhomogeneous solutions, respectively. $\alpha^\mu_{(h)}$ obviously represents a transverse mode, and the second equation gives, in hydrodynamic terms, a balance between rotational and irrotational modes. The existence of such a balance is well documented in the hydrodynamic literature explaining the mathematical description of irrotational motion of a two-dimensional incompressible fluid. Due to the irrotationality of the motion, the velocity vector $(v_1, v_2)$ is expressed in terms of the gradient of the vector potential $\hat{\phi}$, namely, $(v_1 = \partial_1 \hat{\phi}, v_2 = \partial_2 \hat{\phi})$; on the other hand, the incompressibility of the fluid makes its motion non-divergent such that $(v_1, v_2)$ is alternatively expressed as $(v_1 = -\partial_2 \hat{\psi}, v_2 = \partial_1 \hat{\psi})$, where $\hat{\psi}$ denotes a streamfunction. Equating these two, we obtain $\partial_1 \hat{\phi} = -\partial_2 \hat{\psi}, \partial_2 \hat{\phi} = \partial_1 \hat{\psi}$, showing that $\hat{\phi}$ and $\hat{\psi}$ satisfy the Cauchy–Riemann relation in complex analysis. This example serves as a useful reference in proving that *a null vector current $\partial^\mu \phi$ propagating along the $x^1$ axis perpendicular to $F_{\mu\nu}$ can be reinterpreted as the current of the longitudinal ($x^1$-directed) electric field,* of which a detailed explanation is given in reference [21] and the existence of such longitudinally propagating electric field was actually reported by [39]. Thus, based on the above arguments on $\phi$ and $B$, we can say that they are physically meaningful key quantities in formulating the Clebsch dual field.

The orthogonality condition $F_{\mu\nu}(\partial^\nu \phi) = 0$ derived by (32) is mathematically equivalent to the relativistic hydrodynamic equation of motion of a barotropic (isentropic) fluid [40]: $\omega_{\mu\nu}(wu^\nu) = 0$, where $\omega_{\mu\nu} := \partial_\mu(wu_\nu) - \partial_\nu(wu_\mu)$, $u^\nu$ and $w$ are the vorticity tensor, 4-velocity, and proper enthalpy density of the fluid, respectively. This observation suggests that we look into the unknown form of 4-vector potential $U_\mu$ relating to a longitudinal virtual photon that may have space-like momentum by the method of Clebsch parametrisation [41]:

$$U_\mu = \lambda \partial_\mu \phi, \tag{37}$$

where the two scalars $\lambda$ and $\phi$ become canonically conjugate variables in the parametrized Hamiltonian isentropic vortex dynamics. Now, let us determine the $\lambda$ field by referring to the following structures of electromagnetic waves: (1) $\partial^\sigma \partial_\sigma F_{\mu\nu} = 0$ and (2) $F_{\mu\nu}$ is advected along a longitudinal null Poynting 4-vector. Corresponding to these structures, we introduce, with a constant $\kappa_0$ to be determined, a space-like KG equation $\partial^\nu \partial_\nu \lambda - (\kappa_0)^2 \lambda = 0$ (the middle equation of (4)) with the directional constraint $C^\nu \partial_\nu L_\mu = 0$, where $C^\nu := \partial^\nu \phi$ and $L_\mu := \partial_\mu \lambda$. Multiplying this constraint by $L^\mu$ and $C^\mu$ yields

$$L^\mu(C^\nu \partial_\nu L_\mu) = 0 \to C^\nu \partial_\nu (L^\mu L_\mu) = 0; C^\mu(C^\nu \partial_\nu L_\mu) = 0 \to C^\nu \partial_\nu (C^\mu L_\mu) = 0,$$

which shows that $\rho := L^\nu L_\nu$ and $C^\nu L_\nu$ are advected along $C^\mu$. In particular, if $C^\mu$ and $L_\mu$ are perpendicular at the initial time, then they remain so after that. Thus, as an important constraint, we can introduce

$$C^\nu L_\nu = 0. \tag{38}$$

The main results of the Clebsch dual formulation can be summarized as follows by classifying this formulation into two categories: i.e., [I] the light-like ($U^\nu(U_\nu)^* = 0$) case possessing "gauge symmetry (GS)" in the sense of (33) and [II] the space-like ($U^\nu(U_\nu)^* < 0$) case with broken GS.

**Category I.**
(1) The field strength $S_{\mu\nu} := \partial_\mu U_\nu - \partial_\nu U_\mu$ corresponding to $F_{\mu\nu}$ is given by a simple bivector with the important orthogonality condition that cannot be satisfied when $L^\mu$ is a time-like vector:

$$S_{\mu\nu} = L_\mu C_\nu - L_\nu C_\mu, \quad C^\nu L_\nu = 0. \tag{39}$$

(2) $U_\mu$ is a tangential vector along a null geodesic satisfying the following wave equation:

$$U^\nu \partial_\nu U_\mu = -S_{\mu\nu} U^\nu = 0, \iff \partial^\nu \partial_\nu U^\mu - (\kappa_0)^2 U^\mu = 0. \tag{40}$$

(3) The EM tensor corresponding to (3) with the opposite sign can be defined together with its conservation law. In references [5,20–22] referred to at the beginning of subsection 4, this sign change is not properly accounted for, which should be fixed as a typo. The sign change is necessary because we are dealing with the negative energy that can be clearly seen in the $\rho$ field in (41),

$$\hat{T}_\mu^\nu = S_{\mu\sigma} S^{\nu\sigma} = \rho C_\mu C^\nu, \ \rho := L^\nu L_\nu; \ \partial_\nu \hat{T}_\mu^\nu = S_{\mu\sigma} \partial_\nu S^{\nu\sigma} = S_{\mu\sigma}(\kappa_0)^2 U^\sigma = 0. \tag{41}$$

The first equation in (41) clearly shows that the Clebsch wave field has the dual representation of a wave, $S_{\mu\sigma} S^{\nu\sigma}$, and longitudinally moving particles, $\rho C_\mu C^\nu$ with negative "density" $\rho$ ($L^\nu(L_\nu)^* < 0$ because $L^\nu$ is a space-like vector), which corresponds to an unphysical longitudinal mode in QED. Equation (40) proves (6) in subsection 2.2. Thus, we have shown that *the Clebsch dual field given in (6) possessing space-like momentum characteristics carries a longitudinally propagating electric field satisfying* "gauge invariant" condition (33), which implies that the quantization of the Clebsch dual field gives an alternative representation of a $U(1)$ gauge boson that emerges in the perturbative calculations in QED.

**Category II.**
(1) $U_\mu$ that is advected by $U^\mu$ along a geodesic is redefined.

$$U_\mu := \frac{1}{2}(\lambda C_\mu - \phi L_\mu), \implies U^\nu \partial_\nu U_\mu = -S_{\mu\nu}U^\nu + \frac{1}{2}\partial_\mu(U^\nu U_\nu) = 0, \tag{42}$$

$$\partial^\nu \partial_\nu \lambda - (\kappa_0)^2 \lambda = 0, \quad \partial^\nu \partial_\nu \phi - (\kappa_0)^2 \phi = 0, \quad C^\nu L_\nu = 0. \tag{43}$$

The form of $S_{\mu\nu}$ given in (39) remains unchanged by (42).

(2) The EM tensor satisfying the conservation law of $\partial_\nu \hat{T}^\nu_\mu = 0$ is redefined.

$$\hat{T}^\nu_\mu = \hat{S}_{\mu\sigma}{}^{\nu\sigma} - \frac{1}{2}\hat{S}_{\alpha\beta}{}^{\alpha\beta}\eta^\nu_\mu, \quad \hat{S}_{\alpha\beta\gamma\delta} := S_{\alpha\beta}S_{\gamma\delta}. \tag{44}$$

$\hat{S}_{\alpha\beta\gamma\delta}$ defined above has the same antisymmetric properties as the Riemann tensor $R_{\alpha\beta\gamma\delta}$ including the first Bianchi identity, $S_{\alpha[\beta\gamma\delta]} = 0$, which holds well since $S_{\mu\nu}$ is a bivector field given by (39). Thus, $\hat{T}^\nu_\mu$ given in (44) becomes isomorphic to Einstein tensor $G^\nu_\mu := R^\nu_\mu - Rg^\nu_\mu/2$, where the Ricci tensor $R_{\mu\nu}$ is defined as $R_{\mu\nu} := R^\sigma{}_{\mu\nu\sigma}$.

## 6. Conclusions

In this article, we have discussed the important role played by the space-like 4-momentum in electromagnetic field interactions and found that the space-like momentum field is embodied by the Majorana fermion, of which time-like modes are now attracting the attention of scientists in the field of solid-state physics [42]. The investigation of the Majorana field unexpectedly opened up a new dynamic channel through which we have identified the causes of the three enigmatic phenomena of DPs, dark energy, and dark matter. The former are generated by the pair annihilation of unstable time-like Majorana particles, while the two fields in the latter come into existence as the compound ground state $|M3\rangle_g$ of the Majorana field and the revised cosmological term $-\lambda_{DP}g_{\mu\nu}/3$ through the simultaneous conformal symmetry breaking in electromagnetic and gravitational fields.

Our interpretation on dark matter defined as $-\lambda_{DP}g_{\mu\nu}/3$ with (29) is consistent with the fact that it can provide the triggering mechanism of galaxy clustering formation since non-zero $W^2$ in (29) acts as the core stuff of such dynamical processes. If we regard such galaxy clustering formations as the time evolution of material subsystems in the universe, then we can say that the simultaneous existence of the dark matter and energy sustains such subsystems' evolutions, respectively, as the unseen driving forces of attraction and repulsion with different magnitude, both of which are external to the subsystems in the sense that they are not bound to the time-like sectors in the spacetime. Their remarkable abundance ratios in comparison to a negligible one of ordinary matter suggests an extended thermodynamical viewpoint in which the evolution of material subsystems in the universe can be compared to the "heat engines" working between a couple of "heat reservoirs" with higher and lower temperature, which correspond respectively to the dark matter with positive energy and the negative dark energy.

**Author Contributions:** H.S. contributed to the basic structure of this article as well as to the Clebsch dual representation applied to the discussions of dressed photons and cosmology. I.O. contributed to providing the knowledge on fundamental quantum field theory which gives the justification of introducing Clebsch dual representation to electromagnetic field interactions and also to the improvement of the basic structure of the article. M.O. contributed to the experimental achievements on dressed photon phenomena. H.O. contributed to the derivation of Equation (29), which suggests the physical meaning of the cosmological term. All authors have read and agreed to the published version of the manuscript.

**Funding:** This research received no external funding.

**Acknowledgments:** This research was partially supported in the form of collaboration with the Institute of Mathematics for Industry, Kyushu University. We thank the anonymous reviewer for his questions and comments, which helped us to improve the quality of this article.

**Conflicts of Interest:** The authors declare no conflict of interest.

## References

1. Ohtsu, M.; Kobayashi, K. *Optical Near Fields*; Springer: Berlin, Germany 2004; pp. 11–51.
2. Ohtsu, M. *Dressed Photons*; Springer: Berlin, Germany, 2014; pp. 89–214.
3. Kawazoe, T.; Ohtsu, M.; Aso, S.; Sawado, Y.; Hosoda, Y.; Yoshizawa, K.; Akahane, K.; Yamamoto, N.; Naruse, M. Two-dimensional array of room-temperature nano-photonic logic gates using InAs quantum dots in mesa structures. *Appl. Phys. B* **2011**, *103*, 537–546. [CrossRef]
4. Ohtsu, M.; Kawazoe, T. Principles and practices of Si light emitting diodes using dressed photons. *Adv. Mater. Lett.* **2019**, *10*, 860–867. [CrossRef]
5. Ohtsu, M.; Ojima, I.; Sakuma, H. *Progress in Optics*; Visser, T., Ed.; Elsevier: Amsterdam, The Netherlands, 2019; Chapter 1, Volume 62, pp. 45–97.
6. Ohtsu, M. *Silicon Light-Emitting Diodes and Lasers*; Springer: Berlin, Germany, 2016; Chapter 8, pp. 121–138.
7. Kawazoe, T.; Kobayashi, K. Nonadiabatic photodissociation process using an optical near field. *J. Chem. Phys.* **2005**, *122*, 024715. [CrossRef] [PubMed]
8. Kawazoe, T.; Ohtsu, M. Adiabatic and nonadiabatic nanofabrication by localized optical near fields. *Proc. SPIE* **2004**, *5339*, 619–630.
9. Kawazoe, T.; Kobayashi, K.; Ohtsu, M. Near-field optical chemical vapor deposition using Zn(acac)$_2$ with a non-adiabatic photochemical process. *Appl. Phys. B* **2006**, *84*, 247–251. [CrossRef]
10. Ohtsu, M.; Kawazoe, T. Experimental Estimation of the Maximum Size of a Dressed Photon. 2018. Available online: http://offshell.rodrep.org/?p=98 (accessed on 16 February 2018).
11. Sangu, S.; Kobayashi, K.; Ohtsu, M. Optical near fields as photon-matter interacting systems. *J. Microsc.* **2001**, *202*, 279–285.
12. Yatsui, T.; Hirata, K.; Nomura, W.; Tabata, Y.; Ohtsu, M. Realization of an ultra-flat silica surface with angstrom-scale average roughness using nonadiabatic optical near-field etching. *Appl. Phys. B* **2008**, *93*, 55–57. [CrossRef]
13. Hirata, K. Realization of high-performance optical element by optical near-field etching. *Proc. SPIE* **2011**, *7921*, 79210M.
14. Teki, R.; Kadaksham, A.J.; Goodwin, F.; Yatsui, T.; Ohtsu, M. Dressed-photon nanopolishing for EUV mask substrate defect mitigation. In Proceedings of the Society of Photo-Optocal Instrumentation Engineers (SPIE) Advanced Lithography, San Jose, CA, USA, 24–28 February 2013; Paper 8679-14.
15. Yatsui, T.; Nomura, W.; Ohtsu, M. Realization of ultraflat plastic film using Dressed-Photon-Phonon-Assisted selective etching of nanoscale structures. *Adv. Opt. Technol.* **2015**, *2015*, 701802. [CrossRef]
16. Lehmann, H.; Symanzik, K.; Zimmerman, W. Zur Formulierung quantisierter Feldtheorien. *Nuovo Cim.* **1955**, *1*, 425. [CrossRef]
17. Jost, R. *The General Theory of Quantized Fields*; American Mathematical Society: Providence, RI, USA, 1963.
18. Dell'Antonio, G.F. Support of a field in $p$ space. *J. Math. Phys.* **1961**, *2*, 759–766. [CrossRef]
19. Bers, A.; Fox, R.; Kuper, C.G.; Lipson, S.G. The impossibility of free tachyons. In *Relativity and Gravitation*; Kuper, C.G., Peres, A., Eds.; Gordon and Breach Science Publishers: New York, NY, USA, 1971.
20. Sakuma, H.; Ojima, I.; Ohtsu, M. Dressed photons in a new paradigm of off-shell quantum fields. *Progr. Quantum Electron.* **2017**, *55*, 74–87. [CrossRef]
21. Sakuma, H.; Ojima, I.; Ohtsu, M. Gauge symmetry breaking and emergence of Clebsch-dual electromagnetic field as a model of dressed photons. *Appl. Phys. A* **2017**, *123*, 750. [CrossRef]
22. Sakuma, H. Virtual Photon Model by Spatio-Temporal Vortex Dynamics. In *Progress in Nanophotonics*; Yatsui, T., Ed.; Springer: Cham, Switzerland, 2018; Volume 5, pp. 53–77.
23. Ojima, I. A unified scheme for generalized sectors based on selection criteria—order parameters of symmetries and of thermal situations and physical meanings of classifying categorical adjunctions. *Open Syst. Inf. Dyn.* **2003**, *10*, 235–279. [CrossRef]

24. Ojima, I. Micro-macro duality in quantum physics. In Proceedings of the International Conference on Stochastic Analysis: Classical and Quantum, Meijo University, Nagoya, Japan, 1–5 November 2004; World Scientific: Singapore, 2005; pp. 143–161.
25. Ojima, I. Micro-Macro duality and emergence of macroscopic levels. *Quantum Probab. White Noise Anal.* **2008**, *21*, 217–228.
26. Ojima, I. Nakanishi-Lautrup B-Field, Crossed Product & Duality. *RIMS Kokyuroku* **2006**, *1524*, 29–37.
27. Nakanishi, N.; Ojima, I. *Covariant Operator Formalism of Gauge Theories and Quantum Gravity*; World Scientific: Singapore, 1990.
28. Aharonov, Y.; Bohm, D. Significance of electromagnetic potentials in the quantum theory. *Phys. Rev.* **1959**, *115*, 485–491. [CrossRef]
29. Bratteli, O.; Robinson, D. *Operator Algebra and Statistical Mechanics*, 2nd ed.; Springer: Berlin, Germany, 1987; Volume 1.
30. Snyder, H.S. Quantized space-time. *Phys. Rev.* **1947**, *71*, 38. [CrossRef]
31. Aharonov, Y.; Komar, A.; Susskind, L. Superluminal behavior, causality, and instability. *Phys. Rev.* **1969**, *182*, 1400–1402. [CrossRef]
32. Doplicher, S.; Haag, R.; Roberts, J.E. Fields, observables and gauge transformations I & II. *Comm. Math. Phys.* **1969**, *13*, 1–23.
33. Feinberg, G. Possibility of Faster-Than-Light Particles. *Phys. Rev.* **1967**, *159*, 1089–1105. [CrossRef]
34. Liu, H. Available online: https://www.quora.com/What-is-the-best-estimate-of-the-cosmological-constant (accessed on 15 April 2020).
35. Petit, J.P. Twin Universes Cosmology. *Astrophys. Space Sci.* **1995**, *226*, 273–307. [CrossRef]
36. Jezierski, J.; Lukasik, M. Conformal Yano-Killing tensor for the Kerr metric and conserved quantities. *arXiv* **2005**, arXiv:gr-qc/0510058.
37. Sakuma, H.; Ochiai, H. Note on the Physical Meaning of the Cosmological Term. OffShell: 1909O.001.v2. 2019. Available online: http://offshell.rodrep.org/?p=249 (accessed on 15 April 2020).
38. Maldacena, J. The large N limit of superconformal field theories and supergravity. *Adv. Theor. Math. Phys.* **1998**, *2*, 231–252. [CrossRef]
39. Cicchitelli, L.; Hora, H.; Postle, R. Longitudinal field components for laser beams in vacuum. *Phys. Rev. A* **1990**, *41*, 3727–3732. [CrossRef] [PubMed]
40. Landau, L.D.; Lifshitz, E.M. Fluid Mechanics. In *Course of Theoretical Physics*, 2nd ed.; Elsevier: Oxford, UK, 1987; Volume 6.
41. Lamb, S.H. *Hydrodynamics*, 6th ed.; Cambridge University Press: Cambridge, UK, 1930.
42. Kasahara, K.; Ohnishi, T.; Mizukami, Y.; Tanaka, O.; Sixiao, M.; Sugii, K.; Kurita, N.; Tanaka, H.; Nasu, J.; Motome, Y.; et al. Majorana quantization and half-integer thermal quantum Hall effect in a Kitaev spin liquid. *Nature* **2018**, *559*, 227–231. [CrossRef]

© 2020 by the authors. Licensee MDPI, Basel, Switzerland. This article is an open access article distributed under the terms and conditions of the Creative Commons Attribution (CC BY) license (http://creativecommons.org/licenses/by/4.0/).

MDPI  
St. Alban-Anlage 66  
4052 Basel  
Switzerland  
Tel. +41 61 683 77 34  
Fax +41 61 302 89 18  
www.mdpi.com

*Symmetry* Editorial Office  
E-mail: symmetry@mdpi.com  
www.mdpi.com/journal/symmetry